The body of evolutionary theory that currently occupies a dominant position in biological thought is neo-Darwinism. While this theory has considerable explanatory power, it is widely recognized as being incomplete in that it lacks a component dealing with individual development, or ontogeny. This lack is particularly conspicuous in relation to attempts to explain the evolutionary origin of the 35 or so animal body plans, and of the developmental trajectories that generate them.

This book examines both the origin of body plans in particular and the evolution of animal development in general. In doing so, it ranges widely, covering topics as diverse as comparative developmental genetics, selection theory and Vendian/Cambrian fossils. Particular emphasis is placed on gene duplication, changes in spatiotemporal gene-expression patterns, internal selection, coevolution of interacting genes, and coadaptation.

The book will be of particular interest to researchers and students in evolutionary biology, genetics, palaeontology and developmental biology.

Wallace Arthur is Professor of Evolutionary Biology at the University of Sunderland, UK. He is the author of two earlier books on the evolution of developmental systems: *Mechanisms of Morphological Evolution* and *A Theory of the Evolution of Development*.

The Origin of Animal Body Plans

# The Origin of Animal Body Plans

## A Study in Evolutionary Developmental Biology

**WALLACE ARTHUR**
*University of Sunderland, UK*

PUBLISHED BY THE PRESS SYNDICATE OF THE UNIVERSITY OF CAMBRIDGE
The Pitt Building, Trumpington Street, Cambridge, United Kingdom

CAMBRIDGE UNIVERSITY PRESS
The Edinburgh Building, Cambridge CB2 2RU, UK
40 West 20th Street, New York, NY 10011-4211, USA
10 Stamford Road, Oakleigh, Melbourne 3166, Australia
Ruiz de Alarcón 13, 28014 Madrid, Spain
Dock House, The Waterfront, Cape Town 8001, South Africa

http://www.cambridge.org

First published 1997
First paperback edition 2000

Typeset in 10.5/14 pt Melior

*A catalog record for this book is available from the British Library*

*Library of Congress Cataloging in Publication data is available*

ISBN 0 521 55014 9 hardback
ISBN 0 521 77928 6 paperback


Transferred to digital printing 2004

"We have seen that the members of the same class, independently of their habits of life, resemble each other in the general plan of their organisation. This resemblance is often expressed by the term 'unity of type'; or by saying that the several parts and organs in the different species of the class are homologous. The whole subject is included under the general name of Morphology. This is the most interesting department of natural history, and may be said to be its very soul."

Charles Darwin (1859, p. 434)

# Contents

# Preface

A new discipline – *Evolutionary Developmental Biology* as Hall (1992) has called it – is in the process of being born. Its origins are scattered and heterogeneous: from von Baer and Haeckel, through D'Arcy Thompson, Garstang, de Beer and Waddington, to the many recent and current students of comparative developmental genetics and related topics whose work is examined herein. Its potential is enormous: essentially to build a bridge between the mechanisms of population genetics and evolutionary ecology on the one hand and the patterns of comparative anatomy and palaeontology on the other; and thereby to help unify evolutionary theory in general.

This book is intended to be a contribution to this new discipline. It has a focus, though not an exclusive one, on the origin of animal body plans. It probes the question of how the morphological 'designs' of the thirty-five or so animal phyla arose in a burst of creative evolutionary activity in the distant geological past. In particular, it asks the question: were the genetic, developmental and population-level processes involved the same as those occurring in present-day speciations? If not, then what was different about those key early evolutionary events? In attempting to answer these questions, the book ranges from molecules and cells through developing organisms and natural populations to cladograms and ancient fossils.

My own background is in evolutionary ecology and, to a lesser extent, in population genetics. However, to write this book I have had to divert time away from reading in my own field in order to cover all the areas mentioned above, and more besides. In consequence, I have doubtless become 'a jack of all trades and a master of none', with the emphasis firmly on the latter. But, while a position of being adrift in an interdisciplinary sea surrounded by vigorous, more tightly-focused endeavours on all sides feels somewhat vulnerable, I do not regret adopting it. New syntheses between established disciplines are catalysed by individuals adopting uncomfortable positions at interfaces. And the gradual emergence of evolutionary developmental biology as a discipline in its own right will decrease the discomfort as time goes on.

At the beginning of this endeavour, I rebelled against my neo-Darwinian background. By the time the first draft was three-quarters written, however, I had become clear that I was still a neo-Darwinian after all — albeit one who is highly critical of the extreme 'nondevelopmental fringe' of that tradition. Everything in this book is based on the view that key evolutionary changes take place in the context of mutation and selection occurring in localized populations. No 'higher order' or 'emergent' mechanisms are required. However, large-scale evolutionary patterns, such as von Baerian 'deviation', most certainly do occur within and between particular higher taxa; and their explanation in terms of the relative probabilities of different kinds of evolutionary change in developmental pathways represents a major challenge. Also, I argue that neo-Darwinism needs to take more fully on board the concepts of developmental constraint, internal selection and the coevolution of developmentally interacting genes.

With regard to the book's structure: it is designed to be read straight through, and to that end I have tried to keep it short. However, Chapters 2–6 (and Section 9.2) include background information necessary for the main argument of the book (the core of which occupies Chapters 5–11). Readers coming to the book from particular disciplinary bases may wish to omit those bits of background with which they are already familiar (e.g. Sections 2.2 on cladistic methods, 5.2 on some aspects of basic cell biology or 9.2 on simple population genetic models).

A word on style. I have deliberately varied the 'voice' throughout the book, interspersing 'we', 'I' and the passive. Overuse of 'I' can sound arrogant, while overuse of the passive is dull. My own preference is for the 'mathematician's we', representing as it does an author's attempt to meander through the arguments jointly with the reader. However, the 'we' style also has its limitations. At times it is simply inappropriate; and its overuse might be seen to indicate delusions of royalty. The mixed style that I have adopted will not appeal to everyone, especially advocates of the 'pan-passive' (Leather 1996); but then again, what style would be universally liked by all?

And now, some much-felt words of thanks. Draft chapters have been looked at, and critically commented on, by the following: Michael Akam, Helen Arthur, Simon Conway Morris, Gabby Dover, Richard Fortey, Scott Gilbert, Brian Hall, Philip Ingham, Alec Panchen, Linda Partridge, Colin Patterson, Keith Thomson, George Williams and Greg Wray. I have also benefited from discussions with Derek Briggs, Mark Davies, Paul Eady, Malcolm Farrow, Brian Goodwin, John Lazarus and Colin Scrutton. My visits to the geographically-distant of the above were funded through a Leverhulme Research Fellowship. This Fellowship also freed up much of my time for writing: without it, the book might never have come to fruition. I have had secretarial assistance second to none from Pam Giblin and Carolyn Stout. Access to the extensive library collections of both the Literary and Philosophical Society and the University of Newcastle upon Tyne has been invaluable. A special word of thanks goes to Alec Panchen for his help and encouragement throughout the book's gestation. Finally, the year of publication coincides with the year of retirement of my erstwhile PhD supervisor, Bryan Clarke, to whom I would like to say a belated 'thank you' for setting me off on a scientific trajectory which has never been easy but has usually been fun.

Wallace Arthur
Sunderland, UK
November 1996

# Preface to the Paperback Edition

Two years have elapsed since publication of the hardback edition, and many significant advances have been made during this period. I mention several of these below, ordered according to the first chapter on which they impact.

Molecular studies of the age of the animal kingdom, including attempts to date major divergences (e.g. protostome–deuterostome) have continued to proliferate (Ayala *et al.* 1998; Bromham *et al.* 1998; Wang *et al.* 1999). These have produced very varied estimates, and metazoan origins may yet lie anywhere between about 600 and 1200 *my* ago, as indicated in Chapters 1 and 3, or indeed somewhat earlier. Most of the molecular studies favor an early metazoan origin. However, the continuing lack of clear fossil evidence of undisputed metazoans in pre-Vendian times is still a powerful argument for a late origin.

Von Baer's proposed pattern of increasing phenotypic divergence through embryogenesis (Chapters 2, 4) continues to be challenged, yet continues to survive, albeit in modified form and in restricted numbers of taxa. The latest challenge has come from Richardson *et al.* (1997), who show that Haeckel's drawings, which ironically have provided part of the evidence for von Baer's 'laws', are somewhat inaccurate, and tend to gloss over some early embryonic differences. This acts to diminish, but not to eliminate, von Baerian divergence (see Chapters 2, 4 and 11).

There has been a major development in high-level animal phylogeny with the proposal of two new superphyletic groupings: Ecdysozoa and Lophotrochozoa (Aguinaldo *et al.* 1997; de Rosa *et al.* 1999; Ruiz Trillo *et al.* 1999). This is relevant to the phylogenies depicted in Chapters 3 and 7. In Section 3.2, I pose a series of questions regarding the structure of the Metazoa and the relationships between phyla. The last of these is 'What Other Superphyletic Clusters are Discernible?' (after protostomes and deuterostomes). Those pictured are conventional ones, including, for example, an arthropod/annelid/mollusc grouping. However, if the new proposal is correct, then arthropods are in Ecdysozoa along with nematodes and others, while annelids and molluscs are in Lophotrochozoa. It is not yet certain that these new groupings are correct, but the balance of probabilities is swinging in their favor.

Research into the genetic and cellular basis of development itself (Chapter 6) has continued at its characteristically rapid pace. One consequence of this is that the third body axis (left-right) is becoming much better understood (see review by Levin and Mercola 1998). It remains to be seen what evolutionary stories emerge as more types of organism are studied in relation to this axis. Recent work on the molecular developmental genetics of the anteroposterior and dorsoventral axes, and of the proximo-distal axis of limbs, has focused much attention on homology of genes, homology of structures, and the differences between them (Chapter 7; see also Abouheif et al. 1997; Panganiban et al. 1997; Shubin et al. 1997; Arthur et al. 1999).

There have been recent theoretical and empirical studies of the kinds of mutation that can occur in developmental genes (Chapter 8), and in particular the subset of these that can be incorporated into evolutionary change at the population level and beyond (Chapter 9). Orr's (1998) model of an inverse relationship between magnitude of phenotypic effect and frequency of contribution to evolutionary change adds rigor to one of the central proposals of the book. Orr's model is a development from the Fisher principle (Chapter 2), but it is a distinct step forward because it avoids Fisher's extreme 'micromutationist' view. It also looks at the time sequence of mutations fixed as the optimum is approached, which Fisher did not consider. On the empirical side, Stern (1998) has shown that small-effect mutations in *Hox* genes can contribute to interspecific divergence (between *Drosophila* species). This helps to refine our view of *Hox* genes as being not exclusively big-effect 'selector genes', but capable of a range of effects from very minor to very major (Akam 1998a,b).

So, two years on and there has been considerable progress on many fronts, but much remains to be done. The next couple of decades hold much promise as Evolutionary Developmental Biology continues to grow and to mature as a discipline. It is interesting that since publication of the hardback the number of journals dedicated to Evolutionary Developmental Biology, or 'evo-devo', has risen from one to three: to *Development, Genes and Evolution* have been added *Evolution and Development* and also *Molecular and Developmental Evolution* (a section of *Journal of Experimental Zoology*). Still a small drop in the ocean of biological journals, of course; but this 'proportional tininess' is more than compensated for by the huge conceptual excitement that characterizes the field. Long may it continue.

Wallace Arthur
Sunderland, UK
July 1999

# CHAPTER ONE

# Introduction

## 1.1 A Developmental Approach to an Evolutionary Problem

We humans take some fifteen to twenty years to make the developmental journey from conception to our final adult form. A major feature of this journey is that it involves an enormous increase in organismic size – from a single cell to many trillions of cells – and a corresponding increase in complexity, leading to the adult complement of more than 200 different cell types and a wide range of organs and structures. This increase in organismic complexity is, however, far from linear. Indeed, the vast majority of the developmental period is dominated by allometric growth of already-formed parts, leading only to a change in relative bodily proportions. This is true not only of postnatal growth but also of around three-quarters of the time spent *in utero*. The basic body plan, including all of the major organ systems, is established within the first couple of months after fertilization.

This picture of an early 'creative' or morphogenetic phase followed by a much longer phase of allometric growth has long been recognized by embryologists, and is applicable to a wide range of animal taxa. Working outwards from our human starting point, it is certainly true of other mammals, and of birds. It also appears to hold for many invertebrate phyla.

1

The morphogenetic/allometric distinction is more difficult to apply to groups characterized by complex life histories, such as amphibians and insects. However, I would argue that it still *can* be applied, albeit with care, and in some cases only to the development of selected life stages. *Drosophila,* as a representative of the holometabolous insects, would seem at first sight a particularly difficult case. However, the egg-to-larva and imaginal-disc-to-adult systems are to a large extent separate. (For accounts of *Drosophila* development see Demerec 1950 and Bate and Martinez-Arias 1993.) It is apparent that a short morphogenetic phase giving way to a longer phase of allometric growth characterizes the development of the *Drosophila* larva, but not that of the adult.

We have, then, a generalization about the nature of development which applies widely, albeit not universally, across the animal kingdom. Against this background, let us now consider a hypothetical proposition. Suppose that the members of a research group working on postnatal muscle growth in humans claimed to have formulated a complete theory of development based solely on their own study system: would we believe them? Almost certainly not. Of course, at a very general level some of the processes involved in postnatal allometric growth and in early morphogenesis are similar. For example, an interplay between growth factors and gene switching mechanisms is an important part of both. But despite some such similarities, the *differences* are sufficiently pronounced that it seems unreasonable to expect studies on (for example) cell proliferation in muscle tissue to provide a complete explanation of development. We would conclude, then, that our would-be theory makers were in error.

We now need to confront the question of whether the biological community – or at least a large proportion of it – has come to accept a theory of evolution that is based on a broadly parallel error. Our case studies on the action of natural selection all involve microevolutionary changes occurring within particular lineages hundreds of millions of years after the origin of the major body plans of which the species concerned represent variations. Many of these case-studies are well known, especially the evolution of industrial melanism in *Biston* (Bishop and Cook 1980), the evolution of pigmentation patterns in *Cepaea* (Jones, Leith and Rawlings 1977) and the evolution of Batesian mimicry in several lepidopterans (Turner 1977). Many palaeontological case studies are also restricted to particular lineages, with studies on the horse (Simpson 1951; MacFadden 1992) and the molluscs of Lake Turkana (Williamson 1981) being among the best known. While such studies are usually transspecific, and therefore in the realm of 'macroevolution', they are only a very short distance in that direction from an origin-of-body-plans perspective. (Simpson (1944) used the term 'mega-evolution' for the biggest-scale evolutionary events such as body plan origins, but this term has not become widely adopted.)

So, this book is starting with an exhortation to the reader to believe that current evolutionary theory, based on natural selection and adaptation in present-day lineages is, at the very least, incomplete; and this exhortation is based on the drawing of a parallel between the processes of development and evolution. But this begs the question: is there really a true parallel here? In other words, did evolution really have an early 'morphogenetic' phase during which most major body plans originated? It is to this question that we now turn.

## 1.2    The Early History of the Animal Kingdom

The existence of a 'Cambrian explosion' of metazoan fossils has long been known. Numerous phyla were present in the early Cambrian (see Benton 1993), whereas before the base of the Cambrian (dated at 543 *my* ago by Bowring *et al.* 1993) only the enigmatic Ediacaran animals are found. While the taxonomic status of these is not yet clear, it seems likely that Seilacher's (1984) view that the Ediacarans had quite distinct body plans from Phanerozoic Metazoa is broadly correct (see also Buss and Seilacher 1994; further discussion in Chapter 3).

One possible phylogenetic scheme, then, is that multicellular animals arose about 600 *my* ago (Valentine 1994), and that they underwent two successive radiations. The first produced the Ediacaran fauna which persisted from about 580 to 540 *my* ago, and the second (in the early Cambrian) produced the Eumetazoa.

Two recent papers force us to question this picture. Fortey, Briggs and Wills (1996) emphasize the possibility of a Vendian origin for the Eumetazoa. They point out that our broad cladistic picture of the animal kingdom is rather robust: the origins of individual phyla such as annelids and arthropods were preceded by the protostome/deuterostome divergence, which in turn was preceded by the divergence of Cnidaria and Bilateria. Given that, then if Glaessner's (1984) view that the Ediacaran fossils belong to conventional phyla such as

Annelida and Arthropoda is correct for even a single genus (perhaps *Spriggina* – see Chapter 3), with the bulk of the fauna being of an entirely different construction, along the lines that Seilacher (1984) suggested, this would push the whole 'Cambrian explosion' back into the Vendian. Pre-Cambrian Eumetazoa are then represented by 'ghost lineages' leaving no fossil trace, possibly as a result of very small body size.

While Fortey *et al.*'s (1996) picture (Figure 1–1a) shifts the protostome/deuterostome divergence back by 50 *my* or so (see also Valentine, Erwin and Jablonski 1996), the recent molecular study by Wray, Levinton and Shapiro (1996) argues for a much more radical shift in excess of 500 *my* – see Figure 1–1b. Given the potential of this study to double the accepted view of the age of the animal kingdom, it is worth examining it in some detail.

Wray *et al.* (1996) analysed taxonomic patterns of sequence similarity in each of seven genes (and in 18S rRNA). For each gene, they proceeded as follows. First, they took a pair of vertebrate species whose divergence time was reasonably well established from fossil data. Then they determined the degree of sequence similarity. They repeated this procedure for many pairs of vertebrate species, thus giving a large scatter of points. These scatters were approximately linear for all the genes examined, thus suggesting a reasonably constant (or clock-like) pattern of long-term substitutions. They thus provide a calibration line from which divergence times can be estimated from sequence similarities for pairs of species from different phyla, where fossil evidence of divergence is sparse or nonexistent (see Figure 1–2). Although the standard errors are rather large, all genes indicated the 'deep' pre-Cambrian split shown in Figure 1–1b.

It remains a distinct possibility that extrapolation from within-vertebrate quasilinearity over the last 400 *my* to a panmetazoan linearity extending over a much longer period is incorrect. It may well be that gene sequence evolution was in general much faster in earlier evolution, with rates slowing as gene products converged on a near-optimal structure. Indeed, Wray *et al.*'s (1996) data on 18S rRNA, which were calibrated further back in time through the use of echinoderms and molluscs in addition to vertebrates, suggest this very pattern.

While further work is clearly necessary to resolve this issue, a view that the studies by Fortey *et al.* (1996), Valentine *et al.* (1996) and Wray *et al.* (1996) share is that metazoan lineage divergences (which may or may not have been 'explosive') were quite distinct from, and preceded,

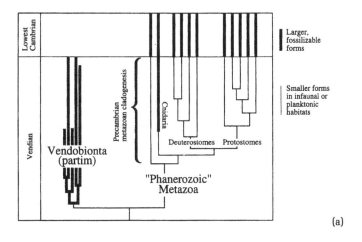

(a)

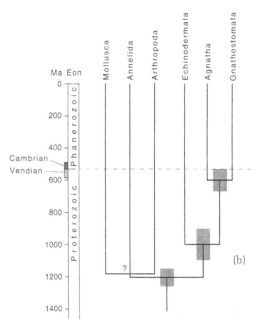

(b)

**Figure 1–1    Alternative views of the timing of major lineage-splitting events**

(a) The "possible model" presented by Fortey *et al.* (1996); and (b) the "estimated divergence times" of Wray *et al.* (1996). Note that the protostome/deuterostome split occurs at around 600 *my* ago in (a) since the Vendian extends from about 650 to 550 *my* ago, but 1200 *my* ago in (b). (Different publications use MYr, *my* and Ma to indicate millions of years, usually measured backwards from the present.) (a) From: Fortey *et al.* 1996. *Biol. J. Linn. Soc.*, **57**, 13–33; (b) from Wray *et al.* 1996. *Science*, **274**, 568–573, with permission.

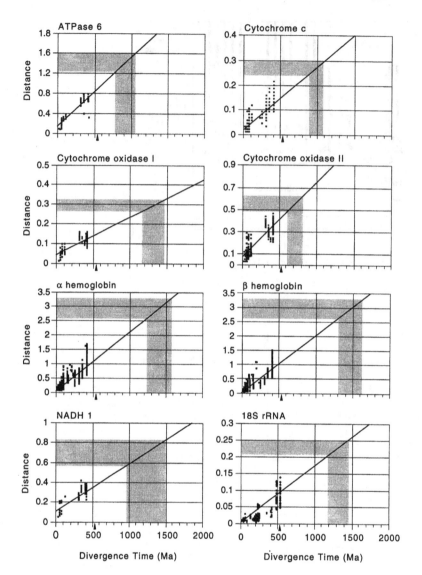

**Figure 1–2  Relationship between estimated lineage splitting times (based on palaeontological data) and extent of sequence divergence in seven genes (and 18S rRNA)**

Points: data for vertebrate species pairs (and others in the case of rRNA).

Line: calibration line, with extrapolation backwards in time on the basis of a presumed linear relationship throughout evolutionary history.

Shaded zone: band of sequence differences and estimated lineage divergence times for all inter-phylum comparisons that were made. From Wray *et al.* 1996. *Science*, **274**, 568–573, with permission.

the explosive appearance of Cambrian metazoan fossils. As Vermeij (1996) states, in commenting on the Wray *et al.* (1996) paper, that study "in no way diminishes the significance of the Vendian–Cambrian revolution, but it does separate the ecological innovations of that episode from the earlier evolution of the basic animal body plans." If this 'decoupling' (as Fortey *et al.* 1996 call it) is correct, then there are three possible explanations for the lack of fossil metazoans in the Vendian:

1. Lineage divergence (Vendian or pre-Vendian) occurred prior to multicellularity (Cambrian).
2. Lineage divergence (Vendian or pre-Vendian) occurred after multicellularity but before body size became macroscopic and/or before the development of fossilizable skeletal structures (Cambrian).
3. Lineage divergence (Vendian or pre-Vendian) occurred simultaneously with body size increase and the development of skeletal structures, but taphonomic conditions resulted in a complete absence of fossils of the lineages concerned.

The similarity of developmental–genetic control mechanisms that is now known to exist between phyla (see Slack, Holland and Graham 1993 and Chapter 7) virtually excludes the first possibility; and the worldwide Ediacaran fossils from the Vendian render the third very unlikely. Our conclusion, then, is *either* that **2** above is correct, in which case the Cambrian explosion is a parallel explosion of body size and morphological innovation in many lineages (perhaps being driven by a coevolutionary 'race') *or* that lineage divergence and morphological disparity exploded all at once in the early Cambrian after all. The evidence is not yet sufficiently clear or extensive to enable a firm choice to be made between these two hypotheses.

Regardless of the timing of early divergences, it appears that no phylum-level body plans have arisen in the animal kingdom in the last 500 *my*. This contrasts with the situation in plants, where the angiosperm body plan arose relatively recently (probably about 130 *my* ago: see Hickey and Doyle 1977; Crane, Friis and Pedersen 1995). Perhaps this difference relates to a difference in developmental–genetic control mechanisms in the two kingdoms, with some genes controlling the determination of animal body axes and other key processes of early ontogeny being more 'generatively entrenched' (Wimsatt 1986) than their nearest equivalents in plants. However, we

7

are a long way from being able to make progress in testing such hypotheses.

I will discuss the structure and history of the animal kingdom in more detail in Chapter 3. For now, the main message to emerge is twofold: (a) animal evolution was more 'creative' before 500 *my* ago than since then, when it has been rather conservative, so there is a broad parallel with development, as suggested at the outset; but (b) there is considerable uncertainty over the number, nature and timing of early creative 'explosions'. This uncertainty has implications for our choice of approach to the question of how body plans arose, and, more generally, how developmental systems evolve. The next section briefly explores these implications.

## 1.3    Alternative Strategies

One possible approach to the study of the origin of body plans would be to take a particular such plan – say the molluscan one – and to attempt what Eldredge (1979) calls an adaptive scenario. That is, we try to explain the origin of molluscan morphology in terms of the environmental conditions prevailing at the time, for example an increase in predation or a change in sea level. We then repeat the process piecemeal for every other body plan origin. Environmental conditions can change markedly in 5,000 years, let alone 50 million or 500 million, which, as we have seen above, represent possible degrees of error. It would seem essential that an alternative strategy for approaching the problem is adopted.

In fact, a pure 'adaptive scenario' approach would be undesirable even if there were no uncertainty over the timing of body plan origins, because it tacitly assumes that such origins may be fully explained in terms of adaptation to external environmental conditions. A recurring theme throughout this book is that, while external adaptation is important, so too is internal coadaptation, and the process of internal selection that generates it (Whyte 1965).

I will now outline the alternative approach to the study of the origins of body plans (and the evolution of development more generally) that will be adopted herein. This outline will serve to elucidate the structure of the 'core' of the book (Chapters 5 through 11: see Preface).

We start by focusing on the development of selected present-day animals, and in particular on the very early developmental stages which represent the ontogenetic (as opposed to evolutionary) origin of the relevant body plan (Chapter 5). We then look at the genes involved in early development, and what they do (Chapter 6). The approach then becomes explicitly comparative, and we address the question of the extent to which equivalent (homologous) genes are involved in diverse taxa (Chapter 7). Next, we ask (a) what kinds of mutation might have been involved in generating the intertaxon differences we observe – gene duplication, exon shuffling, standard point-mutation and so on (Chapter 8); and (b) what kinds of population-level process might have permitted their spread (Chapter 9). We then take a brief look at the relative importance of, and the interplay between, creative and destructive forces (Chapter 10). Finally, we broaden out and look afresh at the overall relationship between ontogeny and phylogeny (Chapter 11).

Clearly, this kind of approach is intended to throw light on the general principles underlying the origin of body plans, rather than provide detailed explanations of individual cases. There is no point in attempting to explain why certain features of 'body plan *x*' arose 530 *my* ago if in fact they may equally have arisen 1060 *my* ago under entirely different biotic and abiotic conditions.

## 1.4    Creation versus Destruction

As I indicated in the opening section, our current (neo-Darwinian) theory of evolution is incomplete. Let us now look at this problem from another angle. In fact, neo-Darwinian theory is incomplete even when assessed against its own internal criteria. The essence of the neo-Darwinian view is that the evolutionary process is of a two-fold nature, involving the production of organismic novelties (of whatever sort) ultimately by mutation and the sieving of these by natural selection. (Optional extras to this world-view, depending on how strict a neo-Darwinian you might wish to be, include micromutation and associated gradualism, and also selection operating at the level of the individual or family rather than at the level of the group or species.)

The main problem with neo-Darwinism in its current form is that its theoretical structure is extremely lopsided. There has been sustained development of quantitative models of the action of selection, from the pioneering work of Fisher (1930), Haldane (1932) and Wright (1931) up to recent work such as that of Charlesworth (1994); while the mutational and developmental production of the variants being sieved by selection has continued to be treated by too many evolutionists as a 'black box', despite the numerous advances that have been made in developmental genetics in recent years (see Chapter 6). Essentially, the individual and population levels have been treated as quasi-independent. The fitnesses of mutant genotypes have been considered to be crucially important in models of selection, while the ways in which fitness effects are produced (for example in terms of altered cell proliferation patterns in development) have been largely disregarded.

This situation should of course be considered undesirable by all evolutionary biologists, including the strictest of neo-Darwinians, but how serious a problem the lack of a mutational/developmental component of evolutionary theory is perceived to be depends on the extent to which the 'perceiver' is a gradualist. If, despite the views put forward herein, *all* evolution proceeds through the accumulation of very minor variations – an extreme view popularized by Dawkins (1986) – then it may not be too much of a deficiency in the theory to simply assume that mutation perpetually generates morphologies that are slight variants on the existing form. But to anyone proposing the existence of one or more radical morphogenetic phases in evolution, the need for an adequate picture of the genetic architecture of development and of the ways in which this is altered by mutation becomes compelling. Hence the feelings of dissatisfaction that many evolutionary developmental biologists have with neo-Darwinism. There is nothing *wrong* with elaborate models of selection, but a detailed quantitative statement of how existing types are sorted and selectively eliminated (or held in a state of stable equilibrium) cannot pretend to be a complete theory of evolution.

Ironically, most of the alternative approaches to evolution that have proliferated in the last few decades have allowed the focus on destructive rather than creative forces to persist. The neutral theory of molecular evolution (Kimura 1983) – arguably *within* a broad neo-Darwinian world view – concentrates on the stochastic loss of neutral and nearly neutral alleles produced in an unspecified way by mutation. Punctuated equilibrium (Eldredge and Gould 1972) is a pattern, not a process, and may simply be a geological reflection of the standard neo-Darwinian mechanism of allopatric speciation, although some authors (e.g. Williamson 1981) have suggested otherwise. The recent interest of some palaeontologists in extinction (see Stanley 1987 for a general account) and the question of whether extinction rates are cyclical (Raup and Sepkoski 1984; Patterson and Smith 1987) continues the tradition of focusing on destructive processes – albeit this time high-level ones, and ones that may have a 'morphogenetic aftereffect' through the freeing-up of niche space for new adaptive radiations.

The *only* approach to evolution that has attempted to focus on creative forces has been that of *Evolutionary Developmental Biology*. I use this label (which is due to Hall 1992) to cover the work of a heterogeneous group of biologists including, among others, von Baer (1828), Thompson (1917), de Beer (1930), Goldschmidt (1940), Waddington (1957), Gould (1977a), Raff and Kaufman (1983), Buss (1987), Arthur (1988), Thomson (1988) and Raff (1996). I will give an account of the structure and history of this discipline in Chapter 4; but now I must deal with another crucial area of the overall evolutionary debate, namely systematics, whose interface with neo-Darwinism is of particular importance in understanding the conflicts within present-day evolutionary biology as a whole.

## 1.5    Systematics and the Concept of Natural Classification

More than a decade ago, a leading cladist declared that there was "no need to placate the ghost of neo-Darwinism; it will not haunt evolutionary theory for much longer" (Rosen 1984). This view appears to be shared by many other cladists (e.g. Platnick 1977) and it is important to enquire into the rationale behind it, because this will help to reveal fundamentally different views of what evolutionary theory 'is for', that is, what an adequate theory of evolution should be able to explain.

The population genetics approach to evolution, which is generally neo-Darwinian, is based on studies of changes in gene frequency within populations – often under the influence of natural selection – and extrapolation of the consequences of these changes (and of instances of reproductive isolation) taxonomically 'upwards' to the divergence of daughter species and beyond. The emphasis is on mechanism, not pattern, and on variation at an intraspecific, rather than transspecific, level (see, for example, Lewontin 1974a).

Clearly, this approach can potentially explain the maintenance of diversity *within* species: see Clarke (1975a, b) and, for the neutralist view, Kimura (1983); and it can also explain the existence of a diversity of species, in terms of their adaptive divergence from a common ancestral stock. But at this point we need to ask whether the population genetics approach can explain the existence of particular *patterns* characterizing the diversity of species. This brings us back to the quotation with which this section started.

Systematists focus their interest on elucidating, and interpreting, what can be variously called Nature's hierarchy (Rosen 1984), the natural hierarchy, or the pattern of natural classification (Panchen 1992). This is the pattern of nested homologies that we see in the animal kingdom (and also in plants, where it is partially obscured by frequent occurrences of 'reticulate evolution': Grant 1981), and which Scotland (1992) includes among the axioms of cladistics. For example, the vertebrates as a group share a homologous internal skeleton, but only a subgroup, tetrapods, are characterized by the dactyl limb, and only a subgroup of tetrapods (birds) have wings with feathers. An abstract diagram of this sort of nested arrangement is shown in Figure 1–3. (Note that the tetrapod dactyl limb is not always *penta*dactyl – see Coates and Clack 1990 for an example of early octodactyly.)

Now if we imagine that a pattern of natural classification, or nested homologies, such as that shown in Figure 1–3, is based entirely on adult characters, and that we have no information whatever to elucidate a developmental dimension to the pattern, it can be argued that neo-Darwinism *does* provide a reasonable causal explanation. Indeed, a branching phylogeny such as that shown in Figure 1–4 will produce the pattern depicted in the previous figure, and the main driving force behind the divergences incorporated in the phylogeny may well be local adaptation (driven by natural selection) coupled with reproductive isolation.

Let us now bring in a developmental dimension. It is generally the case, in patterns of nested homology, that the characteristics involved in the 'outer' homology, of greater taxonomic breadth, also appear earlier in development (von Baer 1828; Patterson 1982; de Queiroz 1985 – further discussion in Chapters 2 and 11). If we consider organisms as 'black boxes' subject to random mutation of their character states, and selection as a force that sieves the resultant variation purely in relation to adaptedness to the external environment, there is no reason why such a pattern should have been predicted, and indeed even in retrospect there is no way that it can be explained without bringing in some additional sort of information.

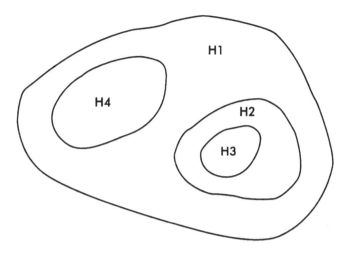

**Figure 1–3  A group of nested homologies, given in the form of a Venn diagram**
The outermost homology (H1) characterizes the greatest number of species. The example given in the text may be represented by H1–H3.

Now this may be an unfairly negative perspective on neo-Darwinian theory. True, neo-Darwinism has, to its detriment, been distinctly 'non-developmental'. Yet there are parts of the theory which, when cast in a more developmental light, may have considerable explanatory power. I would like to focus, here, on an argument developed by Fisher (1930) on the relationship between a mutation's magnitude of phenotypic effect and its probability of being selectively advantageous. Basically, Fisher argued that these two variables were negatively related (see Figure 1–5). While there are some limitations to this approach (Arthur 1984: Chapter 12), the basic idea seems sound, and I will refer to this proposed negative correlation as the 'Fisher principle'.

We need, now, to add a developmental dimension to this principle. It seems reasonable to suppose that genes controlling early developmental decisions, such as which end of the embryo is anterior, will be subject, on average, to mutations with more major phenotypic effects than genes controlling later developmental processes, such as the production of mammalian hair (Arthur 1982a). If so, then under the Fisher principle early genes will evolve more slowly than their later counterparts (Morgan 1932), and, under certain assumptions (Arthur 1988,

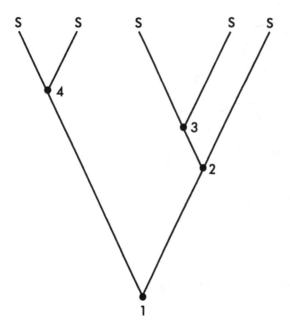

**Figure 1–4 Schematic phylogeny**
Divergences 1–4 correspond to the origins of homologies 1–4 in
Figure 1–3. Extant species are represented by S.

Chapter 2), this will mean that the earlier stages themselves will be
more evolutionarily conserved. In a 'typical' lineage, then, most evolution will concern late developmental stages; and this kind of evolutionary change is capable of explaining a pattern of nested homologies
wherein the taxonomically broadest characters are also the embryolo-
gically earliest. There are many complexities to this argument, for
example the reuse of genes at different stages of development. I will
examine these in subsequent chapters; but essentially what I am proposing here is that Evolutionary Developmental Biology has the potential to form a bridge between population genetic processes and
systematic patterns; and thus to help unify evolutionary biology in
general.

There is, however, a problem. Those genes that control key early
developmental processes are involved in the establishment of the basic
body plan. Mutations in these genes will usually be extremely disadvantageous, and it is conceivable that they are *always* so. (I exclude,
here, phenotypically silent mutations.) But it is clear, for example from
cross-taxon comparisons of a particular group of such genes – the
homeobox-containing *Hox* genes (Garcia-Fernàndez and Holland

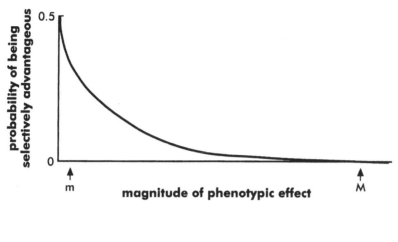

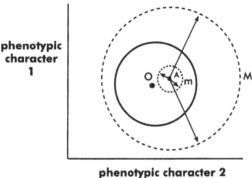

**Figure 1–5  The Fisher principle**
The top diagram illustrates the proposed negative relationship between a mutation's magnitude of phenotypic effect and its probability of being selectively advantageous. The underlying rationale (bottom) is that very small-effect mutations (m) have almost a 50 percent chance of taking the actual phenotype (A) closer to the optimum (O), while large-effect mutations (M) have no chance of doing so. (There is an underlying assumption that mutations are equiprobable in all phenotypic directions. This may well be wrong: see Chapter 10.)

1994) – that homologous 'developmental' genes in different phyla are different. Even where there is an unexpectedly high degree of similarity (e.g. 98 percent between some human and *Drosophila* homeoboxes – Boncinelli *et al.* 1985), there is still a difference. The genes concerned *have* evolved. And comparative embryology shows that early developmental pathways *have* changed in consequence. Does this mean that mutations of these genes are sometimes advantageous? If so, are these in some sense successful macromutations? There is a can of worms here, which has been opened with generally negative consequences on several occasions, most notably by Goldschmidt (1940). But we cannot bypass this issue, however unpopular, if we are to do evolutionary developmental biology; and I will deal with it in the following section.

## 1.6 Micromutation versus Macromutation

"There is no such thing as incipient species. Species and the higher categories originate in single macroevolutionary steps as completely new genetic systems." Thus said Goldschmidt (1940), and thereby excluded developmental considerations from the modern synthesis much more effectively than the essentially non-developmental approach of many previous and subsequent neo-Darwinians. Goldschmidt was pro-development; he dealt explicitly with mechanisms rather than patterns; and he was wrong – or was he?

It now seems clear that Goldschmidt was indeed wrong regarding the nature of the typical speciation event. But it may yet turn out that his much-ridiculed concepts of systemic mutations and hopeful monsters – or ideas not too far removed from them – have a role to play in relation to the origin of major body plans.

Goldschmidt's views on the role of macromutations in evolution were extreme – but at least he stated them clearly. It is much harder to find a clear statement of the neo-Darwinian view of the role of macromutations, beyond a general impression – perhaps deriving ultimately from the Fisher principle – that such mutations are not of importance in evolution. Dawkins (1986) – whose views are equally extreme but in the opposite direction to Goldschmidt's – gives what at first sight seems a clear statement of the neo-Darwinian view. He describes evolutionary theories "that depend upon macromutation" as saltational theories, and goes on to say that there are "very good reasons for rejecting *all* such saltationist theories of evolution" (my italics). Of course, since macromutation is undefined, and the interpretation of "depend on" obscure, these statements are actually rather vague. But they do serve to emphasize the general neo-Darwinian negativity on macromutations.

The position of ecological genetics (Ford 1971), which represents one strand of neo-Darwinism, is particularly interesting. Historically, this stemmed from Fisher and it incorporates his rejection of an important evolutionary role for macromutation. Yet many of the case-studies of ecological genetics, for example those relating to selection on polymorphic variation in *Cepaea* (see Jones *et al.* (1977) for a review), deal with mutations that have individually large effects – usually on pigmentation – and are quite unlike the 'polygenes' of quantitative genetics that underlie such continuously variable characters as organismic size, shape and weight.

What are we to make of this? I would argue that most neo-Darwinians accept that some mutations with large effects on the phenotype do after all contribute to evolution, but only in special cases: particularly pigmentation, but some others too, such as symmetry reversals (see Clarke and Murray 1969). What is denied is both a *general* role in speciation (a denial with which I agree) and an 'important but infrequent' role in relation to the origin of body plans, on which, for the moment, I will reserve judgement.

This is probably a fair reflection of neo-Darwinian views generally, and not just those prevailing within the sub-discipline of ecological genetics. Coyne (1983) concludes from a genetic analysis of morphological differences between *Drosophila* species that the speciation events through which the *D. melanogaster* sibling group arose involved polygenic-type divergences rather than macromutations (but see also Orr and Coyne 1992). He is appropriately cautious about extrapolating this conclusion to the origin of more major body plans, but he nevertheless states that "there is currently no evidence that major and minor taxonomic bifurcations involve different mechanisms of genetic change". I will argue in Chapters 9 and 11 that this view is wrong.

Let us now return to the question of what constitutes a 'macromutation'. This is a difficult issue, and in fact *most* users of the term have failed to define it – it is not just Dawkins (1986) who can be criticized in this respect. My response to this problem, however, is not to attempt a definition but rather to argue that the view of a dichotomy between micro- and macromutation is an unenlightened one that provides a particularly poor basis from which to make progress in understanding how the genetic architecture of development is modified in the course of evolution.

There are three problems (at least) with the micro/macro dichotomy. First, if we were to arrange all known mutations on a scale of 'magnitude of phenotypic effect' we would find a complete spread; the distribution is more or less continuous (as indeed is implicit in Figure

1–5) and it most certainly does not take the form of two discrete clusters, one of small and one of large effect. Second, we need to take on board Bonner's (1974) point that the phenotype is four-dimensional. Rather than focusing specifically on the adult, we need a time-extended view of a mutation's effects – both the span of developmental time over which an effect can be detected at all, and the way in which the magnitude of effect varies over that span. Third, we need to attempt to clarify different *types* of effect. Is a mutation reversing the chirality of a gastropod shell more or less 'macro' than one that turns dipteran halteres into wings? This is a meaningless question. Both mutations have major effects, but they are very different in kind.

These deficiencies in the micro/macro dichotomy are of course linked to the lopsided nature of current neo-Darwinian evolutionary theory, mentioned above. That is, we have been content with rather basic classifications of mutations, while we have developed ever more elaborate conceptions, and corresponding models, of selection. One aim of this book is to correct this lopsidedness by beginning the formulation of a developmental classification of mutations. I will do this in two stages. First, in the following section, I will emphasize the developmental time dimension of the magnitude of a mutation's effect. Second, in Chapter 8, I will examine the *types* of developmental effect mutations have; and also the different molecular causes of mutation: base substitution, transposable element insertion, exon shuffling, gene duplication, and so on. It remains to be seen whether in the end the book will be perceived as reinforcing neo-Darwinism, by contributing to its missing developmental component (which is how I see it), or as being anti-neo-Darwinian because some of the conclusions reached about the processes involved in the origin of body plans are unpalatable to a gradualist/externalist view.

## 1.7    Developing Organisms as Inverted Cones

My starting point for the formulation of a developmental time-extended view of a mutation's effects is the study by Wieschaus and Gehring (1976) of cell lineages in the development of *Drosophila*. This study may initially seem a rather odd starting point, since it involves somatic recombination rather than germline mutation, but its relevance will gradually become apparent.

A developing organism is essentially a proliferating clone of cells all ultimately derived from division of the zygote. At any stage in development, we can identify a particular cell and then look at the descendant cells to which it gives rise in later development and, ultimately, in the adult. This group of cells is a subclone within the overall clone represented by the whole developing organism. (There is a broad parallel here with clades-within-clades in phylogeny; I will return to this point in Chapter 11.)

What Wieschaus and Gehring (1976) did was to use a genetic technique to mark particular clones. Using a stock of *Drosophila* heterozygous for the mutation *mwh* (*multiple wing hairs*), they X-irradiated at different stages of development. Occasionally, an X-ray 'hit' causes somatic recombination, resulting in a single cell with a homozygous genotype (*mwh/mwh*). The 'offspring' cells resulting from proliferation of this single homozygous cell are also of the genotype *mwh/mwh*, and so ultimately this clone is identifiable, in the adult wing, as a dark patch in which each cell gives rise to several hairs as opposed to the usual one (see Figure 1–6).

The size of the visible dark patch reflects the number of cells in the marked clone at the adult stage. By irradiating at different stages of development, Wieschaus and Gehring were able to show that earlier production of a homozygous progenitor cell produced larger marked clones, and this makes intuitive sense. (It is assumed that the *mwh* genotype of a cell does not affect its rate of proliferation.)

The wing is often used in such studies because it reduces morphological patterns from three dimensions to two. I now wish to generalize from these results, however, and this produces the problem that a time-extended view of three-dimensional morphology takes us into four dimensions and, consequently, considerable difficulty in picturing – either mentally or on paper – the processes involved. To get around this, I intend to use an abstract picture of an organism as a two-dimensional disc, with the third dimension being developmental time. A developing organism thus becomes an inverted cone, as in Figure 1–7. This simplification has some costs, but it also has certain advantages, as will shortly become clear.

We can now picture early-induced (large) and late-induced (small) marked clones in a generalized, albeit abstract way rather than in the particular context of *Drosophila*. Now we need to make another transition – from somatic recombination to germline mutation. In a phylogeny of proliferating lineages of 'cone creatures', lineages become

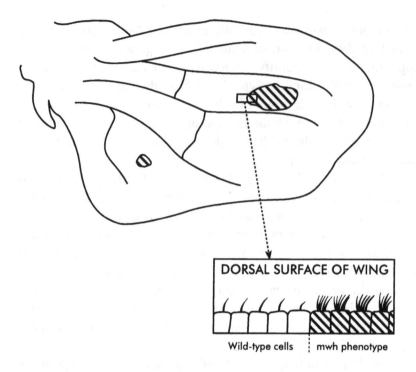

DORSAL SURFACE OF WING

Wild-type cells | mwh phenotype

**Figure 1–6   Genetically marked clones of cells in a *Drosophila*
wing**
Large and small marked clones (top) resulting respectively from
early and late production of clones of cells exhibiting the
multiple wing hair phenotype (see box).

morphologically different only if one (or some) undergo a germline
mutation of a gene controlling development – for example, one
which makes affected cells 'rough' rather than smooth. Now although
the mutant gene will be present in the zygote, it may be inactive. It is
apparent from extensive studies of *Drosophila* (see Ingham 1988;
Lawrence 1992 for reviews) that genes with a controlling role in development have characteristic periods of action: some early, some late;
some brief, some prolonged; some 'single-pulse' and some intermittent
(see Figure 1–8). In general, those genes with earlier onset of activity
can potentially alter the morphology of a larger sub-clone than genes
whose activity begins later – assuming for the moment that, like *mwh*,
the gene concerned is 'cell autonomous' in its effects (meaning that
these effects are restricted to those cells in which the gene is active). In
other words, the same link between magnitude of phenotypic effect in

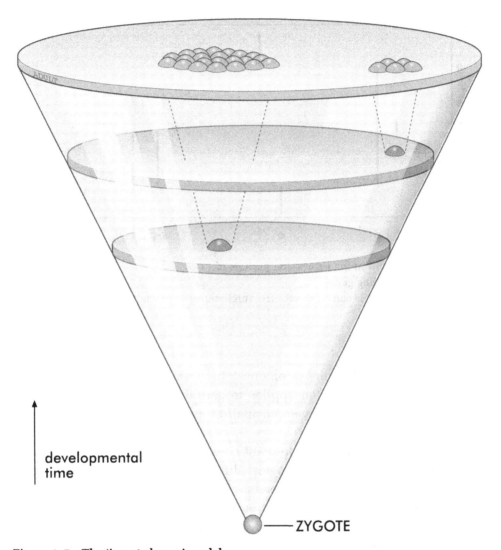

developmental
time

ZYGOTE

**Figure 1–7  The 'inverted cone' model**
A hypothetical organism develops from zygote to a disc-shaped
arrangement of cells which grows in size through cell
proliferation. The size of a morphological change in the adult is
determined by how early in development the mutant gene
controlling the change is switched on: early onset of activity
gives rise to a more extensive change.

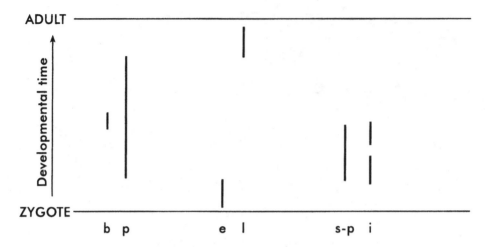

**Figure 1–8  Variation in the expression periods of development-controlling genes**
Brief (b), prolonged (p); early (e), late (l); single-pulse (s-p) and intermittent (i).

the adult and earliness of onset in development that we saw for somatic recombination applies to germline mutation, and hence to the appearance of developmental/morphological novelties in evolution.

This approach has important consequences for the neo-Darwinian view that major morphological differences between organisms from different higher taxa have been produced through the gradual accumulation of very small changes. Specifically, while in a non-developmental approach it seems plausible that many small changes can indeed accumulate to give a larger one, in a developmentally explicit approach it is clear that *many late changes can not accumulate to give an early one.* Thus if taxonomically distant organisms differ right back to their early embryogenesis, as is often the case, the mutations involved in their evolutionary divergence did *not* involve the same genes as those involved in the typical speciation event, where usually the early embryogeneses of the daughter species concerned are virtually identical.

There are, of course, all sorts of complications to the argument, including the fact that some development-controlling genes make diffusible morphogens. The morphological effects of such genes will be much more widespread than those of cell autonomous genes. I will postpone consideration of such complexities to Chapters 5 (morphogens) and 8 (mutations). Other complications to the argument – including the challenges posed by complex life cycles – will be dealt with in Chapters 9 and 11. But even this brief discussion of the time dimension of a mutation's phenotypic effect has been useful in exposing the inadequacy of classifying mutations as 'micro' and 'macro'. To make progress in understanding the evolution of development, including the origin of major body plans, a much more sophisticated view of mutation is essential. A clear view of what is meant by 'body plan' is also essential: this issue will be discussed in the following chapter.

# What is a Body Plan?

## 2.1  Body Plans and Taxonomic Levels

Body plans are easy to exemplify but difficult to define. It is generally accepted that higher taxa such as phyla and classes are characterized by unique body plans, while lower ones, such as genera and species, are not. Frequent references are made to, for example, the vertebrate body plan, the insect body plan, the molluscan body plan, and so on. But few biologists would consider a pair or group of congeneric species as differing in body plans – rather such species typically *share* their body plan with each other and with numerous other close relatives.

Let us consider one example of closely-related species in each of the three above-mentioned higher taxa. The genus *Geospiza* ('Darwin's finches') contains several species and has been the subject of considerable study over many years (see Lack (1947) and Grant (1986) for reviews). These, like other birds, clearly exhibit both vertebrate and avian body plans; we do not generally think of a geospizid 'plan'. The flour beetles *Tribolium confusum* and *T. castaneum* – much studied from an ecological perspective (Park 1948, 1954, 1962) – are characterized by insect and coleopteran body plans rather than a specifically *Tribolium* one. The land snails *Helix aspersa* and *H. pomatia* are typical gastropod molluscs in their body plan (see Kerney and Cameron 1979 for descriptions), and while they belong to the family Helicidae, we would not so readily categorize their body plan as being a helicid one.

Two general observations emerge from consideration of these specific examples. First, the body plan that comes to mind for a particular species is invariably one which characterizes a higher taxon, but the exact level varies. (This variation is hardly surprising; higher taxa are inconsistently used across the animal kingdom.) For the *Geospiza, Tribolium* and *Helix* examples, respectively, we ended up with body plan descriptions that referred to subphylum/class, class/order, and phylum/class, although the levels of taxon in the *Tribolium* case are subject to the continuing debate on the high-level taxonomy of the arthropods (Wheeler, Cartwright and Hayashi 1993; Boore *et al.* 1995; Friedrich and Tautz 1995; Popadíc *et al.* 1996). Second, it is interesting to note that the common names for animals used by the 'layman' are often closely associated with one of the taxonomic levels at which biologists would characterize the appropriate body plan: 'bird', 'beetle' and 'snail' in our examples here. This is no accident; body plans refer to major structural features, and it is often the case that these are particularly obvious to the non-professional observer. There are, however, exceptions; the notochord is central to the chordate body plan but it is certainly *not* an obvious feature.

We need, at this stage, to turn to the problematic issue of how 'body plan' should be defined. The above discussion of selected examples clearly assumed that the reader has some sort of mental picture of what a body plan is – perhaps as a result of previous exposure to high-level homologies (see Figure 2–1). The recent outpouring of work on the relationship between *Hox* genes and body plans (e.g. Molven *et al.* 1990) also makes such an assumption; the authors concerned rarely pause to consider what the term means – though phrases such as "bilateral body plan" (Bartels, Murtha and Ruddle 1993) and "segmented body plan" (Gutjahr, Frei and Noll 1993) give us some clues. There is no generally-agreed definition of body plan (or its synonym *Bauplan*), but a good starting point in discussing this issue can be found in Valentine (1986):

> At the upper levels of the taxonomic hierarchy, phyla- or class-level clades are characterized by their possession of particular assemblages of homologous architectural and structural features...it is to such assemblages that the term Bauplan is applied.

It is clear from this comment that Valentine sees body plans as being essentially morphological in nature. (It is doubtful if there is any behavioural equivalent of the concept; a physiological equivalent is easier to imagine, but has not been elaborated in the literature.) Valentine also makes it clear that he uses *Bauplan* to characterize the morphol-

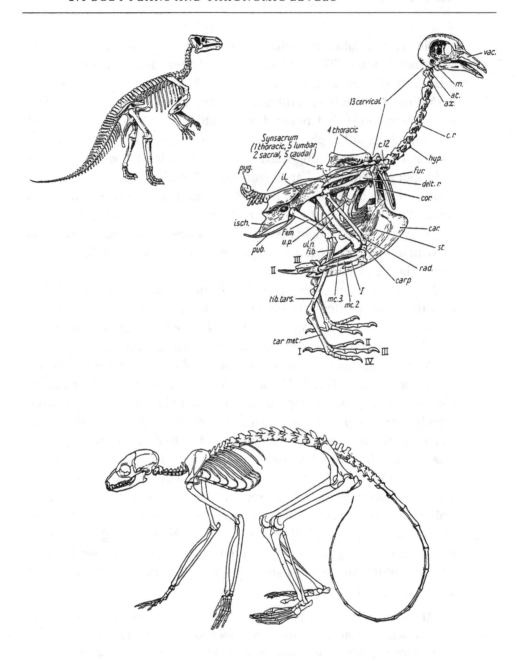

**Figure 2–1   Skeletal homology in the tetrapods**
These skeletons of a dinosaur (*Iguanodon*), a bird (*Columba*) and a
primate (*Lemur*) show numerous homologies of specific bones, e.g.
the radius and ulna of the forearm/wing (marked *rad.* and *uln.* on
the bird skeleton). (From Young 1962.)

Table 2–1. Some characters and character states useful in the delineation of body plans

| Character | Main character states |
| --- | --- |
| Skeleton | Hydrostatic, internal, external |
| Symmetry | Bilateral, radial, asymmetric |
| Pairs of appendages | 0, 2, 3, 4, many |
| Body cavity | Acoel, pseudocoel, coelom |
| Cleavage pattern | Spiral, radial, syncytial |
| Segmentation | Segmented, unsegmented |

ogy of "entire ontogenies", not just the adult stage. Other points that emerge from his working definition are the link with homology (see Section 2.2) and the association with high-level taxa that we have already noted.

It is useful to consider precisely what sorts of morphological characteristics may be considered as major features of particular significance in the characterization of phylum/class-level body plans. In fact, considerable headway can be made using as few as six major characters and only two to five states of each: these are shown in Table 2–1. Many animal body plans are immediately identifiable as unique combinations of these character states. For example, a segmented animal with bilateral symmetry, an internal skeleton, a coelom and two pairs of legs, whose earliest embryogenesis is through radial cleavage, is clearly a tetrapod (a 'superclass' of vertebrates with its origin in the Devonian period: see Ahlberg 1995). Equally, another bilaterally symmetrical, coelomate and segmented animal but this time with an external skeleton, three pairs of legs (in the adult stage) and a syncytial cleavage pattern cannot be other than an insect.

For some phyla – such as the various groups of unsegmented worms – and for classes and orders, it is necessary to invoke other characters and an appropriate array of character states for each. However, the central point remains the same: *characters used in the delineation of body plans are major features of the overall layout of the adult body or of the developmental trajectory giving rise to that adult body.* Relatively superficial characters, such as pigmentation patterns or quantitative variation in size or shape, are *not* useful in this respect.

Ironically, the concept of a body plan also becomes less useful *above* the level of the phylum. It has frequently been pointed out that interrelationships between phyla are usually obscure (Barnes

1984) and that many proposed superphyletic groupings are dubious (Willmer 1990). Even where there is agreement on such groupings, the structural homologies spanning across their constituent phyla are few and relatively unimpressive (Løvtrup 1974): an example is radial cleavage in the deuterostome phyla (Chordata, Hemichordata, Echinodermata).

This leads to another point – that 'body plan' refers to homologies at the 'supercellular' level – that is, to embryological processes and adult structures involving *populations* of cells. All animal phyla share cellular and subcellular homologies (the genetic code, for example), but that is not particularly relevant to discussions of body plans. Also, and related to that point, it must be emphasized that a body plan is an actual spatiotemporal layout of cells, not the genetic information underlying such a layout. So neither the 'zootype' of Slack *et al.* (1993) nor any other pattern of gene activity represents a body plan; rather such patterns may form part of the *explanation* of body plans – a point to which we will return later. As Panchen (1992) has pointed out in a broader context, it is always important, in scientific discussion, to distinguish between the thing for which we seek an explanation (the *explanandum*) and the explanation itself (*explanans*).

The picture that emerges, then, of the link between body plans and taxonomic levels, is shown in Table 2–2. The major body plans of the animal kingdom characterize the 35 or so existing phyla (see Margulis and Schwartz 1988) and perhaps others that were formed in the Cambrian/pre-Cambrian 'explosion' but are now extinct (see Chapter 3). Classes and orders represent what Valentine (1986) calls *Unterbaupläne*. In general, neither low-level taxa nor those above the phylum level can be characterized by their possession of particular body plans.

Of course, body plan characters have to originate somewhere, and ultimately such origins have to be understood within the context of the genetic and phenotypic structure of populations of particular species. So while species themselves are not usually characterized by particular body plans, there must be very occasional instances where they are; and indeed instances where some individuals within a species are characterized by possession of a new body-plan feature. This is rarely observed, both because of its low frequency and because of the concentration of body-plan origins in the distant past. However, it could be argued that the sinistrality of the gastropod family Clausilliidae represents a rare case of a family-level *Unterbauplan*, and that the

**Table 2–2. Relationship between body plans and taxonomic levels**
(Parentheses indicate the less clear-cut cases.)

| Taxonomic level | Characterized by a body plan? | Examples (where appropriate) |
| --- | --- | --- |
| Species | No | – |
| Genus | No | – |
| Family | (No) | – |
| Order | (Yes) | Diptera, halteres |
| Class | Yes | Bivalvia, hinged shells |
| Phylum | Yes | Chordata, notochord |
| Superphyletic group | (No) | – |

polymorphism of chirality in *Partula suturalis* (Clarke and Murray, 1969; Johnson, Murray and Clarke 1993) in recent populations on the island of Moorea (now apparently extinct due to a misconceived biological control programme: Clarke, Murray and Johnson 1984; Murray *et al.* 1988) represents a 'model' of how such an *Unterbauplan* might originate.

I will end this section by summarizing (in Table 2–3) the seven major characteristics that together define the nature of body plans (and their 'opposites' also for increased clarity). This forms a definition of sorts, and those who like their definitions in the form of sentences can easily construct one from the information in the table. As will be apparent, the characterization of the nature of body plans given here is more detailed than, but entirely compatible with, Valentine's (1986) working definition with which we started. Anyone wishing to delve further into the history of the body plan concept should consult chapter 5 of Hall (1992).

## 2.2 Body Plans, Cladograms and Homology

It will be clear from the preceding section that the view of body plans accepted herein renders them a subset of morphological homologies generally. It must be emphasized, however, that this is not a unanimous view. Some authors have considered body plans to be particular morphological layouts, types or grades *regardless of their evolutionary origin*. For example, Willmer (1990) clearly suggests that a particular body plan may arise convergently from two or several

Table 2–3. The seven major characteristics that together 'define' the nature of body plans

| Characteristics of body plans | 'Opposites' |
| --- | --- |
| High taxonomic level | Low |
| Morphological | Behavioural, physiological |
| Homology | Homoplasy |
| Major 'layout' features | Minor 'superficial' features |
| Supercellular | Subcellular |
| Actual structure/process | Coding/information base of |
| Three/four-dimensional | Three-dimensional only |

different groups: "Convergent features range from cellular structures, to whole cell types, to multicellular organs and features of body plans. Even the whole animal *Bauplan* can be convergent".

This kind of usage of the term *Bauplan* (or body plan) is unhelpful. It is true, of course, that the concept of a body plan is a 'grade' or 'type' issue; but it is that *as well as*, not instead of, a clade issue. A particular body plan characterizes a particular major clade. If distantly related clades converge in some major structural feature, then we may find a situation in which unrelated body plans are rather similar – but they will generally not be the same. For example, members of the phylum Brachiopoda and of the molluscan class Bivalvia both possess hinged two-valve shells, but the former are dorsoventrally opposed while the latter are laterally opposed (see Figure 2–2). Here, as in all cases of high-level convergence, the similarity of the convergent forms is incomplete, and it is not profitable to think of them as representing the same body plan.

In his pioneering work, Hennig (1966, 1981) distinguished not only between the commonly recognized categories of monophyletic and polyphyletic groups, but also between both of these and the previously unrecognized category of paraphyletic groups (see Figure 2–3). Examples of these, respectively, are the insect order Diptera, the sponge 'order' Lithistida (Barnes 1984) and the vertebrate 'class' Reptilia (see Figure 2–4 for the rationale behind the last of these designations).

On the basis of the preceding arguments, high-level monophyletic groups will be characterized by particular body plans, but polyphyletic groups will not be. What about paraphyletic groups? For example,

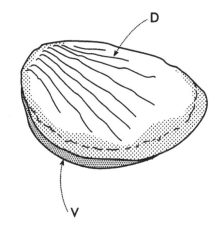

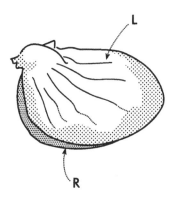

**Figure 2–2  Generalized external morphology of a brachiopod (top) and a bivalve mollusc**
D = dorsal, V = ventral, L = left, R = right

could we reasonably refer to a 'reptilian body plan'? The answer has to be 'yes' – albeit this may not please cladistic purists such as Smith (1994, p.105) who asserts that "paraphyletic groups serve only to mislead in the analysis of evolutionary patterns". In most cases, where we associate a particular body plan with a particular monophyletic group, there are exceptions. There are tetrapods without legs (snakes), snails without shells (slugs) and sponges without choanocytes (carnivorous cladorhizids: Vacelet and Boury-Esnault 1995). The birds, as a group of 'reptiles without scales' (legs excepted, that is) are simply a 'bigger' example of the same phenomenon – and, indeed, such comments on group size are highly time dependent; birds were a tiny group at the time of *Archaeopteryx*.

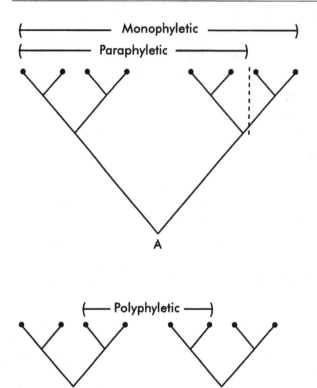

**Figure 2–3  Monophyletic, paraphyletic and polyphyletic groups**
Dots represent taxa; lines indicate evolutionary descent. Note that
group-type can depend on the frame of reference: if we insert a
common ancestor *D* of *B* and *C* then the polyphyletic group
illustrated becomes paraphyletic.

With regard to the nesting of homologies – and in particular the
nesting of *Unterbaupläne* within the major phylum-level body plans –
I should start by making an important distinction between the aims of
systematics and those of evolutionary developmental biology. As we
have already seen, systematists are concerned with elucidating the
pattern of groups-within-groups that constitutes 'natural classification'.
The goal for evolutionary developmental biologists is to explain the
causality of this pattern, through consideration of how natural selection (the domain of population genetics) interacts with the genetic
architecture of development (the domain of developmental biology).
Of course, in the process of doing this, information may well arise that

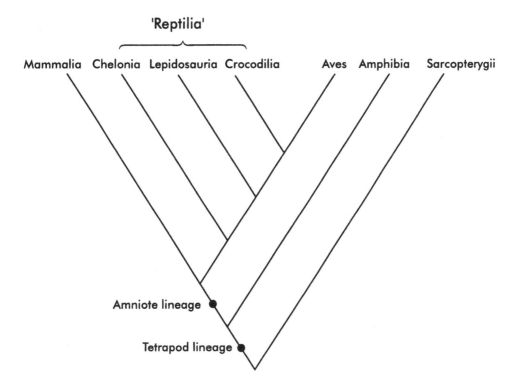

**Figure 2–4  Tetrapod cladogram showing the paraphyletic nature of the reptilian grouping**
The Sarcopterygii (e.g. lungfish, coelacanth) represent the 'sister group' of the tetrapods.

alters our view of some particular systematic/phylogenetic pattern. But it is not the aim of Evolutionary Developmental Biology to attempt to compete with systematics in the latter's own territory.

The ideal way to proceed, then, would be to concentrate on seeking a causal explanation for those nested patterns of *Baupläne* and *Unterbaupläne* that are reasonably well established – those within the vertebrate subphylum, for example. As systematic research progresses, and a broad consensus is reached on the correct patterns for other major groups, then the causal analysis of these too can be tackled. If it were possible to proceed in this way, then it could be argued that evolutionary developmental biologists need not delve into the methods by which nesting patterns are established; they could simply take the resultant patterns for granted.

Of course, reality is somewhat more messy than the above scenario, and consequently it *is* sensible to include here an account of systematic methodology. However, this will be brief; I refer readers wishing to pursue this topic in more depth to Panchen (1992) for a general review. The brief account below focuses on *cladistic* methodology, as it is now clear that, from a phylogenetic perspective, this is superior both to phenetics and to the old-style 'intuitive' approach; although phenetics (see Sokal 1986) remains useful in the analysis of morphological 'disparity' (Chapter 3) and the quantification of morphospace (Chapter 10).

It is difficult for those of us not directly involved in systematic research to put ourselves in the systematist's position. We need to temporarily abandon mental pictures of all forms of branching diagrams, whether cladograms or trees, as these become the goal, not the starting-point. In their place, we need to picture simply a group of species, each with an array of characters and character-states. One essential part of our task, then, is to establish which characters or states are likely to be 'primitive' for the group concerned (and so to have been found in its ancestral stem species) and which are likely to be 'derived' (that is, produced by modification of the stem group character state in one or more lineages of the proliferating clade).

There are four methods for distinguishing primitive and derived morphological (and other) features that are, or have been, in widespread use: the commonality principle (Kluge and Farris 1969; Eldredge 1979); stratigraphy (Fortey and Chatterton 1988); the outgroup method (Watrous and Wheeler 1981); and the ontogenetic method (Nelson 1978). I will describe the first three of these below (with particular reference to body-plan features), but will defer consideration of ontogeny to the following section.

The basic argument underlying the commonality principle is that features that are commoner, in the sense that they are found in more species, will generally be more primitive (i.e. have earlier evolutionary origins) than those that are less common. This approach has been adopted by many systematists, for example Estabrook (1972) and Ekis (1977). Estabrook makes the point as follows: "the relatively more primitive states are likely to be distributed more generally throughout the group under study".

Although there are some cases in which this approach works, it has been heavily criticized by Watrous and Wheeler (1981), and there are clearly situations in which its use would result in making an incorrect

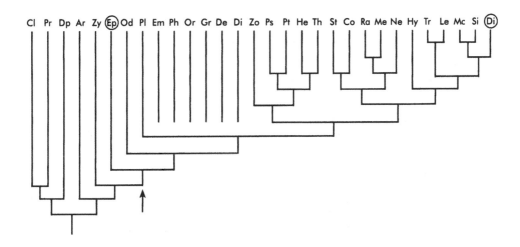

**Figure 2–5   A proposed phylogeny of the insects**
Orders are given in abbreviated form. The pterygote orders include
all those from Ephemeroptera to Diptera (circled). See text for
further explanation. Modified from Kristensen 1981. *Ann. Rev.
Entomol.*, **26**, 135–157.

interpretation of a particular taxonomic dataset. These situations often
arise when we are considering a pair of alternative character states. For
example, consider the distribution of winged and wingless insects, as
shown in Figure 2–5. The arrowed branch-point represents the origin
of winged insects (Pterygota). So the run of twenty-five orders from
Ephemeroptera (Ep) to Diptera (Di) generally includes insects with
wings – although there are, as usual, some cases of secondary loss;
while the wingless insects are represented by Collembola (Cl),
Protura (Pr), Diplura (Dp), Archaeognatha (Ar) and Zygentoma (Zy).

If we were entirely ignorant of insect phylogeny, and used the
commonality principle to deduce which state was primitive and
which derived, we would, of course, come to precisely the wrong
conclusion. We are quite certain that winglessness is the primitive
condition, despite its appearance in only a tiny fraction of current
insect species. Our conclusion, then, is that the commonality principle
can easily mislead, and its use – particularly with alternative character
states – should be avoided.

While the use of 'commonality' is flawed in principle, the use of
stratigraphic information can be troublesome for practical reasons –
stemming from the incompleteness of the fossil record. Given a com-

plete stratigraphic sequence, the direct use of this information to decide which character state was primitive would be the obvious way to proceed. Although there are some cases in which reality comes tolerably close to this ideal situation, they do not include body-plan origins, where stratigraphic information is usually very limited (Fortey and Jefferies 1982). In this context, then, stratigraphy should be regarded as at best providing supplementary information; on its own it cannot be expected to enable us to make sound decisions on which character states are primitive and which derived.

In general, the outgroup method (Watrous and Wheeler 1981; Maddison, Donoghue and Maddison 1984) is preferable to either of the above two, although it requires at least a rough idea of what the pattern of phylogeny is in order to proceed, and so it includes an element of circularity. The basic idea underlying this method (see Figure 2–6) is that to determine which of a pair of character states is primitive within a particular group of species, we examine which state is present in an 'outgroup' that is more distantly related to the group in focus (the 'ingroup') than the ingroup's constituent species are to each other; however within that constraint, the outgroup chosen should generally be the ingroup's closest relative. In general, the state found in the outgroup will be the primitive one.

So far, I have dealt with the overall issue of determining systematic patterns as if, in any one case, we only had access to data on a single type of character. However, in practice we often have access to a whole range of data, including, increasingly, both morphological and molecular data, and potentially many separate characters within each category. Unfortunately, different characters often suggest different systematic patterns as a result of convergent evolution, and the initial naive expectation that molecular data would be sufficiently free from this problem to render phylogenies 'obvious' has been shown to be unwarranted (see, for example, Springer and Krajewski 1989).

Given conflicting suggested patterns from data on different characters, one way to deal with this is to use 'parsimony' (Panchen 1982; Maddison *et al.* 1984; Stewart 1993). Here, we provisionally accept whichever cladogram minimizes the amount of convergence that we are forced to postulate. This can be a highly complex matter when many characters are involved – and, indeed, there are different *kinds* of parsimony (see Harvey and Pagel (1991), chapter 3); but it is greatly facilitated by the wide availability of software packages such as PAUP (Phylogenetic Analysis Using Parsimony: Swofford 1985)

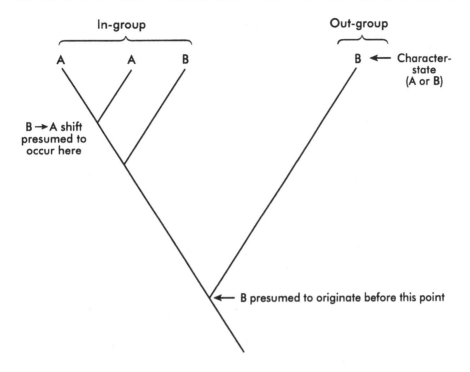

In-group          Out-group

A      A      B          B ← Character-state (A or B)

B → A shift presumed to occur here

← B presumed to originate before this point

**Figure 2–6  The out-group method of determining which of a pair of character states observed within a group of species (the in-group) is primitive and which derived**
In this case, observations on the out-group suggest that B is primitive; note that use of the commonality principle on the in-group alone would have suggested that A was primitive.

I have only been able to devote a small amount of space to methods of cladogram production, and hence phylogeny reconstruction, and have omitted many points of technical detail (multiple versus binary character states, the special case of 'absence' as a character state, the question of how to deal with fossil data (Smith 1994), and so on). However, no one should underestimate the importance of ongoing systematic research in this area. The cases in which we have a clear and agreed picture of the nesting of *Unterbaupläne* within phylum-level body plans, and of the relationship of those major body plans to each other, still represent a small fraction of the overall pattern, even if we only consider the thirty-five or so phyla of the present-day animal kingdom and their first-level *Unterbaupläne*. Vertebrates may be dear to our hearts, and insects may make up about 90 percent of existing animal species, but together they represent just two clades within the overall pattern of animal phylogeny. Much remains to be done.

## 2.3    Body Plans and Embryology

As we saw in Chapter 1, there is a developmental dimension to the pattern of nested homologies. Often, this takes the form of the 'outermost' homologies – those of earliest *evolutionary* origin and most widespread taxonomic distribution – appearing first in *individual ontogeny*. They are then followed, over developmental time, by a series of progressively more taxonomically restricted homologies. This pattern was pointed out (in non-evolutionary guise) by von Baer (1828), and is encapsulated in the first of his four laws, namely: "The general features of a large group of animals appear earlier in the embryo than the special features." (For the first English translation of von Baer's laws, see Henfrey and Huxley 1853; we will revisit von Baer's work in Chapter 4.) The pattern is illustrated, for the case of vertebrate embryos, in Figure 2–7. Clearly, the basic vertebrate body plan is apparent in each case before the more specialized features (fins, beaks, hooves, etc.).

This raises the possibility that in phyla where we are not certain of the correct nesting pattern of high-level homologies around the *Unterbauplan* level, we can in fact establish this pattern by observing the order of ontogenetic appearance of the characters concerned. This 'ontogenetic method' of polarizing homologies (Nelson 1978; Patterson 1982) has been widely advocated and used over the past 20 years, but has also attracted criticism (Kluge 1985; Mabee 1989).

Probably the best-known illustration of the use of the ontogenetic method is Nelson's (1978) consideration of the eyes of flatfish. Suppose that we did not know which came first in evolution: the 'normal' fish morphology with one eye on each side of the head or the 'distorted' arrangement found in flatfish where both eyes are on the same side, that side having rotated so it is now in fact the top. We could resolve this issue in four ways. The commonality principle and stratigraphic method would work here, despite their previously discussed dangers. The outgroup method would also work, as any reasonable choice of outgroup will exhibit the one-eye-per-side state. But we could also use the ontogenetic method: in their development, flatfish go through the

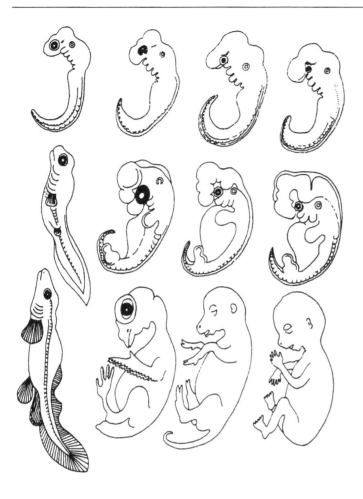

**Figure 2–7   Embryos of (left to right) teleost fish, hen, cow and
human, showing early similarity (top) giving way to later
differences (bottom)**
From Raff and Kaufman 1983. *Embryos, Genes and Evolution.*
Macmillan, New York, with permission.

standard morphological arrangement before they 'flatten', and one eye
migrates to the same side as the other.

Wheeler (1990) argues that the ontogenetic and outgroup methods
are "of about the same efficacy". Kluge (1985) points out that the ontogenetic method, like other methods of producing cladograms, has its
limitations, and that, effectively, we should proceed using a combination of methods. (I will be making this same argument, but for different

reasons to Kluge, below.) In contrast, Mabee (1989) claims an "empirical rejection" of the ontogenetic method. She states that "the ontogenetic criterion is applicable and predicts the correct phylogenetic polarity only when evolution proceeds through terminal addition", and goes on to dismiss the use of ontogeny, as terminal additions constitute a minority of evolutionary changes in her particular case study (on teleost fish of the family Centrarchidae).

Mabee's argument, however, is incorrect. The appropriate comparison to make is *not* between terminal additions and all other evolutionary changes but rather between late changes of *any* kind – including many sorts of modification that constitute neither additions nor deletions – and early changes of any kind. The proportion of cases where the ontogenetic method fails is then considerably reduced, and this puts it on a par with the outgroup method – which brings us back to Wheeler's (1990) 'equal efficacy'. (For further criticisms of Mabee's approach see Patterson 1994.)

It may of course be that the best way to proceed to determine the correct nesting pattern for a series of homologies differs depending on the taxonomic level at which we are operating. I am primarily concerned, herein, with high-level taxa and with *Bauplan/Unterbauplan* characters. But cladistic methods are not focused in this way. Rather, they are intended to be general, though the datasets on which they are employed often relate to a fairly restricted taxon (e.g. a particular group of fish, flies, or orthopterans in the studies by Mabee (1989), Wiegmann, Mitter and Thompson (1993) and Cigliano (1989), respectively). I suspect that the ontogenetic method works particularly well for the resolution of *Unterbauplan* characters (as illustrated by the vertebrate embryos of Figure 2–7). However, the limited embryological similarities between phyla will often preclude its use in the resolution of *super*phyletic groupings; also, it needs to be used with care in groups with complex life histories (see Chapter 11).

We must return, now, to the distinction between elucidating patterns and explaining them. The ontogenetic method is based on comparative embryology. If we produce a high-level phylogeny in this way, and then return to comparative embryology (and in particular to evolution of the genes controlling embryogenesis) to explain the phylogeny, there is perhaps a danger of circularity. It is therefore preferable that the patterns for which we seek an explanation can be derived both from ontogeny *and* from the use of the outgroup method with non-embryological characters.

So for what kinds of pattern, we should now ask, does Evolutionary Developmental Biology seek a causal explanation? Thus far, we have seen that a key pattern of interest is the von Baerian one of progressive divergence from a common early embryonic form. We should be aiming to unite a developmental approach with selected aspects of population genetics so that the resultant body of evolutionary theory actually *predicts* von Baerian patterns (see Arthur 1988, chapter 2) rather than simply permits them. (Current neo-Darwinism permits virtually any pattern, which is one of the criticisms that some cladists (such as Rosen 1984) level against it.)

Von Baerian divergence is by no means the only comparative embryological pattern for which an explanation is needed, nor indeed is it the 'logically prior' one. That is the origin of phylum-level body plans themselves, together with their very different ontogenetic trajectories. Within any one phylum, if the von Baerian pattern applies, as it clearly does in the vertebrates, then that requires explanation, but so too do departures from it and exceptions to it. These include: divergence of *very* early development (pre-pharyngula in the case of vertebrates) and the resultant 'egg timer' pattern (Rieppel 1993; Duboule 1994); heterochrony (Gould 1977a; McKinney and McNamara 1991); different developmental origins for characters that are clearly homologous at the adult stage (de Beer 1971; Panchen 1994); and the evolution of larval forms and complex life cycles (Lillie 1895; Garstang 1929).

Whatever pattern of comparative embryology we seek to explain, the explanation must ultimately involve the evolutionary modification of the genes that control embryogenesis, and the resulting modification of the developmental pathways themselves. Understanding such modification in any depth requires the elucidation of several interconnected aspects of the problem, ranging from the genetic architecture of development (see following section) to the relative importance of (external) adaptation and (internal) coadaptation in driving alterations to that architecture (Section 2.5). The degree to which an existing developmental system exhibits constraint (Arthur 1988) due to the disruptive effects on later development of changes to early stages (Wimsatt's (1986) "generative entrenchment"; Riedl's (1978) "burden") is also of considerable importance.

## 2.4    Body Plans, Genes and Mutations

Our estimates of the numbers of genes in animal genomes are rapidly improving (as are our estimates of the proportion of the total DNA that is genic). It is now clear that the early view that there were about 5,000 genes in a *Drosophila* genome – based on the idea of a 1:1 correspondence between genes and polytene chromosome bands (Judd, Shen and Kaufman 1972) – was a substantial underestimate. The true figure is probably in excess of 20,000. Gene numbers of approximately 18,000 and 65,000 have been suggested for *Caenorhabditis* and *Homo*, respectively (Wilson *et al.* 1994; Fields *et al.* 1994). It is likely that virtually all animal species fall into the 10,000 to 100,000 range.

For our present purposes, we need to divide genes as far as possible into three categories: (1) 'developmental genes': those involved in the determination of body plans and the control of 'downstream' developmental processes, including the determination of the relative size and shape of the organism and its constituent parts; (2) 'terminal target' genes: those that come to be switched on only in certain differentiated cell types, as a result of the activities of developmental genes, but do not themselves have a significant controlling role (e.g. the genes for haemoglobin); and (3) routine metabolic or 'housekeeping' genes (Sang 1984) such as those that manufacture glycolytic enzymes, which are expressed in the vast majority of cells, regardless of type. For an understanding of the evolutionary origin of body plans, the first category is, of course, central. For an overall understanding of the evolution of developmental processes, the second category is important also. But we do not need to devote much attention to the third category in either case. We should not forget, of course, that many genes fall into two or even all three of the above categories, and that within the developmental category the same gene may have both upstream and downstream roles (see Salser and Kenyon 1996). Dual/multiple-role genes clearly *do* need consideration.

What proportion of genes have a developmental role – whether 'body plan', 'downstream', or both – is not yet clear, though it is likely to become more so over the next few years. What proportion of non-genic DNA (e.g. highly repetitive regions) has a developmental role is also unclear. While it will be interesting to discover the numerical values of these proportions – and how they vary taxonomically – the key question is how those genes that *do* have a developmental role interact with each other to produce the correct ontogenetic trajectory;

and how evolution modifies this complex ontogenetic cascade of gene expression and ever-changing phenotypes.

One thing that *is* clear is that we cannot equate 'developmental gene' (as described above) with 'regulatory gene' in the genetic sense, as I have previously pointed out (Arthur 1984). If a particular structural gene makes a routine metabolic enzyme that has no developmental role whatever, then a second, regulatory gene that controls the rate of transcription of this structural gene is not a 'developmental' gene. Equally, another structural gene that simply makes an enzyme and has no gene-regulatory role *in those cells in which it is switched on* may nevertheless control development in distant parts of the embryo if its enzyme product (for example) modifies a small diffusible morphogen molecule from an inactive to an active state. Thus, not all regulatory genes are developmental genes, and not all developmental genes are regulatory (in a within-cell sense). This non-correspondence is illustrated in Figure 2–8. Given this, we should anticipate that 'developmental genes' will turn out to be very heterogeneous in terms of their mode of operation at the molecular level – and indeed some of this heterogeneity is already apparent, since some developmental genes make growth factors, others transcription factors, and so on. What 'developmental genes' have in common with each other is that mutations occurring in them will produce developmentally aberrant phenotypes.

As I have already indicated, there is no clear distinction, within the developmental category, between 'body plan' and 'downstream' genes. Rather, all genes in both categories and those which straddle the two are connected up into an interacting system which we can describe as the genetic architecture of development. Unravelling this is the key task for developmental geneticists; and the picture that they ultimately produce (of which our current view is fragmentary but enticing) is likely to be the most important remaining input into evolutionary theory. Several different features of this genetic architecture need to be elucidated. In general, I will leave this story until Chapters 5–7, but two aspects deserve a brief mention here.

First, what overall form does the pattern of interaction between developmental genes take? Some early workers (e.g. Shumway 1932) seemed to imply a *linear* pattern, but this is hardly likely. My own earlier work (Arthur 1982a, 1984, 1988) has emphasized the importance of *hierarchical* patterns of interconnection (as has that of Slack 1983 and Sang 1984). But simple hierarchies are not enough: *feed-back*

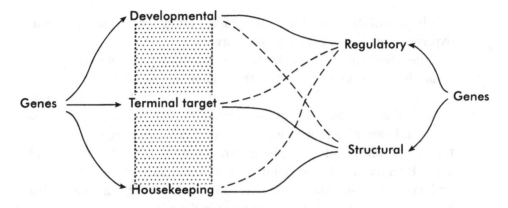

**Figure 2–8  Relationships between two ways of categorizing genes**
Solid lines show often-presumed connections. These *may* represent
the commonest linkages, but linkages indicated by dashed lines are
also possible. Stippling indicates overlap between categories.

*loops* are likely to be common (Kauffman 1993); and some form of
*cross-linking* is probably important between different semi-autono-
mous developmental systems to keep them in phase with each other
(Waddington 1957; Arthur 1988). The high degree of repeatability of
ontogeny implies some sort of *redundancy* (e.g. duplication of controls, as in aircraft). All of this argues for the existence of a highly
complex *network*. However, while this may seem a recipe for despair,
the network is likely not to be completely irregular; rather it will be
patterned in various ways which we must seek to describe. All of this,
of course, has major evolutionary implications. For example, for any
particular developmental gene, the number of other such genes that are
functionally downstream of it will be a factor in determining how
evolutionarily constrained it is. And it may well be that body-plan
genes are the most constrained of all.

Second, what is the chromosomal arrangement of developmental
genes? To what extent are they arranged in clusters or scattered individually across all the various chromosomes of whatever karyotype is
concerned? When they *are* found in clusters, what linear sequence is
found? I must admit that I was initially of the view that the
'chromosomal layout' aspect of genetic architecture was generally
unimportant, but the discovery of 'colinearity' has made that view
untenable (see Lewis 1978; Gaunt, Sharpe and Duboule 1988; Gaunt
1991; Duboule 1994). Colinearity has spatial, temporal and taxonomic

aspects: for example, in the *Hox* gene clusters chromosomal sequence is related to (a) anteroposterior expression patterns; (b) temporal expression patterns; and (c) comparable sequences across diverse taxa. These observations are clearly telling us something, even if it is not yet clear what. This issue will be pursued in Chapter 7.

The molecular mechanisms of gene switching are often classified as *cis* and *trans* (see, for example, Lodish *et al.* 1995). The distinction is between those that rely on regulatory molecules producing their effects by moving along a particular stretch of chromosome (*cis*) and those where the regulatory molecules are more generally mobile and can switch chromosomally scattered genes on or off (*trans*). There is an obvious link here with chromosomal architecture; *cis*-acting control mechanisms are likely to have a greater role in clusters of developmental genes than in their far-flung equivalents. This connects with the interesting question of whether body-plan genes are more clustered (and so more *cis*-controlled) than 'downstream' genes, and if so, why.

No doubt, developmental geneticists working on *Drosophila* would be well satisfied, to say the least, if they were able to unravel the complete genetic architecture (both 'interactional' and 'chromosomal') for that particular organism. Unfortunately, evolutionary developmental biologists have chosen an even more difficult task: understanding how such architectures vary taxonomically and how this variation has arisen through evolutionary change.

In one sense, the view that I will be developing herein (starting from Chapter 7) is a conventional one: I will be arguing that the genetic architecture of development, like other suites of characters, ultimately evolves due to an interplay between mutation and selection. However, I will be giving much more weight to mutation than is usual in evolutionary theory; and will be unable to accept the view expressed by many population geneticists (e.g. Ford 1971) that mutation has no 'directing' role in evolution. Also, it is necessary to depart from old-fashioned 'bean-bag' genetics (wherein genes are seen as independent 'atomistic' units: Mayr 1960, Haldane 1964) and the over-emphasizing of 'point' mutations. We now know the genome to be rather dynamic, subject to all sorts of flux: including exon shuffling, gene duplication, gene conversion and unequal exchange between chromosomes. Some of these processes are likely to be particularly important in the origin and modification of body plans, as will become apparent in subsequent chapters.

Whatever the molecular basis of a novel body plan, the phenotype concerned, at every stage of the process, must be viable and able to spread at the population level. Lack of adequate attention to this problem was the main reason underlying the general rejection of Goldschmidt's (1940) idea of 'hopeful monsters'. But, just as we must depart from bean-bag genetics, so too we must depart from 'ecological panselectionism'. Just as point mutation is a caricature of overall genomic flux, 'ordinary' directional selection caused by the need to adapt to external ecological variables (e.g. the presence of a predator) is a caricature of selection generally. This is not an argument for selection at higher levels of organization: for example, Wynne-Edwards' (1962) group selection or Stanley's (1975, 1979) species selection. Nor is it an argument for relegating selection to a rather minor role (Goodwin 1994a). Rather, it will turn out (in Chapter 9) to be an argument for Whyte's (1965) *internal selection*, my own *n-selection* (Arthur 1984) and for detecting patterns in the probability of different kinds of mutation being selectively advantageous in an 'ordinary' way – patterns which Fisher's (1930) work regrettably discouraged a search for, despite the fact that the 'Fisher principle' (see Figure 1–5), when looked at afresh, *is* one of these very patterns.

## 2.5 Body Plans, Adaptation and Environments

One of Gould's (1977b) "eternal metaphors" of palaeontology concerns the relative evolutionary importance of internal and external factors: essentially developmental and ecological factors, respectively. Of course, both are ultimately involved in evolution, and indeed van Valen's (1974) suggestion that we view the evolutionary process as "the control of development by ecology" has much to recommend it. But nevertheless views have been – and still are – strongly polarized in relation to the internal/external metaphor. Mainstream neo-Darwinism has been primarily externalist in its approach, while many (partially) opposing theories – from Goldschmidt's (1940) systemic mutations to Goodwin's (1984, 1994a) generative paradigm and generic form – have given much greater emphasis to internal (that is, developmental) processes.

Discussion of this issue to date has paid insufficient attention to the distinction between the *origin* and *maintenance* of body plans, and it may well be that these involve a very different balance of internal and external factors. For example, Maynard Smith and Szathmáry's (1995) suggestion that body plans might represent "frozen historical accidents" is essentially a proposal that origination through adaptation to a particular way of life (*external*) is followed by long-term maintenance. This is because the characters involved in the adaptation become subsequently so developmentally networked (or generatively entrenched: Wimsatt 1986) that the need for *internal* coadaptation prevents their removal – in other words, that is what 'freezes' them. While this particular hypothesis may or may not be correct, it does argue for distinguishing between (a) the origin and (b) the stabilization and maintenance of body plans, and I will proceed on that basis. This distinction will come into sharp focus in Chapter 9.

With regard to the origin of the major (phylum-level) body plans, one thing is clear: they do *not* result from invasion of the different major classes of environment: land, sea, freshwater and air. *All* of the major animal body plans are marine in origin. As each has proliferated to form a large clade of species, some have remained exclusively marine (e.g. echinoderms) while some have spread to freshwater and/ or land. Invasion of new environments *has* often been associated with the appearance of new *Unterbaupläne*: for example tetrapods (land) and birds (air) within the chordates; and pulmonate gastropods (freshwater/land) within the molluscs. But even here the picture is far from tidy, for two reasons. First, because impressive *Unterbaupläne* can arise without such invasions (e.g. starfish, sea-urchins, sea-cucumbers and the other main echinoderm types). Second, because where there is an *Unterbauplan*/environment correspondence, it is usually imperfect: whales, as secondarily aquatic ex-tetrapods, are an obvious example; as are the small number of prosobranch land snails.

The fact that all the major body plans are marine in origin does not, of course, mean that their divergence from each other was not externally adaptive. Some body plans may have arisen in connection with occupation of a particular marine microhabitat or niche and/or with adoption of a particular form of locomotion. For example, the chordate body plan based on a notochord and segmented musculature may have arisen partly in connection with sinusoidal swimming (see, for example, Maynard Smith and Szathmáry 1995). However, non-adaptive origins of body plans have also been proposed (e.g. van Valen 1974); and in either case the underlying genetic changes are unspecified. This brings to mind another of Gould's (1977b) eternal metaphors (sudden versus gradual), but I will defer further consideration of that to subsequent chapters.

Let us now turn to the maintenance of body plans, which in some ways is even more important than their origins. All characters originate, but most also disappear again or are fundamentally transformed; that is, they are evolutionarily transient. One of the main features of body plans, in contrast, is their high degree of conservation, extending over hundreds of millions of years and – in the main phyla – across many thousands of species. What causes this extraordinary degree of conservation?

The correct answer to this question is probably 'developmental constraint', but we need to be very careful about precise meanings here (Resnik 1995). In particular, the distinction between developmental and selective constraints, which has been emphasized by Maynard Smith *et al.* (1985) needs to be highlighted, as does the grey area in between these two.

At one end of the spectrum, some variant developmental trajectories are simply never produced within a particular phylum; we assume, then, that the nature of the developmental system is such that it just cannot produce these variants. Limitations on the *production* of variant ontogenies clearly constitute developmental constraints.

At the other end of the spectrum, consider a recent species that has existed unchanged for several million years (as in the punctuated equilibrium 'model': Eldredge and Gould 1972) despite the existence now, and probably throughout its duration, of considerable intraspecific variation. (There are several good examples of this pattern in the bivalve molluscs: see Stanley 1977.) If the variation is there, and available for selection to work on, but the character distribution remains essentially constant, then one interpretation is that stabilizing selection caused by *external factors* (an important proviso, as we shall see) is preventing directional change. If so, then this is clearly selective, not developmental, constraint.

Now consider an intermediate scenario: the developmental system *is* capable of embarking on a variant pathway, but the developing organism dies a short way through its ontogeny because of lack of internal coadaptation. This kind of constraint could reasonably be described as developmental, selective, or both. Because of the definition they use, Maynard Smith *et al.* (1985) consider it selective; but this results in their view of what constitutes a *developmental* constraint being very narrow.

Table 2–4. Developmental versus selective constraints

| Developmental pathway | Reason for non-production or negative selection | |
|---|---|---|
| | Internal/developmental | External/ecological |
| Not embarked upon | D | N/A |
| Initiated, but inviable to adulthood (i.e. embryonic/ juvenile lethal) | D/I | E |
| Initiated, and potentially viable to adulthood, but disadvantageous at | | |
| (a) developmental stage | D/I | E |
| (b) adult stage | I | E |

D = developmental constraint; I = internal selective constraint;
E = external selective constraint.

For the purposes of this book, I will consider 'developmental constraints' in a broader sense (see Table 2–4) to include both (a) limitations to the range of variant ontogenetic pathways that can be embarked upon within a particular higher taxon; and (b) limitations to the range of adult forms that can be produced due to the inviability of organisms with variant pathways at any pre-adult stage, with such inviability arising from lack of internal coadaptation. This sort of inviability is indicative of 'negative internal selection' (Whyte 1965). Internal selection is of crucial importance in the evolution of development; I discuss it further in Sections 5.3, 6.5 and 9.3.

Finally, returning to the different forms of adaptation, the above arguments can be summarized as follows: the origin of phylum-level body plans may include elements of *adaptation* to particular marine environments and forms of mobility within those; *Unterbaupläne* often, but not always, originate with the invasion of a new environment, and this will frequently involve the modification of existing structures for other purposes (sometimes referred to as *pre-adaptation* but preferably called *exaptation*: see Gould and Vrba 1982); body-plan characteristics are highly conserved evolutionarily because they become an integral part of the developmental system and thus have to be retained in order to avoid loss of *coadaptation*. This process of 'internalizing' may be associated with the evolution of increased phenotypic complexity (Chapter 11).

So much for generalities. I hope that by this stage the reader has a clear idea of what I mean by 'body plan' and how it relates to taxonomic levels, homology, ontogeny, genes and adaptation. It is now time to turn to specifics: which body plan is related to which. That will take us, in the following chapter, into metazoan cladograms and thence into the fossil record.

# Patterns of Body Plan Origins

## 3.1 Strategy

In this chapter, I will proceed according to the three-stage strategy
described by Eldredge (1979). That is, I will first consider possible
patterns of relatedness of the major metazoan groups using cladograms; I will then attempt to put approximate times to the important
lineage divergence points, thus producing phylogenetic trees; finally, I
will bring in some relevant palaeoecological data to produce what
Eldredge calls adaptive scenarios – though the treatment of these
will be brief, given the criticisms of such scenarios considered in
Section 1.3. At each shift (cladogram-to-tree and tree-to-scenario) the
proposal being made has a higher information content but also a high
probability of being wrong: there are many possible trees for each
cladogram, many possible adaptive scenarios for each tree.

The 'internal' side of things – that is, the genetic and developmental
architecture of the evolutionary changes involved (as discussed in
Section 2.4) – will for the moment disappear from view. It will re-
surface later (in Chapter 5 and beyond) when I deal with it in more
detail. Ultimately, it is the *connection* between that architecture and
the phylogenetic patterns on which we will now focus that is most
important. I will, consequently, at one stage in this chapter, focus on
the relationships between taxa containing species which are used as
'model systems' for the study of developmental genetics.

One important question that I discuss in this chapter is whether
metazoan groups do indeed connect in the form of a branching tree,
or whether they originated independently from a very distant ancestor

(e.g. a flatworm, or even a protozoan: see Nursall 1962). The latter pattern has been referred to (e.g. by Willmer 1990) as a 'field of grass' rather than a tree. This issue is important, not just from the stance of wishing to understand the actual pattern of phylogeny, but also because it has a bearing on the degree to which divergence of early ontogenetic pathways has occurred over evolutionary time. In the extreme case of independent origins of all metazoan phyla from ancient protozoans, there would be no need to try to explain (for example) how we get from spiral to radial cleavage, because that evolutionary shift would never have happened. As will become clear, however, I do not support an extreme 'field-of-grass' view; and indeed this view looks increasingly untenable as more molecular evidence accumulates.

Other issues that will be dealt with in this chapter include (a) the perennial problem of evolutionary convergence and (b) the question of how developmentally 'experimental' pre-Cambrian and Cambrian evolutionary processes were, and, linked to that, whether some fossils from these periods represent phyla quite different from those found today.

Some readers may wish to pursue the fascinating problem of high-level animal phylogeny in more detail than I can provide within the confines of a single chapter. Important recent books in this area include those by Willmer (1990) and Nielsen (1995); see also Glaessner (1984) on the Ediacaran fauna and Whittington (1985) and Gould (1989) on the Burgess Shale.

## 3.2    Patterns of Metazoan Interrelationships

Depending on exactly how particular animal phyla are delineated, there are somewhere between thirty and forty of them with living representatives, including the recently designated phylum Cycliophora (Funch and Mobjerg-Kristensen 1995). (There may be others – the number of which is very difficult to estimate – whose representatives are entirely extinct.) We now ask: how are these thirty to forty phyla interrelated? What cladogram represents the actual pattern of evolutionary branching, or lineage splitting, that has occurred between the appearance of the first animals and the present?

It is easy to be despondent about our ability to reconstruct phylogenetic history. The evolutionary literature of the past 150 years contains hundreds of different proposed patterns of interrelationship. And while comparative molecular data (only available in the last few decades) have helped, they have not proved as unambiguous as we had hoped. Molecules, like morphologies, vary in their evolutionary rates and are subject to parallel and convergent evolution; and in consequence different molecules often suggest different phylogenies, just as do different morphological characters.

However, I intend, here, to take a positive stance. I will concentrate on what we know rather than what we do not. While there are indeed hundreds of proposed phylogenies or cladograms in existence, the number of possible ways of arranging thirty to forty entities (e.g. the animal phyla) in a branching diagram runs well into the billions. A quick, albeit simplified, calculation is useful here. Suppose that there are thirty-two phyla connected in a bifurcating cladogram whose branches split in synchrony at each level. There are then $32!/2^{31}$ possible patterns of interrelationship, which works out at about $1.2 \times 10^{26}$. So from the viewpoint of a Bayesian statistician (see, for example, O'Hagan 1988) we are not doing too badly. The wide *a priori* range of possibilities has been drastically restricted by the massive body of data on comparative anatomy, embryology, molecular biology and palaeontology now at our disposal. As further data accumulate, perhaps one day our *a posteriori* range of possibilities will be reduced to a single pattern.

I will now survey a selection of suggested patterns of interrelationship from the recent literature, including standard undergraduate texts (Barnes 1987; Barnes, Calow and Olive 1988; Margulis and Schwartz 1988), research-level monographs (Willmer 1990; Nielsen 1995), 'primary' papers (Nursall 1962; Schram 1991; Backeljau, Winnepenninckx and de Bruyn 1993) and review articles (Conway Morris 1993; Raff, Marshall and Turbeville 1994). (The difference between the last two categories is not clear-cut in this area: all novel proposals of metazoan cladograms draw on much past data, and authors of review articles often find it hard to resist supporting particular proposals at the expense of others.)

A proposed pattern of interrelationship is determined by two things: the dataset upon which it is based and the way in which that dataset is analysed, both in terms of broad approach (cladistic versus phenetic versus intuitive) and specific method (e.g. PAUP versus PHYLIP, and different procedures within each: Platnick 1987, 1989). The sources that I have used differ in both respects, so it comes as no surprise that they end up favouring different patterns. But conversely – and in line with my intended positive stance – results on which there is a broad consensus, despite such methodological and dataset differences, must surely be regarded as very robust.

The extremes of cladistic and non-cladistic approaches are represented, among my sources, by Backeljau *et al.* (1993) and Willmer (1990), respectively. Extreme datasets, in the sense of being exclusively morphological and predominantly molecular, are exemplified by Barnes (1987) and Raff *et al.* (1994). It should be noted that the comparative molecular data that are relevant to high-level animal phylogeny concern molecules with wide taxonomic distribution and slow rates of evolutionary change. Many of the early studies in this area, including the pioneering work of Field *et al.* (1988) and Lake (1990), were based on 18S ribosomal RNA, which possesses both of these attributes and is abundant within cells. More recently, study of a wide range of slow-evolving stretches of DNA has become possible, despite low cellular abundance, because of the amplification technique of PCR (polymerase chain reaction: see Mullis, Ferré and Gibbs 1994).

Figures 3–1 to 3–4 illustrate both different suggested phylogenies and different forms of presentation. The complete list of ten sources that I am using for this purpose is given in Table 3–1, along with a summary of the views expressed in each of these in relation to a series of important phylogenetic questions that we now need to address.

## Is the Animal Kingdom Monophyletic?

As can be seen from Table 3–1, the majority view here is 'yes'. That is, all the animal phyla arose by the repeated splitting of an originally singular lineage which was initiated when a particular unicellular group – probably either flagellate or ciliate 'protozoans' – evolved towards multicellularity. Nursall (1962) departs most dramatically from animal monophyly in suggesting that virtually all phyla arose independently from different unicellular ancestors. Willmer (1990) argues for multiple origins of flatworm-type body plans from different protozoans. Barnes *et al.* (1988) support monophyly of the Bilateria but suggest independent origins for Placozoa, Porifera, Cnidaria and Ctenophora. Gradually accumulating molecular evidence strongly supports animal monophyly (see Slack *et al.* 1993).

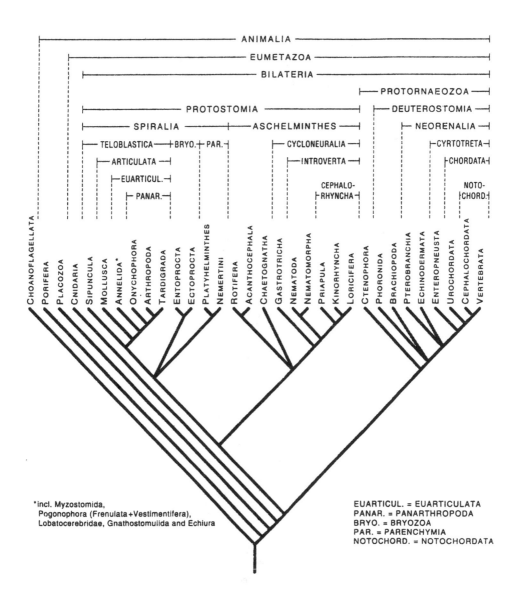

**Figure 3–1    Interrelationships between phyla according to Nielsen (1995)**
From *Animal Evolution*. Oxford University Press, Oxford, with permission.

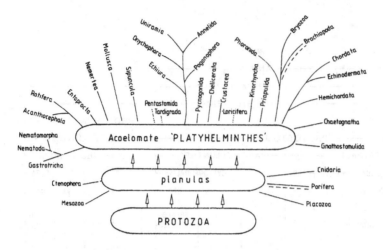

**Figure 3–2 Interrelationships between phyla according to Willmer (1990)**

As can be seen, this view incorporates multiple origins of both 'planulas' and 'platyhelminthes'.

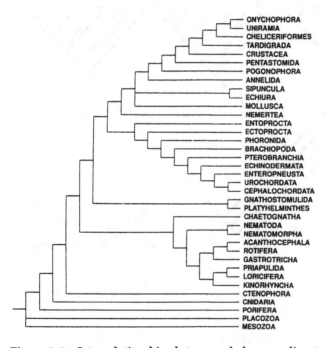

**Figure 3–3 Interrelationships between phyla according to Schram (1991)**

This particular presentation of Schram's views, which is arguably clearer than the original, is from Backeljau *et al.* 1993. *Cladistics*, **9**, 167–181, with permission.

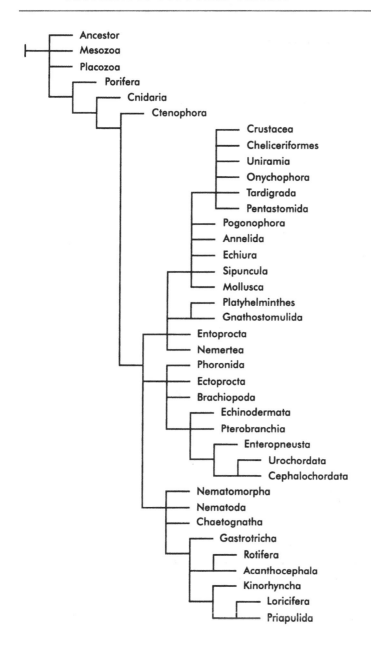

**Figure 3–4  Interrelationships between phyla according to Backeljau *et al.* (1993)**

These authors present several alternative cladograms. This corresponds to their figure 4D. Some of their others are so 'flat' (i.e. characterized by extensive polytomies) that they contain very little information about interrelationships. From Backeljau *et al.* 1993. *Cladistics*, **9**, 167–181, with permission.

**Table 3–1. Alternative views of metazoan phylogeny**

| Source | Monophyletic Animalia? | Early off-shoots? | Monophyletic Bilateria? | Protostomes and deuterostomes? | Monophyletic Arthropoda? | Other 'clusters'? |
|---|---|---|---|---|---|---|
| Nursall (1962) | N | N/A | N | N | Y | –M– |
| Barnes *et al.* (1988) | N | Y | Y | D | N | L– – |
| Margulis and Schwartz (1988) | Y | Y | Y | D | Y | –MA |
| Willmer (1990) | N | Y | N | D | N | L– – |
| Barnes (1987) | Y | Y | Y | P + D | Y | –M– |
| Schram (1991) | Y | Y | Y | D | Y | –MA |
| Backeljau *et al.* (1993) | Y | Y | Y | D | Y | –MA |
| Conway Morris (1993) | Y | Y | Y | P + D | Y | – –A |
| Raff *et al.* (1994) | Y | Y | Y | P + D | Y | LM– |
| Nielsen (1995) | Y | Y* | Y | P + D | Y | –MA |

*Yes for Cnidaria, Placozoa, but No for Ctenophora.
Y, N, N/A = Yes, No, Not applicable
D, P = Deutrostome monophyly, protostome monophyly
L, M, A = Lophophorates; a Mollusc/Annelid/Arthropod group; and Aschelminthes.

## Which Groups Split Off Prior to the Main Bilaterian Radiation?

The groups at issue here are the Porifera, Cnidaria, Ctenophora, Placozoa and Mesozoa. However, there is some uncertainty over the taxonomic status of the last of these, which may in fact constitute a diphyletic grouping of two distinct phyla (Dicyemida and Orthonectida: Nielsen 1995). Many authors omit them from consideration, and I will not discuss them further here.

As can be seen from Table 3–1, there is near unanimity on the status of the other four groups. All ten sources argue that all of these groups represent early side-branches, with a single exception: Nielsen (1995) argues for a close relationship between ctenophores and deuterostomes, based on an unusual view of ctenophore embryology. Molecular data (based on 18S and 28S rDNA) do not support Nielsen's view (Raff *et al.* 1994).

## Do All the Other Animal Phyla Represent a Monophyletic Bilaterian Radiation?

There is a strong consensus here that the answer is 'yes', with Nursall's (1962) view again being the most divergent. In his 'field-of-grass' scheme there is *no* bilaterian radiation. In Willmer's (1990) 'meadow' there is a radiation of most bilaterian phyla from *different* groups of flatworms, so the Bilateria are certainly not a monophyletic clade in her scheme. I suspect that molecular data will shortly prove them to be wrong.

## Are Protostomia and Deuterostomia Monophyletic Groups?

The reality of a deuterostome grouping of chordates, hemichordates and echinoderms is virtually unchallenged. There is some debate about the internal structure of this clade (see Jefferies 1986) and about the position of conodonts and chaetognaths. The former are almost certainly well within the clade (Purnell 1995; Gabbott, Aldridge and Theron 1995) while the latter may not be in it at all (Telford and Holland 1993; Conway Morris 1993; Nielsen 1995).

The evidence for a monophyletic protostome clade is less convincing, and one possibility is that Protostomia is a paraphyletic group which includes the ancestors of the deuterostomes (Backeljau *et al.* 1993). At any rate, this seems a less important question. The two groups (protostomes and deuterostomes) are not really equivalent. Protostomia is an enormous collection of very different phyla, and it is doubtful if a grouping of this kind is of much use in understanding the evolution of development, even if it is indeed monophyletic. It is arguably more important to be able to recognize 'tighter' superphyletic groupings *within* the Protostomia.

## Do the Arthropods Represent a Monophyletic Grouping?

The view that different arthropod groups – in particular Uniramia, Crustacea and Chelicerata – may represent convergent lineages was forcibly argued by Manton (1974, 1977). Increasingly, though, this view looks untenable. The majority of the ten sources used in Table 3–1 support arthropod monophyly, as do the recent cladistic analysis of Wheeler *et al.* (1993) and several molecular studies (e.g. Akam *et al.* 1994a). Some of the most recent studies, while again supporting arthropod monophyly, question the validity of the Uniramia (insects and myriapods) and propose that crustaceans, not myriapods, represent the sister group of Insecta (Boore *et al.* 1995; Friedrich and Tautz, 1995; Popadic *et al.* 1996).

## What Other Superphyletic Clusters are Discernible?

This is where the going gets difficult. The main proposals at issue are:

1. Lophophorates (phoronids, ectoprocts and brachiopods)
2. An annelid/mollusc supergroup (possibly closely related to the arthropods)
3. Aschelminthes (comprising rotifers, gastrotrichs, nematodes, nematomorphs, acanthocephalans and a variable number of others).

As can be seen from Table 3–1 there is considerable disagreement on these issues, both with respect to whether these groups exist as monophyletic entities and, if so, what other groups they are most closely related to – for example, where the lophophorates fall in relation to the protostome/deuterostome split. Molecular studies tend to ally the lophophorates with protostomes (Raff *et al.* 1994) but morphological approaches often place them closer to deuterostomes (Willmer 1990; Nielsen 1995).

For the purposes of this book – and of making progress in understanding the evolution of development generally – what is most important is to be able to construct a cladogram of interrelationships which incorporates those higher taxa containing species that represent 'model systems' for the study of comparative developmental genetics. We need to be reasonably confident that this cladogram is correct, for otherwise we will be wrong in our identification of the evolutionary transitions that have taken place in developmental mechanisms.

Starting from the plethora of views summarized in Table 3–1, I will attempt to produce such a cladogram; and I will get there in two stages, both of which are simplifications. First, I will drop from consideration many of the 'minor' phyla and will allow unresolved polytomies. (These indicate uncertainty in the order of branching events, not an active advocacy of simultaneous events.) This leads to the pattern shown in Figure 3–5, which is based largely on a consensus of the ten sources used. This cladogram of twenty-two higher taxa is less informative, but also less risky, than those involving more than thirty

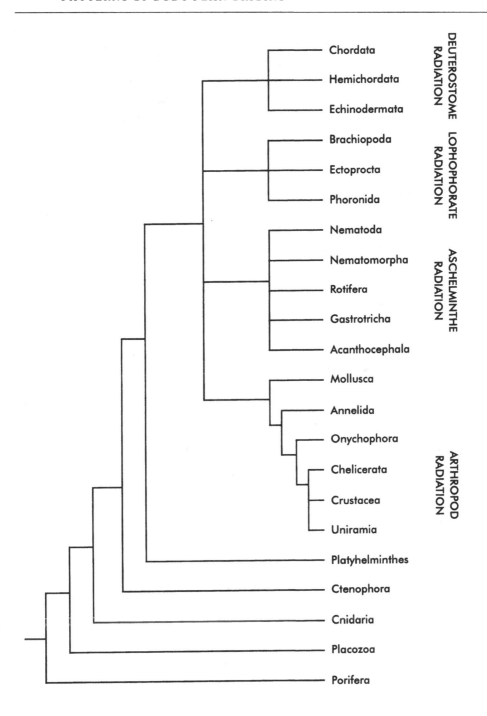

**Figure 3–5  Cladogram of twenty-two major phyla, based on a consensus of the ten sources used**

taxa that we have already seen. (Although it is less risky, it will almost certainly turn out to be wrong in some respects.)

The second simplification is to drop even 'major' phyla (i.e. those with high current biodiversity, such as Mollusca) if they do not contain species which have been used intensively as developmental model systems. To do this, we need to identify which species constitute such systems; this information is given in Table 3–2. It must be emphasized that the selection of species to be included here is somewhat arbitrary. The classics of developmental genetics are all included, as are the best-known species of what might be called experimental embryology; but the 'tail' of the distribution – representing species about whose embryology we know progressively less – has to be

**Table 3–2. Species used as model systems for developmental genetics and experimental embryology**

In some cases the specific name given is somewhat arbitrary as several related species are used. This is particularly true of sea urchins – see chapter 1 of Stearns (1974). Higher taxa given are inconsistent in rank but chosen so as to be useful to readers whose taxonomic expertise is limited. The species listed include a plant 'outgroup' (*Arabidopsis*). The references given here are mostly to general accounts – these complement the more specific references to many developmental systems given in Chapter 5.

| Species | Common name | Higher taxa | Reference |
|---|---|---|---|
| *Drosophila melanogaster* | fruitfly | Insecta: Diptera | Lawrence (1992) |
| *Caenorhabditis elegans* | 'worm' | Nematoda: Rhabditidae | Stern and DeVore (1994) |
| *Danio rerio* | zebrafish | Vertebrata: Osteichthyes | Kimmel *et al.* (1995) |
| *Branchiostoma floridae* | amphioxus | Chordata: Cephalochordata | Holland *et al.* (1994a) |
| *Xenopus laevis* | clawed toad | Vertebrata: Amphibia | Hamilton (1976) |
| *Mus musculus* | mouse | Vertebrata: Mammalia | Monk (1987) |
| *Gallus domesticus* | chick | Vertebrata: Aves | Patten (1971) |
| *Paracentrotus lividus* | sea urchin | Echinodermata: Echinoidea | Hörstadius (1973) |
| *Hydra viridus* | polyp | Cnidaria: Hydridae | Tardent and Tardent (1980) |
| *Arabidopsis thaliana* | thale cress | Angiospermae: Cruciferae | Meyerowitz and Somerville (1995) |

chopped off somewhere. So while *Hydra* is included, *Helix* (whose embryology has been studied to a reasonable degree), is not. I have tried to use paucity of genetic or experimental data as my criterion, rather than a lack of descriptive embryological information.

A cladogram representing the pattern of interrelationships among our selected 'model system' species is given in Figure 3–6. Since this is a further simplification, involving fewer taxa even than in the cladogram of Figure 3–5, it is again both of lower information content and lower probability of being wrong. My three hopes of this cladogram are that (a) the phylogenetic information retained is the most important for making progress in understanding the evolution of development at the present time; (b) the pattern of interrelationships shown is in fact correct – that is, it is a true representation of the actual pattern of lineage splitting, over phylogenetic history, that has led to the taxa concerned; and (c) the 'cautious trichotomy' will be resolved in the direction expected on the basis of an early protosome/deuterostome divergence.

## 3.3    Early Fossils: from Cladograms to Trees

So far, we have concentrated largely upon cladograms, which represent the most abstract way of looking at interrelationships. In a typical cladogram, neither the vertical nor the horizontal axis has a clear meaning (except in as much as both are used to allow a nesting pattern to be depicted), and neither is calibrated using physical units. The vertical axis does not represent time, nor the horizontal one morphology. Bifurcations can be rotated through 180° with no effect. Indeed, a cladogram can be replaced with a Venn diagram (see Figure 1–3, for example). In the view of some 'transformed' or 'pattern' cladists, even the connection with phylogeny is unnecessary (Platnick 1979), evolutionary lineage divergence being regarded only as a *possible* explanation of the pattern observed in the cladogram, and one that should not be assumed at the outset, despite the difficulty in imagining an alternative explanation.

What all this means is, that while cladograms are a good starting point in our attempts to discern patterns of relationship among organisms, they are certainly not our ultimate goal. For those of us interested in the evolution of development (and the origin of body plans as a component of that) it is particularly necessary to bring both time and morphology explicitly back into the picture, and thus, in a sense, to

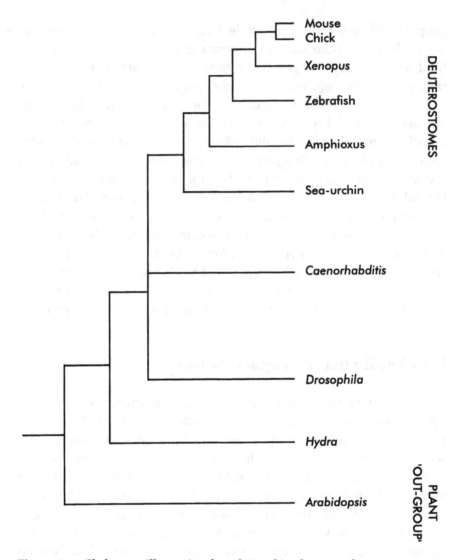

**Figure 3–6   Cladogram illustrating the relationships between the main developmental 'model systems'**

calibrate the cladograms. I will consider timescales here, morphology in the following section.

From a body-plan perspective, the most pressing task is to attempt to date the origins of the various animal phyla and, associated with that, to determine the degree to which body-plan radiation was 'explosive' or gradual and sequential. As we saw in Chapter 1, there is considerable uncertainty at present over this issue, partly as a result of the recent molecular study by Wray *et al.* (1996), which suggests that existing views need to be questioned.

From the database compiled by Benton (1993), it appears that the following groups were present in the early Cambrian (the Caerfai epoch in Table 3–3): Porifera, Cnidaria, Mollusca, Annelida, Crustacea, Chelicerata, Brachiopoda, Echinodermata. It now seems likely that Chordata (Chen *et al.* 1995; Shu, Conway Morris and Zhang 1996) and Hemichordata (Shu, Zhang and Chen 1996) were also present. Mapping these taxa on to the cladogram shown previously as Figure 3–5 produces an interesting result (Figure 3–7). It is apparent from the taxonomic distribution of known early Cambrian groups that all but one of the major radiations of phyla shown in the figure had already occurred by this time. The one possible exception is the Aschelminthes, but these have a very poor fossil record because of their lack of hard body parts, and they may well have been present much earlier than current evidence suggests. The apparent absence of Uniramia is more likely to be real, given (a) their fossilizable exoskeletons and (b) their present-day link with terrestrial environments.

The conclusion that may apparently be drawn, then, is that most of today's animal phyla originated *at least* as early as 520 *my* ago. How much earlier than this they may have originated is a more difficult question, involving, as it does, interpretation of the sparse fossil record of the Vendian period, and in particular the Ediacaran fauna.

Despite its name (after Ediacara, in South Australia, where it was first discovered (see Sprigg 1947, 1949)) this is in fact a world-wide fauna, as can be seen from Figure 3–8. Much of the early description of the Ediacaran fauna was carried out by Glaessner (1958, 1959, 1969, 1971, 1976, 1984), who interpreted this fauna as a range of genera belonging to Cnidaria, Annelida and Arthropoda, apart from one species, *Tribrachidium heraldicum*, of unknown taxonomic position (see Cloud and Glaessner 1982; Glaessner 1984, pp. 52–3).

At least two alternative views have been expressed on the nature of the Ediacaran fauna. Seilacher (1984, 1985, 1989) has suggested that forms such as *Charniodiscus, Dickinsonia, Spriggina* and *Praecambridium*, attributed by Glaessner (1984) to present-day phyla, in fact represent an extinct high-level taxon (phylum, or even kingdom) of organisms with a 'quilted' body plan which he calls Vendozoa or Vendobionta. (A similar suggestion was made by Pflug (1972a, b) for Ediacara-type fossils from south-west Africa.) More

**Table 3–3. Epochs and stages of the Cambrian and Vendian periods**
The Table spans the period from around 500 to around 600 *my* ago. The base of the
Cambrian was previously considered to be about 570 *my* ago, but recent evidence
suggests a figure of about 543 *my* (Bowring *et al.* 1993). Data extracted from
Harland *et al.* 1990 and modified by the insertion of the Manykaian.

| Era | Period | Epoch | Stage |
|-----|--------|-------|-------|
| Early Palaeozoic | Cambrian (Cbn) | Merioneth (Mer) | Dolgellian (Dol) |
| | | | Maentwrogian (Mnt) |
| | | St David's (StD) | Menevian (Men) |
| | | | Solvan (Sol) |
| | | Caerfai (Crf) | Lenian (Len) |
| | | | Atdabanian (Atb) |
| | | | Tommotian (Tom) |
| | | | Manykaian (Man) |
| Neoproterozoic | Vendian (Ven) | Ediacara (Edi) | Poundian (Pon) |
| | | | Wonokan (Won) |
| | | Varanger (Var) | Mortensnes (Mor) |
| | | | Smalfjord (Sma) |

recently, Retallack (1994) has made the rather bizarre suggestion that
the Ediacaran 'animals' are in fact marine lichens – on the basis of a
claimed lack of compressibility.

The difficulty of interpretation is clear, and is well stated by
Seilacher (1985): given the limited morphological resolution that is
possible in these ancient fossils, "any investigator will be highly influenced by previous experiences and preconceptions", i.e. will be
biased! My own inclination is to give qualified support to Seilacher's
concept of Vendozoa. The morphology of most Ediacarans is generally
unlike that of present-day animals, with two possible exceptions. First,
some genera (for example, *Charniodiscus* (Figure 3–9)) resemble the
modern cnidarian group known as sea-pens. Second, it is possible that
*Spriggina* is a primitive bilaterian, perhaps even an early annelid, as
Glaessner (1984) suggested. The other Ediacarans – including
*Tribrachidium* and *Dickinsonia* shown in Figure 3–9 – seem genuinely
unique, when compared to today's phyla, and are probably best interpreted as a range of body plans that became extinct in the late Vendian
and early Cambrian.

One possible evolutionary tree incorporating both Ediacaran and
Burgess Shale (see below) fossils is given in Figure 3–11. Here, the
Bilateria only existed in Ediacaran times as a stem-group, which either

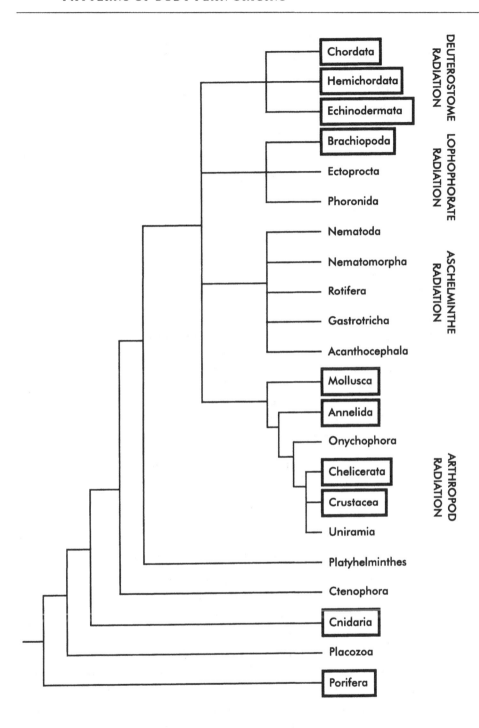

**Figure 3–7** The cladogram given in Figure 3–5, but with the phyla known to have been present in the early Cambrian highlighted

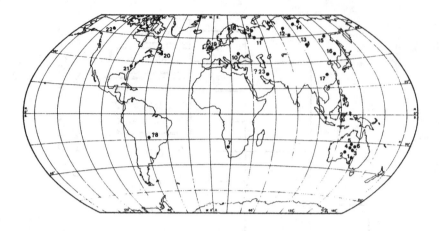

**Figure 3–8   Worldwide occurrences of Ediacaran fossils**
*Australia:* 1, Ediacara, Flinders Ranges; 2, Punkerri Hills, Officer
Basin; 3, Deep Well, southeast of Alice Springs; 4, Laura Creek,
southwest of Alice Springs; 5, Mt Skinner; 6, Jervois Ranges,
southwest Georgina Basin. *Africa:* 7, southwest Africa/Namibia.
*South America:* 8, Mato Grosso, southwest Brazil. *Ex-USSR:* 9,
Northern Russia; 10, southwest Ukraine (Podolia); 11, western
Urals; 12, Yenisey River (Igarka-Turukhansk); 13, Irkutsk (Lake
Baikal); 14, Anabar and Olenek (northern Siberia); 15, River
Maya. *China:* 16, northeast China (Longshan); 17, Yangtze Gorge.
*Northern Europe:* 18, Lake Torneträsk (Sweden); 19, English
Midlands   (Leicester).   *North       America:*   20,   southeast
Newfoundland; 21, North Carolina; 22, northwest Canada
(Mackenzie Mountains); *Middle East:* 23, central Iran. From
Glaessner (1984).

left no fossil trace or is represented by *Spriggina*. If *Spriggina* were a
genuine annelid, this would dramatically alter the structure of the tree,
as it would push the bilaterian radiation back well into the Vendian
(see Fortey *et al.* (1996) and Figure 1–1a).

Prior to the Ediacaran fauna (located in time at approximately 550–
580 *my* ago according to Bowring *et al.*'s 1993 timescale; see also Knoll
1996) there was a series of glaciations (the Varanger ice age: see next
section), and prior to those the only evidence of animal life takes the
form of trace fossils such as horizontal 'burrows'. However, these may
yet turn out to be pseudofossils (McMenamin 1989); and molecular
work suggesting a very 'deep' pre-Cambrian origin of animals (Wray
*et al.* 1996) may yet turn out to be based on an incorrect picture of

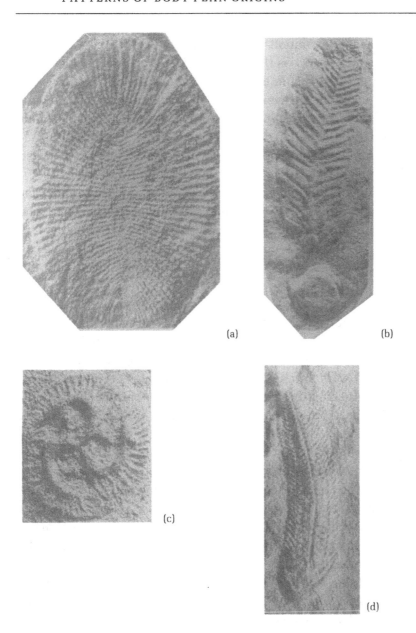

**Figure 3–9   Selected Ediacaran fossils**

(a)   *Dickinsonia*;   (b)   *Charniodiscus*;   (c)   *Tribrachidium*; (d) *Spriggina*. From Glaessner (1984). My own interpretation is that *Charniodiscus* may be a Cnidarian and *Spriggina* a primitive bilaterian, while the others represent a now-extinct Vendozoan grouping. Alternatively, it remains entirely possible that *all* Ediacaran fossils represent this grouping. From Glaessner (1984).

constant molecular evolutionary rates. It is therefore still entirely possible that Valentine (1994) is right that multicellular animals originated as late as 600 *my* ago, first in the form of Vendozoa, which either were subjected to an extinction event in the late Vendian caused by competition with "more progressive lines of Metazoa which extended into the Phanerozoic" (Fedonkin 1985), or, more likely, underwent a dramatic decline in the late Vendian, but with some Vendozoan lineages persisting until at least 510 *my* ago (Crimes, Insole and Williams 1995).

Whether or not this view is correct, it seems almost certain that there was rapid proliferation of animal morphological characteristics around the base of the Cambrian. We now need to enquire whether this proliferation produced not only body plans still represented in today's fauna, but also others, whose 'bearers' all became extinct before the end of the Ordovician.

Opinion on this matter – much of it centred around the famous Burgess Shale fauna from British Columbia – has almost (but not quite) come full circle. The discoverer of the Burgess Shale, Charles Walcott, described all the fossils he found as belonging to phyla still represented today, including annelids and arthropods (Walcott 1911a, b, c, 1912). Subsequently, many of these fossils were re-studied and considered to be representatives of other phyla, now wholly extinct, whose body plans were unlike any characterizing current metazoans (see, for example, Whittington 1975). Gould (1989, pp. 210–211) provides a list of all the redesignations of Walcott's specimens up to 1985.

Ironically, at about the same time as Gould's book *Wonderful Life* appeared, Briggs and Fortey (1989) performed a cladistic analysis of selected Burgess Shale fossils, along with other Cambrian fossil material from Wisconsin and Orsten (Sweden) and a few extant arthropods. One result was that two of the Burgess Shale arthropod fossils that had been thought to be unrelated to existing forms (*Yohoia*, Whittington 1974 and *Burgessia*, Hughes 1975) ended up, according to the cladogram, as being closely related to the chelicerate *Limulus*. In a follow-up study, Briggs, Fortey and Wills (1992) confirmed this close relationship of *Yohoia*, *Burgessia* and *Limulus*, and generalized this and similar results into a view that the Burgess Shale does not include a wide range of 'bizarre' or experimental body plans. A similar view, based on studies of the early Cambrian Chengjang fauna, has been advanced by Ramsköld and Xianguang (1991). More recently, Conway Morris and Peel (1995) have argued that *Wiwaxia*, which had previously been thought to be a unique body plan, is in fact a stem-group Annelid.

It will now be clear that there is a parallel between the debates on the nature of the Ediacaran and Burgess Shale faunas. In both cases an initially conservative description was challenged by a radical one; and the interplay between these conflicting views has, in both cases, seemed at least in part ideologically driven. The truth, as is often the case in such conflicts, is likely to lie somewhere between the extremes. It is difficult to dispute Briggs *et al.*'s (1992) rigorous cladistic analyses, which do indeed suggest a 'conservative' placing of *Yohoia* and *Burgessia*. Yet equally, it is interesting to note that these authors omitted from consideration *Anomalocaris*, whose extraordinary morphology (see Figure 3–10) does indeed argue for its being attributed to a new higher taxon, as suggested by Whittington and Briggs (1985). A review of diverse opinions on the taxonomic status of *Anomalocaris* has been provided by Collins (1996); and the varying interpretations of the four above-mentioned genera over time are summarized in Table 3–4.

Although there is clearly a healthy ongoing debate about the taxonomic status of many Burgess Shale and other early/mid Cambrian fossils, there is general agreement on the loss of some Cambrian body plans prior to the Ordovician. Even Wills, Briggs and Fortey (1994), who have argued strongly that overall morphological disparity was similar in Cambrian and Recent arthropods, conclude that "many Cambrian body plans have been lost". This conclusion is built into Figure 3–11, which represents a hypothesis of the general form that early animal phylogeny may have taken. Here the lineage divergence times illustrated correspond to the *latest possible* such times (cf. Figure 1–1a and b). Future fossil discoveries, and further molecular analyses like that of Wray *et al.* (1996), will no doubt eventually enable a consensus to be reached on when lineages actually diverged; but such a consensus is still some way off.

## 3.4 Bringing Back Morphology

At the start of this chapter, I indicated that I would embark upon Eldredge's (1979) three-stage strategy: from cladograms to trees to scenarios. The conventional view, at least within cladistics, is that to achieve the first of these transitions, we only need to bring geological time into the picture. That is: cladogram + timescale = tree; or, as Smith (1994) puts it: "cladogram + biostratigraphy → tree".

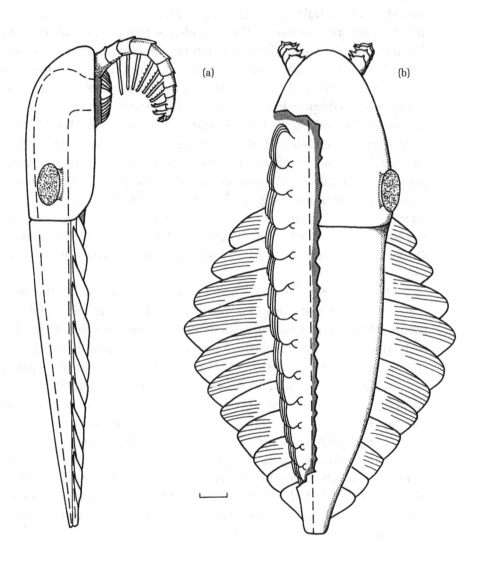

(a)

(b)

**Figure 3–10    Reconstruction of *Anomalocaris nathorsti*,
appendage partly flexed, lateral lobes in still position**
(a) Right lateral view, outline of alimentary canal dashed; (b) dorsal
view, portion of external cuticle on left side cut away to show gills
(diagrammatic) below, left margin of alimentary canal dashed.
Scale bar, 1 cm. From Whittington and Briggs 1985. *Phil. Trans.
Roy. Soc., Lond.*, **B309**, 569–609, with permission.

**Table 3–4. Changing taxonomic interpretation of selected Burgess Shale fossils**
Question-marks indicate tentative designations.

| Genus | Original interpretation by Walcott | Re-interpretation (and re-interpreter) | Third interpretation (and source of) |
|---|---|---|---|
| *Yohoia* | Branchiopod Crustacean | Unique Arthropod? (Whittington 1974) | Chelicerate (Briggs and Fortey 1989) |
| *Burgessia* | Branchiopod Crustacean | Unique Arthropod? (Hughes 1975) | Chelicerate (Briggs and Fortey 1989) |
| *Wiwaxia* | Polychaete Annelid | New phylum (Conway Morris 1985) | Stem Annelid (Conway Morris and Peel 1995) |
| *Anoma-locaris* | Branchiopod Crustacean | New phylum (Whittington and Briggs 1985) | Various (see review by Collins 1996) |

While evolutionary time clearly *must* be incorporated into the '*y* axis' – and I began to do this in the previous section – I would argue that in addition some explicit measure of morphology must be incorporated into the '*x* axis'. This is especially true when the trees concerned relate to the evolution of development, for in that context we are not just interested in lineage divergences, but also in what is happening to the developmental programmes within each lineage.

While this point may seem obvious to many evolutionary biologists, it may encounter some resistance within cladistics, and it is thus worth developing it in a little more detail: I will do this in the context of the two evolutionary trees shown in Figure 3–12. The patterns of lineage divergence in these are the same, and consequently so are the corresponding cladograms, which can be represented as ((A + B) + (C + D) + E) – leaving aside for the moment the ancestral/fossil species F.

Now consider the effects of calibrating the *x*-axis in morphological terms, as shown in Figure 3–12. The first, and most obvious, point is that in the upper tree, species E has diverged much more from the rest (about six 'units') than it has in the lower tree (where it has only diverged by a single unit). A second point worth noting is that we can no longer rotate any divergence through 180°; in either tree, D is morphologically closer to E than is C, for example. Finally, in the upper tree, because the fossil ancestor F is morphologically much

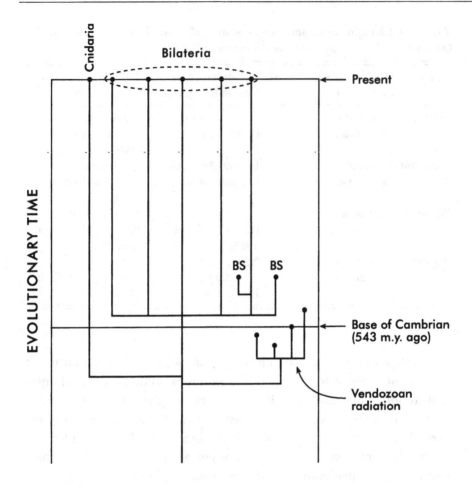

**Figure 3–11  An outline animal phylogeny, showing the most recent possible lineage divergence times**

BS = extinct animals present in Burgess Shale. Vendozoa = Seilacher's (1984) 'quilted' body plan. Fewer phyla of present-day Bilateria are shown than exist: for more detail see Figures 3–1 to 3–4. Also, contrast with the earlier proposed divergences shown in Figure 1–1.

closer to descendant species A–D than is the remaining descendant E, a morphological grouping of (A, B, C, D, F) makes sense, even though, from a cladistic perspective, such a group is paraphyletic.

We now turn to the difficult issue of what kind of units to use to calibrate morphology – or, as Gould (1991) has put it, to "quantify morphospace" (see also Chapter 10). Clearly, for consideration of body-plan divergence, measures of the size or shape of particular struc-

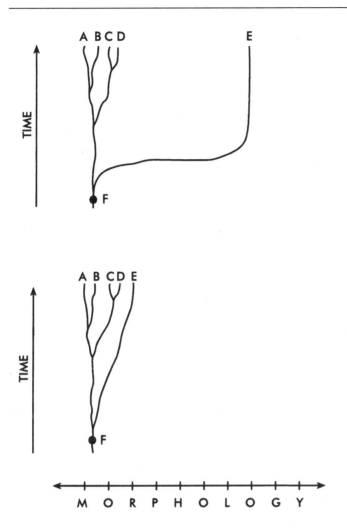

**Figure 3–12  Two trees which are identical in terms of cladistic structure and time calibration but different in the pattern of morphological divergence**

It is not intended to imply, here, that morphology can be satisfactorily measured along a single dimension – this is a limitation imposed by the two-dimensional page. See text for discussion of how to measure morphological divergence.

tures are inappropriate, since many organs or structures will be unique to particular body plans. Also, any measure of morphology *per se* is inappropriate in that evolutionary convergence will be capable of producing cross-overs or reticulations. For example, consider the case of gastropods evolving in shell shape as measured by the ratio of size of aperture to overall size of shell. (This is not a body-plan example, of course, but it is easier to make the point with a simpler, smaller-scale sort of divergence.) It would be quite possible in such a case to get a pattern involving morphological cross-overs, but this is unhelpful in two respects. First, if there are many cross-overs an uninterpretable tangle of lineages will result. Second, it is apt to be confused with the genealogical reticulation that results from speciation by allopolyploidy.

One way to solve this problem is to measure morphological divergence by the number of unique derived characters (autapomorphies) in each lineage. Thus, a very divergent group, such as E in the upper tree of Figure 3–12, is one possessing many autapomorphies. Another way to proceed is to use earliness of divergence of embryological pathways (i.e. earliness in developmental, not evolutionary, time). These two approaches should be broadly in agreement, since earlier divergence in development should generally lead to a greater number of autapomorphies – though of course there will inevitably be exceptions to this general correspondence.

This approach, then, allows us to recognize different degrees of morphological change associated with lineage divergences, in addition to the temporal order of these divergences; and it avoids the problem of 'tangled trees' mentioned above. If morphological separation is calibrated in this way (*number* of autapomorphies and/or *earliness* of developmental divergence), then it would in theory still be possible to rotate lineages through 180° around divergence points; but of course information on the *nature* of the developmental pathways and the resultant adult characters might make one layout look more sensible than its rotated alternative.

Three final points are in order. First, there is no necessary link here with Mayr's (1974) concept of 'genes-in-common'. Major morphological changes can be caused by a few key genes; and equally, changes in many less important genes may have only minor morphological effects. Second, divergence in morphological *type* does not necessarily mean divergence in *grade* – in the sense of one form being more complex or 'advanced' than another. Sometimes, of course, divergence does lead to a change in phenotypic complexity, but consideration of that issue is not necessary here and will be postponed until Chapter 11. Third, the re-injection of morphology into trees should not be taken to indicate that I am advocating a general return to phenetic methods (see

Sneath and Sokal (1973) and Sokal (1986) for details). Despite Hennig's (1981) contrasting of genealogical and morphotypical history, there is no reason why the two cannot be combined, with cladistically-based trees being supplemented with explicit information on morphological, and indeed developmental, distances.

As will be apparent, the suggestions made in this section are very preliminary. Much remains to be done before rigorous, developmentally explicit trees are as commonplace as cladograms are today. But this task is essential if we are to understand the way in which developmental systems have been modified in the course of evolution. Time-calibrated cladograms alone ('semi-trees', if you like) are an insufficient description of the patterns that evolutionary developmental biologists seek to explain.

## 3.5    Palaeoecology and Possible Adaptive Scenarios

Although there is considerable uncertainty over when the main animal lineages diverged, as we saw in Sections 1.2 and 3.3, it is clear that the period from about 600 to 500 *my* ago (essentially the Vendian and Cambrian) was a time of intense morphological innovation. On one view (Valentine 1994; Valentine *et al.* 1996) the main lineage divergences *and* body-plan origins occurred during this period (though not necessarily simultaneously). On another (Wray *et al.* 1996), the main lineages diverged pre-Vendian, but initially comprised animals of small body size and limited morphological complexity which then underwent parallel 'explosions' into a range of more elaborate forms in Vendian/Cambrian times. On either view, it makes sense to examine the ecological conditions that prevailed between 600 and 500 *my* ago – as far as they are known – to see if they give any clue to the reasons for the bursts of morphological innovation that occurred. However, we should keep firmly in mind the limitations of the 'adaptive scenario' approach (see Section 1.3).

Strangely, given the length of time involved, many environmental conditions were broadly similar in the Vendian/Cambrian to those prevailing in the present-day biosphere. It appears that by 600 *my* ago plate tectonic movements were occurring much as they do now (Kröner 1981), and the earth's surface was a mixture of marine areas and emergent continental land masses – the latter having existed since very much earlier times (Buick *et al.* 1995). Of course, the form and positions of the continents were different: at around 600 *my* ago there was a single supercontinent (Murphy and Nance 1991) which broke up over the following 50 *my* or so (Bond, Kominz and Devlin 1989). But the global sea level, which is arguably of more importance to marine animals than the exact positions of the continents, was broadly the same, at the Cambrian/pre-Cambrian boundary, as it is today (Figure 3–13). (See Hallam (1992) for a discussion of this highly complex issue.)

The atmosphere, too, was rather similar in composition to our own. The level of oxygen was approximately the same as it is now (Berner and Canfield 1989; see Figure 3–14), largely as a consequence of extensive Proterozoic photosynthesis by both prokaryotic and eukaryotic unicells – the latter having originated prior to 1.4 billion years ago (Walter, Du and Horodyski 1990), and possibly much earlier (Zhang 1986). Carbon dioxide levels around the base of the Cambrian were somewhat higher than at present (Berner 1991), but even so $CO_2$ appears to have been a relatively minor constituent of the atmosphere, just as it is now.

Global temperatures, also, were not dissimilar, at the base of the Cambrian, to those prevailing today. It is now generally agreed that there was a world-wide Varangan ice age extending from about 610 to 590 *my* ago, immediately prior to the Ediacaran radiation (Brasier 1992; Knoll 1992, 1996). There were probably no further global glaciations until the end of the Ordovician (Harland *et al.* 1990).

Turning to biotic aspects of the environment, there were again many similarities to the current biosphere. By around 600 *my* ago, many prokaryotic and protistan lineages had undergone considerable diversification (Knoll 1991; Lipps 1993), as had multicellular algae (Butterfield, Knoll and Swett 1990). More advanced plant groups (Bryophytes, Pteridophytes, Gymnosperms and Angiosperms) did not originate until much later (see Benton 1993). However, they are absent from most marine habitats today; and we should recall that the Ediacaran and Cambrian animal radiations were entirely marine.

Given all the above observations, it is probably true to say that for a newly-evolved macroscopic Ediacaran or early Cambrian animal, the aspect of their environment that differed most from that of present-day metazoans was the lack of other large animals. There were few pre-evolved direct competitors or mobile predators. There may also have been few animal parasites, although pentastomid parasites had appeared by the late Cambrian (Walossek and Müller 1994).

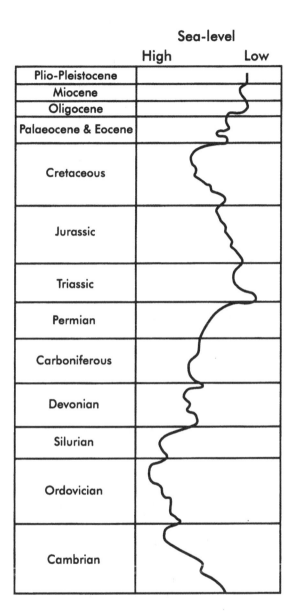

**Figure 3–13 Changes in sea-level from the Cambrian to the present**

Note that at the base of the Cambrian the level was very little higher than now. From Hallam 1992. *Phanerozoic Sea-Level Changes.* Columbia University Press, New York, with permission.

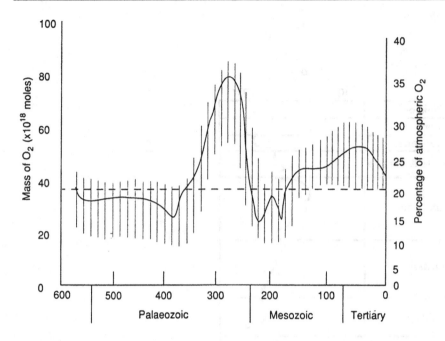

**Figure 3–14    Changes in atmospheric oxygen levels over the last 600 *my***
From Conway Morris 1995. *TREE*, **10**, 290–294, with permission of Elsevier Science Ltd.

The broad adaptive scenario that inevitably suggests itself, then, for the Ediacaran and Cambrian radiations, is one of proliferation into empty "niche space" (Arthur 1987a), "adaptive space" (Valentine and Erwin 1987; Valentine 1991) or "morphospace" (Gould 1991). We need to be careful in our usage of these terms, and in our elucidation of the underlying concepts. I will return to explore this issue in more detail in Chapters 9 and 10; but regardless of the fine tuning, the idea of an explosion of new body plans in relatively uncluttered ecospace is an exciting one. Animal evolution around the Vendian/ Cambrian boundary may have been very different to the much more constrained and conservative pattern that has characterized the later reaches of the Phanerozoic era.

# CHAPTER FOUR

# Evolutionary Developmental Biology

## 4.1 From Pattern to Mechanism

It is clear from the previous chapter that a spectrum of views exists on the pattern of early animal evolution and the way in which body plans originated. This spectrum can be characterized by describing its extremes.

At the 'radical' extremity: there was no multicellular animal life prior to 600 *my* ago; there was an explosion of body plans in Ediacaran times, with many becoming extinct, and a second body-plan explosion in the early Cambrian, again followed by many extinctions; evolution in Vendian and Cambrian times was much more 'experimental' than it is now; and internal factors such as developmental constraint (or early lack of it) are important in evolution as well as considerations about niche space and external adaptation.

At the opposite 'conservative' extremity: multicellular animals have existed since more than a billion years ago; body plans have come into being very gradually since then, and their origins require no special explanation; any appearance of 'explosiveness' is illusory and caused by differing degrees of preservation (due to both presence/absence of hard parts and changes in the prevailing geological conditions); the Ediacaran and Cambrian faunas are composed entirely of members of present-day phyla; early evolution was just as conservative as its later counterpart; and external adaptation is the overwhelming driving force throughout evolution, with internal, developmental considerations being of limited consequence.

Of course, the spectrum of views extending between these extremes is not a simple, unidimensional one. Many 'pick and mix' combinations can be imagined – for example a view that most Ediacaran creatures had bizarre, experimental body plans, while the Burgess Shale animals were all quite standard annelids, arthropods, and so on. But just as in politics, so in science composite radical and conservative counter-views tend to emerge.

My own view, at the time of writing, is that the truth lies closer to the radical than the conservative extreme. But, like any other evolutionary biologist working in the late 1990s, I have to admit that my view may turn out to be wrong. Suppose for a moment that it *is* wrong and that early evolution was really rather conservative. What then can be said about the origin of body plans?

This is an appropriate stage at which to recall the point made in the Preface – and several times since – that studies on the origin of body plans form part of the wider nascent discipline of Evolutionary Developmental Biology. Practitioners of this discipline seek, among other things, to understand the developmental–genetic causality of evolutionary divergence in morphological characters: both major, body-plan characters and lesser ones representing 'variations on a theme'. The fact that mammals have four *Hox* gene clusters while amphioxus has only one (Garcia-Fernàndez and Holland 1994) appears to be telling us – given the key developmental role of those genes – that gene duplication is important in the origin of at least some body plans, as suggested by Ohno (1970). Such a conclusion may well be independent of (for example) whether multicellular animals appeared 600 or 1200 *my* ago (see Valentine 1994 and Wray *et al.* 1996, respectively). It is also independent of many cladistic details, except those characterizing the main divergences within a cladogram of the chordates.

This leads us to an interesting conundrum regarding the interplay between pattern and mechanism. Specifically, let us focus on the following question: to what extent do we need to establish phylogenetic pattern before being able to make progress in understanding the evolution of development at a causal level?

If evolutionary developmental biology is to avoid the criticism levelled against neo-Darwinism by Rosen (1984) and others that it explains everything and therefore nothing, it clearly needs to be more tightly linked to systematic patterns than are the models of population genetics. Yet the above example concerning *Hox* genes reminds us that not all aspects of systematic pattern are important; and in general the resolution of lower-level patterns of cladistic relationship may well be unimportant with regard to establishing the main features of developmental evolution. As Medawar and Medawar (1977) have said, it is often the case that "nothing of any importance turns on the allocation of one ancestry rather than another".

High-level cladistic relationships *are* important, and that is why I concentrated on these in the previous chapter. If amphioxus turned out to be a protostome rather than a primitive chordate (rather unlikely!), then the interpretation of Garcia-Fernàndez and Holland's (1994) study would be very different. If we became uncertain of the monophyly of the vertebrates (equally unlikely), then doubt would be cast on the validity of the main example of von Baer's law. But changes in our view of phylogeny within, for example, particular groups of mammals would have no effect on either.

What of the additional information on pattern that is present in evolutionary trees: namely, morphology and time? The former is clearly crucial, for it is ultimately morphological patterns that we are trying to explain. The latter, time, is sometimes important, sometimes not, depending on our precise aim. And we should now examine in detail exactly what the aims of our new discipline are.

## 4.2    The Aims of Evolutionary Developmental Biology

In Chapter 2, I drew attention (following Panchen 1992) to the important distinction between the thing for which an explanation is sought (the *explanandum*) and the explanation itself (the *explanans*). The *explanandum* of evolutionary developmental biology can be put as follows:

> *The pattern of phylogenetic variation in (a) developmental pathways and (b) their underlying genetic architecture.* This overall pattern includes: the origins of body plans; general high-level trends at the phenotypic level (e.g. von Baer's 'progressive divergence', the statistical trend towards increased complexity, and so on); and variation in the degree to which the genetic architecture of development is hierarchically (and otherwise) organized.

It is also useful to phrase our aims in terms of specific questions for which we seek answers. While the list of three such questions below is not exhaustive, it nevertheless does include the most crucial issues:

1. How do developmental *mechanisms* and their *genetic control systems* map to the phylogenetic tree of descriptive ontogenies?
2. What does this mapping tell us about the sorts of mutational changes (and consequent cellular mechanism changes) involved in the evolution of development?
3. What implications, if any, do data collected to answer questions 1 and 2 have for the mechanism of spread of the new variant ontogenies at the population level?

The multiple points of connection of these questions with diverse areas of endeavour spread through molecular, developmental, population and systematic biology are readily apparent, and give some idea of the magnitude of the task at hand.

To some extent, our three questions form a logical sequence, and I will deal with them in that order in the chapters that follow (after a necessary digression, in Chapters 5 and 6, into the nature of developmental mechanisms). There is currently much interest, as will become apparent, in questions 1 and 2. In contrast, Whyte's (1965) hypothesis of internal selection and my own hypothesis of *n*-selection (Arthur 1984, 1988) are among the relatively few attempts so far to address question 3 – though of course we can envisage a possible answer of 'none' to that question from some neo-Darwinians.

Although I have looked at the systematic side of the overall pattern to be explained, in Chapter 3, I have not included any information on descriptive embryology. Readers wishing to pursue that side of things before going into causal explanations should consult any of the embryological texts available: both the purely descriptive older ones such as Kerr (1919) and also those of the more recent ones that still give plenty of descriptive detail, e.g. Counce and Waddington (1972), Torrey and Feduccia (1979), Balinsky (1981), Gilbert (1994) and Gilbert and Raunio (1997).

The final question that I wish to direct readers' attention to at this stage is: what is the background to modern evolutionary developmental biology or, to put it another way, what relevant work has been done over the last 200 years? While my answer to this will be much too short to satisfy historians of science, it nevertheless is long enough to require a separate section.

## 4.3    A Brief History

I will now give a brief, and inevitably incomplete, history of Evolutionary Developmental Biology: brief because my main intention in this book is to develop the subject in the present, not to analyse its past; and incomplete partly because of my own limitations as a historian, and partly because the discipline is not delineated by clear boundaries, so it is a matter of opinion what is within it and what is not. For those who like their history even briefer, the milestones in Table 4–1 represent a sort of 'potted summary' of this section.

Two points are worth making in advance about this history. First, the various contributions are heterogeneous and disconnected. Haeckel (1866), Thompson (1917), Goldschmidt (1940) and Waddington (1957), for example, are all key figures, but they all took different starting points and built theoretical structures in different directions. This disconnectedness relates to the lack of 'discipline' status – until recently – of Evolutionary Developmental Biology. Second, although much nineteenth-century work was concerned with body plans, particularly the vertebrate one (e.g. Owen 1848), many of the best known studies do not directly connect with the question of body plan origins. For example: Von Baer's laws are usually exemplified by comparisons *within* the vertebrate body plan (as in Figure 2–7); Goldschmidt's 'systemic mutations' are often thought of as changing one sort of insect into another; and Thompson's transformations take us from one crustacean carapace to another (e.g. Figure 4–1), one mammalian skull to another, and so on – although he also was aware that transformations only operated within limited domains, and that 'jumps' between these would sometimes be necessary. Current evolutionary developmental biology clearly needs both to become more coherent and to renew its focus on the body-plan issue.

Without meaning to imply that no one thought about these issues prior to the end of the eighteenth century, I will begin with the period 1790–1830 during which the German school of Naturphilosophen flourished (key figures including Kielmeyer, Autenrieth, Oken and Meckel), as did their equivalent French school, led by Geoffroy Saint-Hilaire and (later) Etienne Serres (for details see Russell (1916) and Gould (1977a, chapter 3)). Theirs was a romantic, metaphysical view of the world, and a central pattern within this view was what we would now call recapitulation: embryos of 'higher' animals (such as ourselves) pass through stages resembling the more primitive animals

**Table 4–1. Milestones in the history of evolutionary developmental biology**
The table incorporates the most influential books; papers are too numerous to receive individual mention. Most of the entries are expanded upon in the text. Good historical sources for additional detail are Russell (1916), Gould (1977a) and Hall (1992). Some book titles are given here in abbreviated form: see References for full detail.

**Phase I: pre 1800–1900: 'Comparative embryology'**

| | |
|---|---|
| ~ 1800 | *Naturphilosophen*: pre-evolutionary romantic recapitulation |
| 1828 | von Baer's *Entwicklungsgeschichte der Tiere*; the principle of divergence or deviation |
| 1848 | Owen's *Archetype and Homology*: evolutionary concepts in non-evolutionary guise |
| 1859 | Darwin's *Origin of Species*: comparative embryology taken on board but not given a central role |
| 1866 | Haeckel's *Generelle Morphologie*: evolutionary recapitulation and the biogenetic law (ontogeny repeats phylogeny); see also Haeckel (1896) |

**Quiescent period: 1900–1940\***

| | |
|---|---|
| 1917 | D'Arcy Thompson's *Growth and Form*: the theory of transformations and the correlation of parts |

**Phase II: 1940–1957: 'Excluded from the synthesis'**

| | |
|---|---|
| 1940 | Goldschmidt's *Material Basis of Evolution*: chromosomal repatterning, systemic mutation and hopeful monsters |
| 1940 | de Beer's *Embryos and Ancestors* (previously published as *Embryology and Evolution*): emphasizes the role of heterochrony |
| 1942 | Huxley's *Evolution: the Modern Synthesis*: development claimed to be central to the synthesis, yet features little, with the exception of allometry |
| 1949 | Schmalhausen's *Factors of Evolution*: stabilizing selection favours robust, buffered morphogenetic systems |
| 1957 | Waddington's *Strategy of the Genes*: epigenetic landscapes, canalization and genetic assimilation |

**Quiescent period: 1957–1977**

| | |
|---|---|
| 1965 | Whyte's *Internal Factors in Evolution*: emphasizes the importance of internal selection for coadaptation |

**Phase III: 1977–Present: 'Current resurgence'**

| | |
|---|---|
| 1977 | Gould's *Ontogeny and Phylogeny*: like de Beer, a focus on heterochrony |
| 1980s | A plethora of books on various topics (listed in Section 1.4) |
| 1992 | Hall's *Evolutionary Developmental Biology*: a name for the 'new' discipline; and a focus on hierarchical epigenesis |
| 1993 | Kauffman's *Origins of Order*: inputs from complexity theory |
| 1994 | Akam *et al.*'s *Evolution of Developmental Mechanisms*: reveals the massive potential impact of 'comparative developmental genetics' on evolutionary biology |

\*But see Chapter 10 for discussion of the mutationist/selectionist debate.

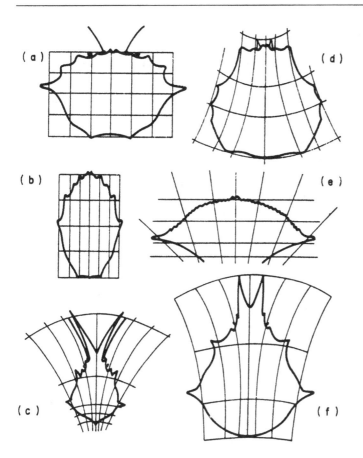

**Figure 4–1  Two-dimensional outlines of the carapaces of several crabs**

(a) *Geryon*; (b) *Corstes*; (c) *Scyramathia*; (d) *Paralomis*; (e) *Lupa*; (f) *Chorinus*. Appropriate transformations of the grid on which the outline of one genus is plotted give good approximations to the carapaces of the other genera. Numerous, apparently unrelated, minor differences in shape are thus seen as part of the same overall transformation. From Thompson (1917).

as they develop; though of course this was not seen in the light of evolutionary causality – it was simply a pattern.

Von Baer (1828) rejected this pattern in his fourth law (Table 4–2) and put in its place an alternative one, which can be called divergence or deviation (see Figure 2–7). Here, common embryonic forms gradually diverge, in different taxa, leading to dissimilar adults. So embryos pass through stages where they resemble other embryos, but never

**Table 4–2. Von Baer's laws (from Gould 1977a)**

1. The general features of a large group of animals appear earlier in the embryo than the special features.

2. Less general characters are developed from the most general, and so forth, until finally the most specialized appear.

3. Each embryo of a given species, instead of passing through the stages of other animals, departs more and more from them.

4. Fundamentally therefore, the embryo of a higher animal is never like [the adult of] a lower animal, but only like its embryo.

through stages where they resemble other adults. This view has gained general acceptance at least among vertebrate workers, though it is now acknowledged that *very* early stages are also dissimilar due partly to the differing amounts of yolk in different kinds of egg. This modification has led some authors to think of development as being the shape of an egg-timer (e.g. Slack *et al.* 1993; Rieppel 1993; Duboule 1994); but of course it is a very asymmetric egg-timer with the point of constriction close to one end. The extent to which von Baer's laws apply to invertebrates, and particularly those with complex life histories, remains to be resolved (see Chapter 11).

When Darwin (1859) changed our world view to an evolutionary one, von Baer's divergence became potentially explicable in mechanistic terms (as did Owen's (1848) homology), and indeed Darwin attempts such an explanation, though he relegates it to the penultimate section of his penultimate chapter, in contrast to his four full up-front chapters on variation and natural selection. He thus (perhaps unwittingly) set the scene for the dominance of external (ecological) over internal (developmental) factors in neo-Darwinism. Despite this relegation, what Darwin says is, as usual, very sensible. He suggests that embryonic similarity giving way to adult divergence in cross-taxon comparisons may be caused by the evolutionary incorporation, in some lineages, of modifications that only take effect part-way through embryogenesis. One goal of evolutionary developmental biology is to flesh out this explanation with developmental genetic detail, and to make it more quantitative (see Arthur 1988, chapter 2; and later chapters herein).

The leading figure in the evolutionary interpretation of comparative embryological patterns in the period immediately following publication of *The Origin of Species* was Haeckel (1866), whose views on pattern were influenced both by the Naturphilosophen and by von Baer. Haeckel's 'biogenetic law' – ontogeny recapitulates phylogeny – is often said to claim that the embryo of a higher animal goes through a stage resembling the *adult* of a lower one (see, for example, Gould 1977a). However, (a) Haeckel was too good an embryologist to believe this (Mayr 1994); (b) some of his statements of the biogenetic law (e.g. Haeckel 1896, p.5) make no explicit mention of adults; and (c) his descriptions of the actual relationships between the embryos of different species sound decidedly 'von Baerian'. Consider, for example, his remarks on mammalian embryos (1896, p.18): "The fact is that an examination of the human embryo in the third or fourth week of its evolution [= ontogeny] shows it to be altogether different from the fully developed Man, and that it exactly corresponds to the *undeveloped embryo-form* presented by the Ape, the Dog, the Rabbit, and other Mammals, at the same stage in their Ontogeny." (my italics)

Towards the end of the nineteenth century, the study of both development and evolution underwent considerable change. In embryology, the comparative approach of von Baer, Owen, Haeckel and others gave way to the more mechanistic, experimental approach of *Entwicklungsmechanik* (Roux 1894). In evolutionary biology, a dispute raged between the Mendelians and the biometricians on the relative importance of continuous and discontinuous variation. While this *could* have included a very explicit developmental dimension, the form the dispute took (see Chapter 10) tended to down-play this 'fourth dimension' of the phenotype, and to concentrate on the other three.

Out of the resultant void, then, came Thompson's *On Growth and Form* (1917), and in particular his theory of transformations, wherein the morphologies of related species are seen to differ not in countless independent respects but rather in a coordinated way, which can be represented by a generalized deformation of one form to produce another (Figure 4–1). Thompson's focus was, as is well known, on elegant mathematical description rather than on causality; but a causal interpretation of the patterns he revealed must be an integral part of evolutionary developmental biology. Thompson was critical of Darwin, arguing that "Our geometrical analogies weigh heavily against Darwin's concept of endless small continuous variations". Yet Darwin (1859) devoted a section to "Correlation of Growth" and was well aware of what quantitative geneticists now describe as the responses of correlated characters to selection on a particular 'target' character (see Figure 4–2 and Falconer (1989) for examples). But what produces (a) correlation of characters within a developing organism, and (b) correlated character changes in the course of evolutionary divergence? Pleiotropic effects of developmental genes, and long-range morphogens such as growth hormones are two of the ingredients – but a detailed answer still eludes us.

One of the foundations of the 'modern synthesis' came in 1930 with the publication of Fisher's *Genetical Theory of Natural Selection* (which was followed by important works by Wright (1931) and Haldane (1932)). While the 'synthesis' has largely served to *exclude* developmental considerations from mainstream evolutionary thought, one of Fisher's arguments – when seen in a different light – may in fact make an important contribution to evolutionary developmental biology. I have already dealt with this (the 'Fisher principle': Section 1.5). In a historical context it serves to illustrate the difficulty, previously alluded to, of deciding what to include in the history of a discipline and where interdisciplinary boundaries lie.

One of the most important, but also most controversial contributions to the evolution of development debate was Goldschmidt's (1940) *Material Basis of Evolution*. In it, Goldschmidt argued that intraspecific variation was irrelevant to transspecific evolution – a view we now know to be wrong. Ironically, the extremity of Goldschmidt's views may have done more to exclude developmental considerations from the modern synthesis than the rather non-developmental tendencies of many of its architects, as noted in Section 1.6. However, while Goldschmidt was mistaken about the nature of the typical speciation event, some of his other ideas may yet turn out to be important.

De Beer's work on comparative embryology (1940) deserves a mention both for what it did and did not do. It did indicate how selection might generate interesting evolutionary patterns, such as 'clandestine evolution' involving divergent larval adaptation followed by paedogenesis (accelerated reproductive maturity) and loss of the adult stage. But in its generally placatory tone it did *not* highlight the difficulties that might lie ahead in integrating development into the synthesis more fully. In a sense, de Beer's message was 'there's no problem'; but this is not the case, as we saw in Chapter 1.

The following two quotations are taken from the Preface of a book written in 1942:

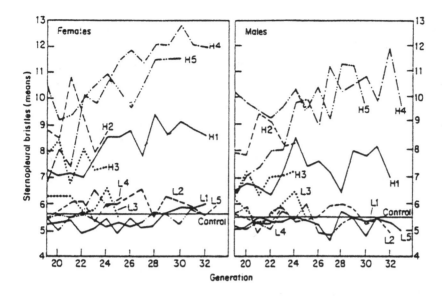

**Figure 4–2   Correlated responses of sternopleural bristles in lines
of *D. melanogaster* selected for high (H) and low (L) numbers of
abdominal sternital bristles**
From Clayton *et al.* (1957).

Any originality which this book may possess lies partly in its ... stressing the fact that a study of the effects of genes during development is as essential for an understanding of evolution as are the study of mutation and that of selection.

Equally obvious is my debt to the Morgan school and to Goldschmidt: but clearly this would apply to any modern book dealing with evolution.

It might not be surprising if the book concerned had been an anti-synthesis monograph by one of Goldschmidt's students. But it *is* surprising that the quotations are in fact from Huxley's (1942) *Evolution: The Modern Synthesis.* It is interesting that Huxley emphasizes so strongly the need for, effectively, a discipline of evolutionary developmental biology. His own contribution of allometric growth (1924, 1932) was valuable, but was not of sufficient importance to make development central in the synthesis as it would seem that he had intended.

One of the most creative evolutionary developmental biologists was Waddington, who in a series of publications (notably 1940, 1953, 1975 and especially *The Strategy of the Genes* (1957)) developed a network of interrelated ideas. These include an abstract picture of developmental pathways (the epigenetic landscape), a dynamic equivalent of homeostasis applicable to these pathways (homeorhesis) and a related concept associated with the idea of developmental constraint (canalization). Waddington also developed, and experimentally tested, the idea of genetic assimilation, wherein selection can make constitutive a phenotype that was previously only produced in response to an environmental stimulus – an apparently, but not actually, Lamarckian process (Waddington 1953, 1956). With the exception of genetic assimilation, Waddington's ideas have not been accepted by neo-Darwinians, one of whom (Jones 1984) describes them as "a posy of analogies". Yet *all* of Waddington's ideas are of potentially wide application, and developmental constraint may be of particular importance.

Many of Schmalhausen's (1949) and Waddington's ideas, though developed independently, had much in common, as Schmalhausen himself remarks in the preface to his book. One of his main themes is the ability of natural selection to make developmental programmes less dependent on environmental cues and more self-reliant, or autonomous. This would be expected, for example, where short-term environmental variation is sufficiently pronounced to cause problems for those genotypes whose ontogenetic direction is environment-dependent. Other genotypes which are capable of a repeatable ontogenetic sequence to the 'correct' adult form *despite* environmental variation would be favoured in such circumstances. There is a clear link here with Waddington's genetic assimilation. Overall, Schmalhausen thought of selection very much as a creative force in evolution. The degree to which selection is truly creative is a complex issue; I will discuss it further in Chapter 10.

Normally, 'selection' is thought of as being related to external factors (predators, resources, climate), while 'constraint' is related to internal ones. But this pairing of the two dichotomies is false. In particular, selection may favour variants which have improved internal co-adaptation – in contrast to improved adaptation to a particular environment. This message has been stressed on several occasions, but has been put particularly clearly by Whyte (1965), who also discusses possible reasons why internal/developmental considerations have so far remained undeservedly peripheral to mainstream evolutionary biology.

Gould's (1977a) *Ontogeny and Phylogeny* is partly an exposition of the importance of heterochrony in evolution. While I agree that it is indeed important, I doubt if heterochrony is the key to understanding the origin of body plans – rather it relates to the proliferation of variants of these plans. New structures and organs have to be produced before they can be accelerated or retarded developmentally in relation to each other. Heterochrony, like allometry, is important throughout evolution, but probably has no special role in the early 'morphogenetic' phase. Gould's book is also historical; and it goes into a good deal more detail than I have done here, so readers interested in the development of the subject would do well to read it.

Not only does Gould's (1977) book include some history – it may also have been instrumental in influencing history, in that it appears to have helped to usher in the "current resurgence phase" (see Table 4–1) in which we now find ourselves. I do not intend to give an account of this here (and so my 'history' effectively ends at 1977). The reason for this apparent omission is that the whole of the book effectively constitutes an account of evolutionary developmental biology as it is now, and as it has developed through the 1980s and 1990s.

## 4.4    Is There a Theory of Development?

It has sometimes been suggested that in order to achieve an expanded evolutionary synthesis we need to incorporate not just piecemeal developmental ideas or mechanisms, but rather some overall 'theory of development' comparable to Mendel's laws of genetics, which of course were incorporated into the modern synthesis at the outset. For example, Waddington (1975, p.11), with a healthy disregard for the predominantly non-evolutionary ethos of the 1820s, states: "It has been clear at least since von Baer's day that a theory of evolution requires, as a fundamental part of it, some theory of development."

This view, however, begs the question posed in the section heading, namely: *is* there a theory of development awaiting discovery? Looking across the various branches of biological science, it is clear that some are very much organized by the presence of a central 'core' theory. This is true of evolutionary biology itself (natural selection) and of transmission genetics (Mendel's laws). It is much less true of, for example, biochemistry or community ecology (see Arthur 1987b). Clearly, we do not *at present* have a comprehensive theory of development.

Berrill's claim (1961, p.1) that "no general theory of development has emerged, in spite of the mounting mass of observational and experimental information" is as true now as it was then. Because we have no theory, we cannot be certain that one is possible (or that it is not).

This fact, however, has not prevented various authors from adhering to strong views one way or another. Waddington was clearly of the opinion that a theory of development will eventually emerge. In contrast, Weiss (1940) makes the following assertion: "The revelation of the multiplicity of developmental processes and mechanisms has been a sad disappointment, for it has removed all hope of a general, comprehensive and universal formula [= theory] of development."

Weiss's logic, however, is imperfect. The existence of multiple mechanisms through which an overall process operates does not of itself necessarily preclude the emergence of a general theory of that process. Transmission of characters from parents to offspring may be sex-linked or autosomal, may exhibit a 'normal' or maternal pattern of inheritance, and may be unifactorial or polygenic – but this diversity does not detract from Mendelian theory.

Also, theories may operate within varied domains and at different levels. For example, it can be argued that the 'variable gene activity theory' (see Davidson 1986, chapter 1) is indeed a general – albeit insufficiently detailed – theory of cell differentiation (the counter-theory of selective loss of blocks of genetic material having been disproved). However, it is clearly *not* a theory of pattern formation; different muscles (for example) with different characteristic shapes are composed of a similar range of cell-types. How they come to be different in overall form requires some quite distinct 'theory of morphogenesis'.

In some cases, a theory emerges from the realization that things are in some sense the opposite of what they were previously thought to be. Transmission genetics is a classic case of this: a theory of particulate inheritance upon which the whole subject is now based triumphed over the previous view of blending inheritance, wherein germ materials from the two sexes were thought to merge irretrievably, and there was no equivalent of Mendelian segregation. (However, pre-Mendelian views were more heterogeneous than is now sometimes thought, and phenomena such as reappearance of traits after a generational gap, which could be taken as suggestive of segregation, were known (see Sturtevant 1965, chapter 1 for a brief review).) In relation to this particular advance, Fisher (1930, p.7) makes the following point:

It is a remarkable fact that had any thinker in the middle of the nineteenth century undertaken, as a piece of abstract and theoretical analysis, the task of constructing a particulate theory of inheritance, he would have been led, on the basis of a few very simple assumptions, to produce a system identical with the modern scheme of Mendelian or factorial inheritance.

Now this may well be true, but there is an important caveat. It was only *after* the devising of Mendelian theory that it became clear that the way in which the old, incorrect 'theory' and the new correct one differed was in their blending versus particulate natures. Fisher's mid-nineteenth century thinker had no reason to go for a particulate theory rather than (say) a 'grandparental theory' (in which inheritance skips a generation: $F_2$ characteristics are determined by the P generation, $F_3$ by the $F_1$ and so on). This also would have differed from the existing ('parental') theory but along a different axis.

In this context, it may be helpful to look at various dichotomies that are already recognized within development. While this can hardly be expected to lead 'eureka-style' to a theory of development, it does assist us to focus on the nature of the overall developmental process, and to separate those dichotomies where we are now certain that one alternative is correct, the other incorrect, from those where there is a messier 'some-of-each' conclusion, in line with Weiss's (1940) pessimistic picture of a plethora of mechanisms.

A list of important 'dichotomies' is given in Table 4–3, although I should stress that some of these are really just opposite extremes on a spectrum of possibilities. I have used superscripts to indicate: cases where we now know one alternative to be correct (1), the other wrong; and cases where we know or strongly suspect one alternative to be much commoner (2) than the other. In all the other cases (no superscript) both are known to be important and of frequent occurrence in developmental systems. In these cases, there may be patterns in the occurrence of the two alternatives – both those that we already know (like the ontogenetic time trend from maternal to zygotic action of development-controlling genes), and those that await discovery.

The most striking conclusion from Table 4–3 is that in only one case can we rule out one of the alternatives. It is now apparent that the old idea of a preformed 'homunculus' inside the egg or sperm is incorrect; rather, all information is heavily encoded at that stage, and development proceeds by a complex epigenetic process of translating this information into actual structures.

**Table 4–3. Developmental dichotomies**

There is no significance to the order of the list given in the table. Relationships between different entries vary: some dichotomies are independent of each other, but some would be expected to interact.

| | |
|---|---|
| Short-range signals<br>  cell surface molecules<br>  induction via contact | Long-range signals<br>  diffusible morphogens,<br>  hormones, growth factors |
| Short-term effects | Long-term effects |
| Cell differentiation<br>  within-cell changes of shape,<br>  size, content | Pattern formation/morphogenesis<br>  processes involving whole cell<br>  populations |
| Tendency to become different<br>  regionalization of embryo<br>  possibly hierarchical<br>  leads to quasi-independence | Tendency to stay the same<br>  includes canalization,<br>  coordination (e.g. of sides),<br>  repeatability (between individuals) |
| Single control systems<br>  high risk of error | Dual-multiple control systems[2]<br>  redundancy: lower risk |
| Gene switching → gene products<br>  control by nuclear factors | Gene products → gene switching<br>  control by cytoplasmic factors |
| Positional information<br>  cell 'knows where it is'<br>  see Wolpert (1969, 1971) | Historical information<br>  e.g. cell registers mitoses since<br>  some 'milestone event' |
| Preformation<br>  zygote contains miniature adult | Epigenesis[1]<br>  zygote contains coded information |
| Systematically determined axes[2]<br>  factors used include maternal<br>  axes, gravity, point of sperm entry | *De novo* stochastic axes<br>  process of axis determination<br>  starts from slight concentration<br>  fluctuations (see Turing 1952) |
| Single gradient systems<br>  applies to mid-range signalling<br>  as in 'French flag' model | Dual/multiple gradient systems[2]<br>  again, mid-range signalling<br>  see Meinhardt (1982) |
| Compartmentalized in space[2]<br>  developmental–genetic evidence<br>  cell communication evidence | Spatially holistic<br>  embryo undivided<br>  found in very simple organisms? |
| Maternally-acting genes<br>  true of many genes controlling<br>  early development | Zygotically-acting genes<br>  true of most genes controlling<br>  later development |
| Continuous time development[2] | Discrete-stage development |
| Morphogenesis via cell migration<br>  cell populations move around<br>  within embryo | Morphogenesis via *in situ* proliferation<br>  cell populations 'stay put' but<br>  divide in spatially controlled way |
| Mosaic development<br>  lack of flexibility<br>  particular cells give rise to<br>  particular structures | Regulative development<br>  more flexible<br>  cell fates can be altered without<br>  derailing development |
| Chemical signals<br>  e.g. hormone | Electrical/other signals<br>  see Hotary and Robinson (1992) |
| Signals causing activation<br>  positive control | Signals causing inhibition<br>  negative control |

For interpretation of superscripts (1, 2) see text.

With regard to the dichotomies where one alternative is known or thought to be predominant, we can say the following: development is normally continuous in time but compartmentalized in space to varying degrees, and its repeatability is ensured by the existence of multiple control systems. The *beginning* of development requires no special explanation as there is normally a spatial carry-over from the previous generation or a mechanism for using environmental information for initiating important early events such as determination of major body axes.

Within these few constraints, then, development can be seen as a very heterogeneous process involving many different mechanisms – in this, Weiss (1940) was certainly right. But was his pessimistic inference of the lack of a general theory also right? Does the existence of multiple mechanisms really indicate that there is no over-arching theory or principle of development?

It is at points like this that scientists resort to what MacArthur (1972) calls "insight" and Thorpe (1978) calls "faith". The phrases 'gut feeling' and 'hunch' also spring to mind. All very disreputable sounding, it has to be said, but then the myth of the totally objective scientist has long ago been laid to rest. My own gut feeling is that an important general principle about development will eventually emerge, and I will permit myself a couple of paragraphs on the sort of thing it is likely to include. Readers who dislike speculation can skip over these two paragraphs.

One of our main problems is to connect within-cell and among-cell processes. Patterns of gene switching within a cell cannot readily be used to answer questions such as why a finger is different from a thumb. To make this connection, we need, among other things, to bring cell surface molecules into the picture, as some of these (e.g. cadherins: see Chapter 5) constitute a sort of 'glue' or 'velcro' that helps cells stick together into characteristically shaped blocks of tissue. As Gilbert (1994, p.88) states:

> The local patterns of expression of these cell surface molecules are thought to provide a major link between the one-dimensional genetic code and the three-dimensional organism. By modulating the appearance of these molecules, the genetic potential of the genome can become manifest in the mechanical process of morphogenesis.

My guess is that the general principle of development, when it emerges, will have three main molecular strands: genes, cell surface molecules and mobile morphogenetic agents. Perhaps pattern formation within regions or compartments will be largely explicable through the first two of these along with mid-range morphogens, while signalling agents with long-distance effects will be seen to aid coordination among regions and between the opposite sides of the body. Perhaps the way in which these three strands interact is, at some level, truly general, and thus applicable to worms, flies, mice and humans equally, despite some differences in molecular detail across different taxa.

So much for speculation. But we ought to ask, at this stage, how it is possible to advance evolutionary developmental biology at all in the absence of the much-sought theory of development.

The answer is that a knowledge of detailed mechanisms is itself helpful even in the absence of a general theory. Moreover, the mechanistic details that have become understood over the last couple of decades are largely *genetic* ones; and genes form a common language for both development and evolution. Mechanisms such as induction, revealed by the elegant experiments of the early embryologists (see Spemann 1938) were rather difficult to translate into evolutionary terms because their genetic basis was obscure (though the role of genes in some inductive processes is now emerging: Chapter 5). In contrast, mechanisms such as genetic cascade systems involving different classes of development-controlling genes (see Ingham 1988) should be less difficult to 'translate' in this way.

So we proceed armed with (a) 'partial theories' (like the variable gene activity theory of cell differentiation), (b) 'old' information on developmental mechanisms (induction, gradients and so on), and (c) relatively recent discoveries on the developmental roles of certain groups of genes, including those containing homeobox sequences (which I will consider in Chapter 6). But before moving from a historical and philosophical perspective into the details of developmental–genetic mechanisms, I would like to highlight opposing views on the usefulness of actively striving to construct a theory of development, expressed by two leading early figures.

According to Spemann (1938, pp. 367–8):

> It is not my intention to conclude these experimental contributions to a theory of development with the attempt to construct such a theory myself. I do not wish to devise hypotheses as long as exact knowledge is attainable by experimental work. Besides, I believe that if the facts are not compiled at random, but gained in a logical proceeding, they

will by themselves join together to build up a genuine theory in the original meaning of this word, i.e., a comprehensive view of all facts afforded by experience.

I should like to work like the archaeologist who pieces together the fragments of a lovely thing which are alone left to him. As he proceeds, fragment by fragment, he is guided by the conviction that these fragments are parts of a whole which, however, he does not yet know. He must be enough of an artist to recreate, as it were, the work of the master, but he dare not build according to his own ideas. Above all, he must keep holy the broken edges of the fragments; in that way only may he hope to fit new fragments into their proper place and thus ultimately achieve a true restoration of the master's creation. There may be other ways of proceeding, but this is the one I have chosen for myself.

Thus I wish to emphasize that I have never constructed, as far as I know, an "organizer theory". I coined the term "organizer" to describe some new and very remarkable facts which I ran across during my experiments. I have, however, from the beginning considered this conception a preliminary one and have, more than once and in a formal manner, characterized it as such. I have nothing to do with any attempts to make it the foundation of a theory.

In contrast, Needham (1934, p.221) is of the view that

there can be no doubt that a plethora of observation and experiment is...bad for scientific progress. Modern biology is the crowning instance of this fact. What has been well called a "medley of *ad hoc* hypotheses" is all that we have to show as the theoretical background of a vast and constantly increasing mass of observation and experiments. Embryology in particular has been theoretically threadbare, since the decay of the evolution theory as a mode of explanation.... The unfortunate thing is that nothing has so far been devised to put in its place. Experimental embryology, Morphological embryology, Physiological embryology, and Chemical embryology form today a vast range of factual knowledge, without one single unifying hypothesis, for we cannot yet dignify the axial gradient doctrines, the field theories, and the speculations on the genetic control of enzymes, with such a position. We cannot doubt that the most urgent need of modern embryology is a series of advances of a purely theoretical, even mathematico-logical, nature.

It will be interesting to see whether the general theory of development, if and when it emerges, is produced by someone adopting Spemann's 'wait and see' approach or by someone operating more proactively, as advocated by Needham. It will also be interesting to see what impact it has on evolutionary biology, over and above the injection of developmental–genetic details which is already happening and is described in subsequent chapters.

# CHAPTER FIVE

# Developmental Mechanisms: Cells and Signals

## 5.1    Strategy

It is important, at this stage, to look at some of the ways in which developmental mechanisms work. Otherwise, the phylogenetic comparisons of development-controlling genes, to be presented in Chapter 7, will have little meaning. Differences between the DNA sequences of homologous developmental genes in different lineages are of some interest in themselves; but they are of much *more* interest if we can connect them with differences in the processes over which the genes concerned have a controlling influence. Ultimately, we want to be able to understand the whole chain of events from altered DNA sequence to altered gene product to altered developmental mechanism to altered morphology.

I will now proceed as follows:

1. In this chapter, my emphasis will be on cellular processes; genes do feature at various points, but they are not the primary focus of attention. In Chapter 6, the complementary gene-centred approach is adopted. These should be seen as convergent routes to a common goal: understanding developmental mechanisms as fully as possible.

2. Throughout this chapter and the next, I will be very selective, using only a few examples to illustrate key points. This approach allows me to deal fairly intensively with each example, without my overall account of developmental mechanisms becoming excessively long. However, it carries a potential danger. Within any one category of mechanism (e.g. induction) there is some variability in the details from one example to another. Much of this variability is 'invisible' in a selective approach, and readers ought to keep this in mind. For a more comprehensive treatment of development, see Gilbert (1994).

3. I have concentrated on between-cell processes. In most cases of intercellular signalling, there is also an intracellular mechanism for getting the incoming signal to the genes, where it has a switching effect of some sort. These intracellular processes – signal transduction pathways – have been reviewed by several authors (see, for example, Lauffenburger and Linderman (1993, Chapter 5), Hoch and Silhavy (1995), Irvine, Michell and Marshall (1996)) and I will devote little space to them here.

4. In the final section of both this chapter and the next I will examine how individual developmental mechanisms happening in particular parts of the embryo at particular times connect up to produce the overall developmental programme. This is of course a very complex issue about which we still know rather little. But its importance is such that we should at least attempt to identify the key questions, and to produce some testable hypotheses by way of answers.

## 5.2 Cellular Processes and Architecture

No developmental stage of any organism is spatially homogeneous. Even the simplest, unicellular stages – such as the zygote – contain cytoplasmic localizations of many things, most importantly of morphogenetically active molecules. So one way of looking at development is as a temporally ordered sequence of spatial heterogeneities or patterns. Often, the time sequence is from relatively simple in spatial terms (with most developmental information present in coded form) to more spatially complex (with much developmental information decoded and 'actualized' in the form of physical structures). However, in organisms with complex life histories there is a within-generation cyclic element to the ratio of 'coding to actualization', as well as the between-generation cycling of this ratio that applies to all organisms.

This temporally ordered sequence of morphological heterogeneities that we call development generates adult tissue patterns that, in some taxa, can be highly complex, involving very precise and repeatable arrangements of billions, even trillions, of cells. Yet such patterns are achieved through the control of just a few major cellular phenomena. Most notable among these are: plane of division; rate of division; tendency to move; direction and rate of movement; differentiation into a particular cell type; and cell death (apoptosis).

Development is possible only if cells 'know' what to do in all these respects. So the key question becomes 'how *do* they know?', and the whole of developmental biology could be regarded as an attempt to answer this question.

Let us focus in on a single cell within some arbitrarily chosen stage in the developmental sequence of an unidentified animal (see Figure 5–1). Any such cell receives a variety of signals from its surroundings, some of which originate from adjacent cells, while others have traversed longer distances before being received by the cell in question. As Sternberg (1993) has commented: "the fate of a single cell depends on multiple intercellular signals". That is, a cell's behaviour, in terms of division, movement and differentiation, is determined by the totality of incoming intercellular signals, in conjunction with its cell-lineage 'history', through which certain combinations of genes will already have been switched on or off.

We will now examine some aspects of cell architecture which form the physical context for intercellular signalling (see Figure 5–2). As far as possible, I will concentrate on aspects of cellular architecture that are fairly general, but at the end of this section I will make a brief mention of some unusual but morphogenetically important departures from the picture presented in the figure.

Cells are surrounded by an extracellular matrix, whose exact composition varies between tissues, as does the size of the intercellular space within which this matrix is found. Some combination of protein and glycoprotein molecules forms the bulk of the matrix. These molecules are in some cases quite loosely aggregated, in others more compacted – as in the case of the basal lamina often found underlying sheets of epithelial cells.

Several types of protein molecule project through the cell membrane, often in both directions so that the molecule has extracellular,

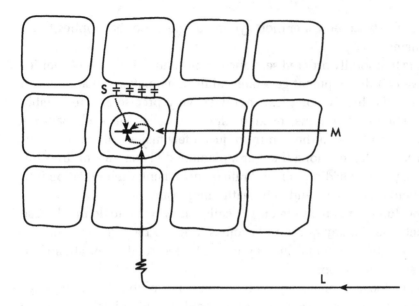

**Figure 5–1  Three types of intercellular signals**
S, short-range e.g. cell–cell contacts; M, mid-range e.g. morphogen
gradients (operating over distances of several cell diameters); L,
long-range e.g. hormones produced by a distant organ and
reaching the target tissue via the transport system. Dashed arrows:
signal transduction mechanisms through which incoming signals
are able to affect gene expression.

transmembrane and cytoplasmic domains. These include: cell–cell
adhesive molecules such as cadherins (Takeichi 1991); molecules
that bind to the extracellular matrix proteins (e.g. integrins: Tamkun
*et al.* 1986); and receptor molecules such as receptor tyrosine kinase
(RTK: see, for example, Aroian *et al.* 1990) which receive signals (e.g.
growth factors) circulating in the intercellular spaces and act as the
initiation of a multistep signal transduction system by which gene
switching is ultimately effected. Often, adjacent cells are connected
by gap junctions which permit the passage of small molecules (up to
a molecular weight in the region of 1,500 – equivalent to a peptide of
approximately fifteen amino acids).

Where signals received by a cell cause it to change in shape (often
as part of differentiation) or to move, these activities are brought about
partly through the cytoskeleton – a network of microtubules and

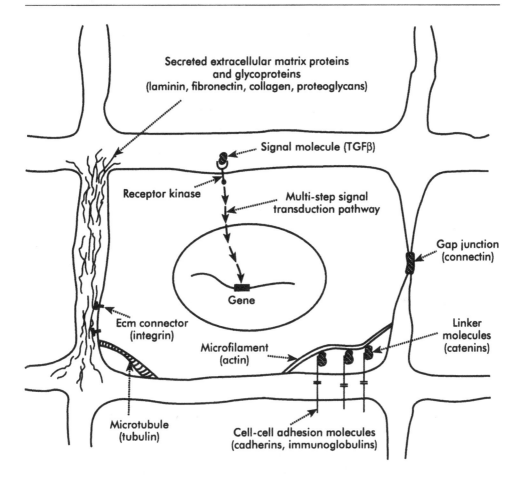

**Figure 5–2  Selected details of cell architecture**
Examples of proteins involved are given in brackets. Cell periphery features illustrated are normally found all around the cell, not just in a few isolated places. TGFβ = transforming growth factor β. Ecm = extracellular matrix.

microfilaments spanning anchorage points in the cell membrane. Also, cell movement in some cases involves a gradient of integrin receptors, with a higher concentration at the leading edge (Lawson and Maxfield 1995). Such facts serve to remind us that morphogenesis is ultimately a mechanical process, and we need to understand its cellular mechanics as well as its genetic control.

Some hormones and other morphogens enter the cell and influence the genes more directly than through a signal transduction pathway.

But even here receptor molecules are involved. In some cases, these bind to the incoming hormone when it reaches the nucleus; in others, a hormone–receptor complex first forms in the cytoplasm and migrates into the nucleus subsequently. The receptor's specificity for particular DNA sequences enables the activation or repression effects of the hormone to be restricted to the 'correct' genes.

Every single molecule and cellular structure mentioned in this brief account has a role in developmental processes. It should definitely *not* be thought, for example, that cadherins are only important for adhesion and not for intercellular signalling (see next section). Cells at the edge of a piece of tissue will be adhering to similar cells on five sides and an entirely different type of cell (or in some cases *no* cell) on their sixth side. This in itself conveys positional information to the cell which is different from that received by an internal cell (i.e. one residing somewhere in the middle of a block of tissue).

In some cases, the typical cell architecture shown in Figure 5–2 does not apply. One of the best-known examples of this is the insect syncytial blastoderm. Here, nuclear division is not initially accompanied by cell division. Cell membranes only form after the outwardly-migrating nuclei reach the periphery of the fertilized egg. Syncytial arrangements, of course, constitute a very different environment from a cellular one for the movement of morphogenetic substances, and this applies regardless of where they are found. (Other examples include syncytial slime moulds, trophoblast syncytia in mammals and syncytial 'cables' in sea-urchin embryos.)

Intermediate between standard cellularization and syncytia are cases where cells exhibit unusual numbers and/or sizes of physical connections to their neighbours. An example of this occurs in the *Drosophila* egg-chamber, where the fifteen nurse cells are connected to each other and to the oocyte (whose cytoplasm they fill with mRNAs and proteins) by a series of so-called ring canals (Koch, Smith and King 1967). Here the link between physical 'bridges' and key morphogenetic processes is particularly clear.

The following three sections deal with short-, mid- and long-range signals. These are sometimes described, respectively, as juxtacrine, paracrine and endocrine. (The remaining term in this series – autocrine – relates to secreted signalling molecules that affect the cell actually secreting them. Such effects are relatively rare and I do not cover them here, but for an example see Goustin *et al.* 1985.)

## 5.3    Short-range Signals: Cell–Cell Contacts

Some of the signals passing between cells that are in contact with each other can be thought of as conveying rather general positional information, while others carry very precise messages, such as 'now differentiate into cell-type X'. I will discuss one example of each below: the cadherin family and the *sevenless* system respectively.

### Cadherins

The group of adhesion glycoproteins known as the cadherins represent a good example of 'generalized' short-range developmental signalling (see reviews by Takeichi (1991), Takeichi *et al.* (1993) and Behrens (1994)). The cadherins collectively constitute a family of proteins, related to each other both in amino acid sequence and functional role. They are calcium dependent and adhesive (hence the name). They are typically rather large (e.g. around 700 amino acids: Tooi *et al.* 1994) and in some cases enormous (with more than 5,000 amino acids: Mahoney *et al.* 1991). Despite this size variation, all appear to have extracellular, transmembrane and cytoplasmic domains.

The system of naming different 'family members' is somewhat disorganized. In several cases, a particular cadherin is given a letter representing the tissue from which it was first identified, for example E-cadherin (epithelia) and R-cadherin (retina). However, many cadherins are found in multiple tissue types, so this kind of label can be misleading. Other cadherins are given numbers (e.g. cadherin-11: Kimura *et al.* 1995) and yet others are given more complex designations, such as XTCAD-1 (*Xenopus* tailbud cadherin: Tooi *et al.* 1994). The overall lack of consistency in the naming system should not distract us from the essential point – that we have a *family* of related proteins. And it is a family of wide taxonomic distribution: as well as the many studies of cadherins in various vertebrates (humans, mice, chicks and frogs), there have been demonstrations of their presence in diverse invertebrate groups, including insects (Mahoney *et al.* 1991). It seems likely that cadherins are a universal intercellular 'velcro' throughout the animal kingdom.

The role of the cadherins in cell–cell adhesion has been demonstrated by 'knock-out' experiments, using either mutation of the cadherin genes or antibodies directed against their products (e.g. Kintner 1992; Oberlender and Tuan 1994; Daniel, Strickland and Friedmann

1995; Dahl, Sjödin and Semb 1996). Such studies have shown very clearly that the cadherins are necessary for tissue integrity, and thus for developmental processes involving the tissues concerned. They have also shown that it is not only the extracellular domain that is necessary. The cytoplasmic domain binds to the cytoskeleton via another protein group – catenins – and is also essential for normal intercellular adhesion.

Many individual cadherins exhibit homophilic binding. That is, cells expressing 'cadherin-*x*' will bind to other cells that also express 'cadherin-*x*'. Thus cadherins are frequently involved in maintaining the cohesion of homogeneous blocks of tissue. Also, these homophilic binding properties may be conserved across evolutionary lineages. Matsunami *et al.* (1993) have shown that cells expressing chick and mouse R-cadherin will co-aggregate, while conspecific but cross-cadherin-type cells do *not* co-aggregate. However, in a few cases heterophilic binding of cadherins has also been observed (Murphy-Erdosh *et al.* 1995).

While the expression of particular cadherins is now recognized to be a property of particular differentiated cell types, it is much less clear what causes this particular aspect of differentiation. It seems that the effects of cadherin genes are cell autonomous – meaning that their effects are restricted to those cells in which the genes concerned are expressed (see, for example, Mahoney *et al.* 1991); and indeed this might be expected, given the immobile nature of their products. Cadherin gene expression can be affected by proteins of the transforming growth factor beta (TGF$\beta$) family, including activin (Miettinen *et al.* 1994; Brieher and Gumbiner 1994). Also, it appears that some transmembrane protein tyrosine kinases may colocalize and interact with cadherins and integrins. But these isolated pieces of information hardly provide us with a satisfactory picture of the overall control of cadherin gene expression.

Some of the cadherin genes have been mapped, and it is becoming apparent that in at least some cases – and perhaps in many – there are pairs or clusters on particular chromosomes. This sort of arrangement has been found in both *Xenopus* (Simonneau, Broders and Thiery 1992) and humans (Bussemakers *et al.* 1994). As the latter authors remark, it may well be that the close proximity of the genes concerned is important in their regulation. Clearly, this is also true of the well-known *Hox* genes, whose developmental role is very different (see Chapters 6 and 7). It is tempting to speculate that chromosomal clustering will turn out to be important for developmental gene families in general, and that there is a common reason for such clustering (connected with coordinate control) that applies in all cases.

## The *sevenless* System

Turning now to very specific short-range intercellular messages, the *sevenless* system in the *Drosophila* eye provides a particularly clear example. Much work has been carried out on this system over the past few years, following the pioneering study by Tomlinson and Ready (1986). Reviews have been provided by Yamamoto (1994) and Dickson (1995).

A *Drosophila* compound eye is made up of many individual units called ommatidia. Each of these consists of eight photoreceptor cells, designated R1 to R8, and four associated cone cells. Of the photoreceptor cells, R7 is the last to differentiate. Before it does so, its precursor cell is part of an 'equivalence group' with the four cone cell precursors: all five of these cells *could* differentiate into R7, but only one does so.

The signal that is of central importance in this system passes from a transmembrane protein in the R8 cell (Bride-of-sevenless or Boss) to the Sevenless (Sev) protein of the R7 precursor cell. Sevenless is a receptor tyrosine kinase, and its activation by Boss is transmitted into the nucleus by a signal transduction pathway; the cellular changes associated with overt differentiation then follow.

Cone cell precursors also have Sevenless protein incorporated into their membranes. The reason they fail to respond to Boss is that they have no direct contact with the R8 cell on whose membrane the Boss molecules are located. In *sevenless* mutants, the Boss signal from R8 is not received despite physical contact, and the R7 precursor cell thus differentiates into an extra cone cell (Tomlinson and Ready 1986).

The main similarity and difference between this system and 'generalized signalling' by cadherins are readily apparent. Both systems are dependent on cell–cell contact. However, the Boss–Sevenless interaction conveys a highly specific instruction to differentiate in a particular way; while cadherin–cadherin interaction, for example in the form of homophilic binding, conveys positional information of a rather general nature.

## Evolutionary Considerations

Although I intend to leave detailed treatment of evolutionary mechanisms to Chapter 7 and beyond, a brief comment may be useful at this stage regarding the coevolution of developmentally interacting genes, and in particular genes whose products have a receptor–ligand relationship. These include both cases of membrane-bound ligands like Boss (see above) and Delta (of the *Notch-Delta* system: see Artavanis-Tsakonas, Matsuno and Fortini 1995) and cases of secreted ligands such as proteins of the Wnt family (one of whose receptors has recently been identified as Frizzled: Bhanot *et al.* 1996).

In many cases, ligand–receptor systems have a wide taxonomic distribution, with homologous genes in vertebrates and insects (see Chitnis *et al.* 1995 for a study of the vertebrate homologues of the *Drosophila Notch-Delta* system). The evolutionary changes in these genes that have accumulated during and since lineage divergence must be strongly interconnected. That is, any change in a ligand is likely to set up a selective pressure for a change in the corresponding receptor, and vice versa. In the long term, the genes whose products interact *coevolve*. One way of looking at this situation is to picture each gene (or its product) as the selective agent acting on the other. We then have a process of 'internal selection' (Whyte 1965). I will consider this process in more detail in Chapter 9.

An obvious question to ask of ligand–receptor coevolution is why, if the interaction between them is efficient at any particular stage in evolution, either should evolve any further. Perhaps the answer to this puzzle lies – at least in part – in the lack of simple 1:1 correspondences. If, for example, two ligands (A and B) both interact with one receptor, then a change in the receptor that improves its interaction with ligand A more than it worsens its interaction with ligand B may be selected for, leading ultimately to selection for a new variant of ligand B. Since real ligand–receptor systems are often more complex than having a mere three components, the likelihood of such knock-on coevolutionary effects is considerable, and the possible patterns they could take are very varied.

## 5.4  Mid-range Signals and the Nature of 'Morphogens'

For the purpose of exemplifying mid-range signals, I will concentrate on the process of embryonic induction in vertebrates (see brief review by Lemaire and Gurdon 1994). In a typical inductive system, one tissue type or region induces some form of developmental patterning – usually including cell differentiation – in an adjacent region of cells. (Homeogenetic induction, where a tissue effectively induces more of itself, is also known, but I will not consider that here.) Examples of tissues or structures that have long been known to have an inductive role in vertebrates include the dorsal lip of the blastopore (i.e. the 'Spemann organizer': see Section 4.4), the notochord and the dermis.

Some inductive effects are dependent on cell–cell contact, and must therefore involve short-range signals; others are not dependent on such contact, and thus necessarily involve a more mobile morphogen of some sort. As I have already considered short-range signals above, I will concentrate here on the mobile (diffusible or otherwise) agents involved in induction.

Before going into the details of a selected system, it is worth considering the general concept of a 'morphogen'. Literally, this is anything that has developmental effects, and it would thus be perfectly reasonable to consider selected ions, small and large molecules, electrical impulses, genes, cells and blocks of tissue all to be different kinds of morphogen. However, such a broad and heterogeneous category is hardly useful, so in practice the term morphogen has generally been restricted to molecules that move between cells (either by active transport or passive diffusion) and influence developmental processes in the cells they traverse or accumulate in. Also, 'morphogen' has generally only been used to describe molecules that traverse a distance of at least a few cells, and has not been used for molecules that only move from one cell to its immediate neighbours. Even with these restrictions, 'morphogen' still represents a heterogeneous category of substances, including cyclic adenosine monophosphate (AMP) (Bominaar et al. 1991), retinoic acid (Chen, Huang and Solursh 1994; Hill et al. 1995), peptide growth factors (Gurdon et al. 1994; Gurdon, Mitchell and Mahoney 1995), steroid hormones (Section 5.5) and other molecular types. It is also likely that some morphogens await discovery, for example because some already-known mobile molecules will be found to have unsuspected morphogenetic effects, or because new members of 'families' of morphogenetic molecules will be character-ized.

In many cases, the effects of a morphogen have been thought to derive from a gradient in its concentration across a block of tissue. Such gradients were proposed several decades ago by Child (1941); and an attractively simple picture of gradient systems is provided by the French flag model (Wolpert 1968; Crick 1970). Here, a concentra-tion gradient causes differentiation into three different cell types: high levels of morphogen produce 'blue' cells, moderate levels produce 'white' cells and low levels produce 'red' cells.

Endless variations on this theme have been proposed: the source and sink can be localized or dispersed; there can be a single morpho-gen, two or several; morphogens can diffuse between cells via gap junctions or move by other means, and so on. It is likely that many *different* sorts of gradient operate in developmental systems, and indeed that in some cases the patterns of spatiotemporal morphogen concentration are sufficiently complex that the idea of a 'gradient' is too simplistic.

General accounts of induction (both in vertebrates and inverte-brates) are widely available (see, for example Ede (1978, chapter 9), Slack (1983, chapter 9) and Gilbert (1994, chapter 18)). For my present purposes, it makes most sense to concentrate on the nature of inducing substances and on the role of genes in the inductive process. I will use a particular system to exemplify these aspects of induction: embryonic patterning of the mediodorsal region of the vertebrate trunk through the inductive influence of the notochord.

Following gastrulation, which results in the standard triploblastic arrangement of ectoderm, mesoderm and endoderm, the embryo elon-gates, and undergoes a series of further changes. In contrast to gastru-lation, this later stage – neurulation – involves little cell migration and operates more through localized processes of shape modification and cell commitment/determination/differentiation. One of the most important processes in the embryo at this stage (and the one which gives rise to the name 'neurulation') is the formation of the rudimen-tary central nervous system, together with other associated patterning events in the mediodorsal region.

Initially, ectodermal tissue destined to become neural forms a dor-sal sheet of cells which narrows and concentrates at the midline, pro-ducing the neural plate. The lateral edges of this plate bend upwards, meet and fuse, forming the neural tube. At about the same time, bands of 'paraxial' mesoderm aggregate on either side of the neural tube, and these become segmented, giving rise to mesodermal blocks called somites. There now follows a series of determinative events in which the notochord plays a central role (see Figure 5–3). In particular, signals from the notochord (itself a mesodermal derivative) govern the fates of cells in three areas, as follows. (a) Cells in the ventral region of the neural tube become 'floor plate cells' (Placzek *et al.* 1990). (b) Cells in the ventrolateral part of the tube become motor neurons – under the influence of both notochord *and* floor plate cells (Yamada *et al.* 1991). (c) Cells in the ventromedial regions of the somites become sclerotome cells (Dietrich, Schubert and Gruss 1993; Pourquié *et al.* 1993). These will later migrate medially, surround the neural tube, become chondrocytes, and ultimately produce the vertebrae.

Although some processes involving cell–cell contact may also be involved (Placzek, Jessell and Dodd 1993), it has become clear that all these inductive signals involve one or more mobile morphogens. We now need to enquire into the identity of these. In the last few years, one molecule in particular – the Sonic hedgehog protein – has emerged as being a key morphogen in notochord-generated inductions.

*Sonic hedgehog* (*Shh*: see Table 5–1 for conventions on naming genes and proteins) is a member of the *hedgehog* (*hh*) family of genes. *hedgehog* was first discovered in *Drosophila* (Nüsslein-Volhard and Wieschaus 1980) and given its name because of the prickly appearance of the mutant larva, which derives from deletion of the denticle-free (or 'naked') sections of the larval segments. Several vertebrate *hedgehog* family genes were described by Krauss, Concordet and Ingham (1993), Riddle *et al.* (1993) and Echelard *et al.* (1993). Subsequent work on one of these, *Sonic hedgehog*, has revealed it to have a crucial inductive role.

Sonic hedgehog is a secreted protein that is produced in notochord and floor plate cells at critical times for the specification of floor plate, motor neurons and sclerotome (see Roelink *et al.* 1994; Johnson *et al.* 1994; Fan and Tessier-Lavigne (1994); review by Ingham 1995). This makes it an obvious candidate for being an 'inducer'. *In vitro* studies have confirmed Sonic hedgehog protein as the inducer: not only can notochord and floor plate cells producing Sonic hedgehog cause inductive effects without cell–cell contact, but other cells producing Sonic hedgehog can mimic this effect. While there will probably turn out to be other molecules involved in notochord-generated inductions

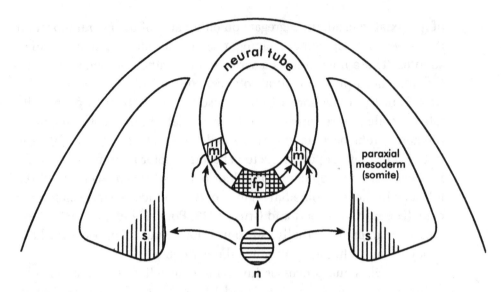

**Figure 5–3  Inductive effects of the notochord in a vertebrate embryo**

n, notochord; fp, floorplate; m, motor neuron; s, sclerotome. Arrow, inductive effect; horizontal shading, inducing tissue; vertical shading, induced tissue.

(including Echidna hedgehog: Currie and Ingham 1996), the role of Sonic hedgehog as the main inducing substance is no longer in doubt.

Sonic hedgehog is not only responsible for mid-range signalling, as discussed above, but is also instrumental in the local 'contact' signals involved in floor plate induction. With the discovery that Shh undergoes autoproteolytic cleavage into a smaller amino-terminal fragment (ShhN: molecular weight ~19 kDA) and a larger carboxy-terminal fragment (ShhC: molecular weight ~26 kDa) came the obvious hypothesis that one was involved in contact-dependent signalling and the other in diffusible/mid-range signalling. However, it has been shown by Porter *et al.* (1995) , Roelink *et al.* (1995) and Fan *et al.* (1995) that it is ShhN that is involved in *both* types of signalling. (Whether ShhC has a role beyond assisting in the autoproteolytic cleavage process itself, and the modification of ShhN, is not yet clear.)

At present, it appears that ShhN is at high concentration around the edges of cells expressing the *Shh* gene, and at progressively lower concentrations at greater distances from these cells. It seems that we have here a gradient of ShhN (Concordet and Ingham 1995) with high

**Table 5-1. Some rules on the naming and abbreviating of genes and proteins**
These rules, although well known to developmental geneticists, may not be familiar to readers from other fields, for whose benefit they are displayed here.

| | |
|---|---|
| *Rule 1* | Gene names are italicized; protein names are not |
| Example | *Sonic hedgehog* = the gene; Sonic hedgehog = the protein |
| *Rule 2* | Initial letter of gene name is upper case if the mutant form of the gene is dominant (i.e. causes the mutant phenotype in heterozygotes); lower case if the mutant is recessive. |
| Example | *Ultrabithorax* = dominant; *ultraspiracle* = recessive |
| Complication | Each gene can have many mutant forms, some of which may be dominant and some recessive; and all of which may be pleiotropic (i.e. have multiple phenotypic effects). Often, the gene will have been discovered through observation of a particular mutation and named after its most obvious – but not necessarily most important – effect. |
| *Rule 3\** | The initial letter of a protein name is generally upper case, regardless of whether the corresponding gene is dominant or recessive. |
| Rationale | In some cases, this helps to distinguish protein names from other uses of the same word. For example: 'Dorsal (= the protein produced by the *dorsal* gene) is not translocated into dorsal nuclei'. |
| *Rule 4* | Gene names are generally abbreviated to three letters, but some abbreviations use two or four. |
| Examples | *wg* = *wingless*; *ftz* = *fushi tarazu*; *Antp* = *Antennapedia* |
| Note | The initial letter rule applies to the abbreviation as well as to the full gene name. |
| *Rule 5* | Abbreviated protein names are the same as the abbreviated gene names (but are not italicized and generally begin with a capital according to rules 1 and 3). |
| Example | Dpp = the protein produced by the *dpp* gene (decapentaplegic) |
| Complication | Processing of proteins may be reflected in their name/abbreviation. For example, Shh undergoes autoproteolytic cleavage into amino-terminal and carboxy-terminal fragments, called Shh-N and Shh-C respectively. |
| *Rule 6* | If a gene name/abbreviation begins with a lower-case letter, this applies regardless of whether the context might suggest otherwise – notably at the start of sentences. |
| *Rule 7* | Most genes are named after an obvious mutant effect; but other names/ abbreviations have a more complex basis. |
| Example | *hedgehog* = prickly appearance (i.e. named after mutant effect); in contrast, *DWnt* is a mixture of *Drosophila*, *wingless*, and *int*egration site. |

*Not all authors/journals observe rule 3: lower-case initial letters of protein names are frequently encountered.

levels giving rise to floor plate cells and low levels causing differentiation into motor neurons.

Clearly, this represents an outstanding example of progress in understanding regional cell differentiation and pattern formation. Early work identified the inducing tissues, and subsequent work pinpointed not only the inducing molecule but even the active part of it. And the importance of *hedgehog* family genes does not end there: they are involved in developmental processes in the limbs of vertebrates (Riddle *et al.* 1993) and insects (Basler and Struhl 1994), and also in vertebrate brain (Ericson *et al.* 1995), avian feathers (Nohno *et al.* 1995) and insect eyes (Heberlein *et al.* 1995). This is clearly a gene family of major importance in animal developmental systems generally, and further studies on it are likely to be highly productive.

## 5.5 Long-range Signals and Panorganismic Coordination

Readers who refer to the paper by Porter *et al.* (1995) mentioned above will find that those authors describe as long-range the kind of diffusible (e.g. motor neuron-inducing) signals that I here call mid-range. The reason for the difference in labelling is that I wish herein to retain a whole-organism view of development. From that perspective, the sort of diffusible signals we have just examined clearly *are* mid-range, and there is a quite distinct array of much longer-range signalling also going on in the embryo.

The need for very long-range communication in a developing organism is clear. How do our right and left sides come to be approximately the same size? (notwithstanding the minor differences known as fluctuating asymmetry, and reviewed by Markow 1995). How do our heads grow at the right rate relative to our trunks so that the head: trunk ratio takes its expected (decreasing) course from infant to adult? In general, how are growth and patterning coordinated in a panorganismic way so that we end up with a correctly proportioned adult?

Once we get past the earliest developmental stages of gastrulation and neurulation, and get part-way into the phase known as organogenesis, various transport systems come into being – their precise nature of course depending on the taxon concerned. As well as carrying nutritive substances around the growing organism, these are also used for transporting morphogenetic substances over long distances. One example of such a morphogenetic substance is the insect 'moulting hormone', ecdysone.

Ecdysone has a variety of roles in the developing insect, but I will concentrate here on just one: its role in initiating pupation in dipterans such as *Drosophila*. Ecdysone is produced by the paired prothoracic glands at the anterior end of the larva, and is transported through the haemolymph to all the larval tissues. Its importance has long been known (see Wigglesworth 1935). Many studies have been conducted over the last few decades on the developmental role of ecdysone, and among them investigations of its gene-switching effects are of particular importance.

Many of these studies have made use of the presence, in dipteran larval tissues, of polytene chromosomes. Microscopically visible 'puffing' of these at particular loci is known to indicate active, RNA-producing genes. By examining puffing patterns over developmental time, we can effectively visualize patterns of changing gene activity – for example those accompanying (and indeed causing) pupation (see Figure 5–4).

Becker (1959) showed that in larvae that had been ligatured posteriorly to the prothoracic glands, the anterior part 'pupated' and exhibited the expected changes in puffing pattern; while the posterior part remained larval, both in puffs and in morphology. This experiment did not of course unambiguously identify ecdysone as the pupation-inducing agent – anything produced in the anterior end might have been responsible. However, Clever and Karlson (1960) showed that injected ecdysone could induce pupation, and Ashburner (1971) demonstrated ecdysone's effects on puffing patterns *in vitro*.

The way in which ecdysone controls gene switching was captured in a beautifully simple model proposed by Ashburner *et al.* (1974). Having considered a whole range of experimental results including dose dependence of responding genes, the effects of antibiotics that block protein synthesis and the consequences of ecdysone 'wash-outs', these authors considered that the model shown in Figure 5–5A fitted all the facts. Subsequent work has confirmed the essential validity of this model (Fletcher and Thummel 1995; Fletcher *et al.* 1995), and elaborated it further to take into account recent findings, especially on the nature of the ecdysone receptor (Figure 5–5B; Huet, Ruiz and Richards 1995).

As will be clear from Figure 5–5B, the functional ecdysone receptor is now known to be a heterodimer formed from the products of two genes: *EcR* (*Ecdysone receptor)* and *usp* (*ultraspiracle*). This heterodimer acts as a transcription factor, showing high-affinity binding to both hormone and DNA (Yao *et al.* 1992, 1993). One way in which Ashburner *et al.*'s (1974) positive effect of early response genes on late ones may work is through the former coding for hormone receptors that will act as transcription factors for the latter (Stone and Thummel 1993).

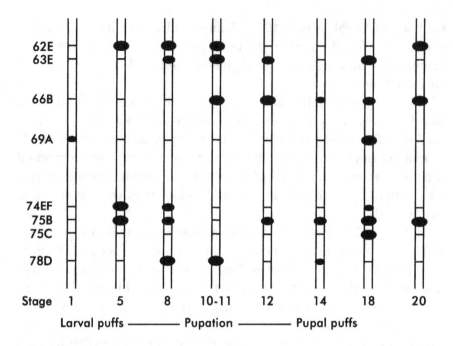

**Figure 5–4 Simplified picture of changing puffing pattern over developmental time in *D. melanogaster***
The chromosome stretch shown is from the left arm of autosome 3. Band subgroups are shown on the left-hand side (62E to 78D). Developmental stages are shown in sequence from left to right. (The first represents active third-instar larvae, the last is around 12h after puparium formation.) From Sang 1984. *Genetics and Development.* Longman, London, with permission.

The overall process of ecdysone-induced pupation is, of course, much more complex than the outline given above. This was indicated clearly by Ashburner *et al.* (1974) when they proposed their model. And recent studies have confirmed this complexity – for example Karim, Guild and Thummel (1993) note that the 'early' genes activate more than 100 later ones. Nevertheless, it seems that the central features of the ecdysone response are now established, even if many additional details remain to be documented.

As might be expected, many aspects of the ecdysone gene-switching system described above are evolutionarily conserved: between species (Jones, Jahraus and Tran 1994), between families (Bienz-Tadmor, Smith and Gerbi 1991) and possibly also between orders and classes.

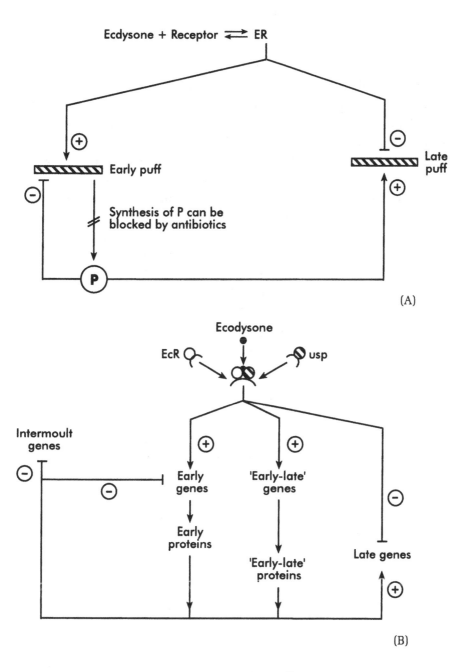

Figure 5–5 Model of ecdysone's gene-switching effects, as proposed by Ashburner *et al.* 1974 (A) and modified by Huet *et al.* 1995 (B)

p = protein; usp = ultraspiracle; EcR = ecdysone receptor

Developmental systems, once established by evolution, are hard to dispense with, but easy (at least in relative terms) to elaborate and modify.

Returning to a more general perspective: hormones and other morphogenetic substances transported around the body can act as co-ordinators of developmental activity in different regions. In a sense we have here the 'opposite' of a gradient: all regions are effectively bathed in a reasonable concentration of the substance concerned. Although the concentration may vary somewhat between tissues or regions and this may sometimes be important, the specificity of responses is often dependent on previous gene-switching events in the exposed cells affecting their 'competence' to respond (e.g. the Broad-Complex in the case of ecdysone: see review by Thummel 1996) rather than on the precise concentration of the hormone. And where effects are general – such as in the cessation of growth throughout the human body at some point in the teenage years – this is due to a change (in this case cessation of hormone production) that is interpreted similarly by *all* dividing cells, despite their individual histories, though perhaps some may respond directly, and others indirectly.

## 5.6  Patterns of Interconnection: Developmental Programmes

Readers may have noticed that the description of particular developmental mechanisms given in the last three sections are, like other such descriptions given by other authors, unsatisfactory because they are intrinsically incomplete. I do not wish to focus here on that aspect of their incompleteness which stems from lack of internal detail (though such details often *are* deficient) but rather on that other aspect of incompleteness which derives from their being considered in isolation from the rest of the developing organism – both in time and space.

There was an example of this in Section 5.2, although I did not emphasize this aspect of it at the time. If some cells are caused to differentiate in response to hormone influx, while other cells receiving the same hormone do not respond, then the cells had *already* differentiated in some way (e.g. by having a hormone receptor gene switched on/off and therein being differentiated in 'competence'). But what caused this prior differentiation? Perhaps it was some other hormone or morphogen suffusing the tissue at an earlier stage. If so, then how did *that* work – yet prior differentiation of some other kind? As will readily be apparent, there is a feeling of 'infinite regress' here: in each case we appear to be explaining little and merely pushing the problem back to some earlier event on which we are not focusing. Ultimately, this regress leads all the way back to the egg.

For fuller consideration of this crucial point, let us return to the notochord inductive system discussed in Section 5.4, an abstract version of which is given in Figure 5–6. Here, each physical entity which sends and/or receives developmental signals is represented simply by a dot, so we lose all literal structural descriptions such as cell numbers or the shapes of blocks of tissue. This simplification allows us to focus on another aspect of the system, which may otherwise be overlooked: namely, the pattern of interconnection of causal links.

Each arrow in the figure is a causal link – that is, a morphogenetic influence of some sort flowing from one 'entity' to another (Arthur 1988). We can see that the particular pattern that applies in this example includes a *divergence* (where more than one effect radiates out from a source, in this case the notochord), a *convergence* (where more than one signal is received by a particular responding entity, in this case the motor neurons) and a *loop* (where a signal bypasses a stage in the main 'route', in this case the direct stimulation of motor neuron differentiation by notochord). These three are superimposed on the basic *cascade* (i.e. a chain of causal effects: in this case the sequence notochord → floor-plate → motor neuron) which can be regarded as intrinsic to all developmental systems. (There are also likely to be negative feedback systems at work: see below.)

All sorts of questions immediately arise. What is the ratio of convergences to divergences? Does it change systematically over developmental time? Do convergences represent redundancy and hence increased repeatability? Do multiple (as opposed to merely two-fold) divergences and convergences occur, and if so, how frequently? There is a whole body of work waiting to be done here, at the interface between abstract complexity theory (Kauffman 1993) and the practical details of real developmental systems.

All the above questions arose from considering a system which involves a mere four 'dots' and four 'arrows'. But this system was itself just an arbitrary starting-point for our discussion. We have already seen that a lot of molecular and cellular detail underlies our dots and arrows (see 'windows' in the figure). If we now head in the other direction – that is, towards a more extensive picture rather than a more intensive one – then we can begin to get away from the 'isolation'

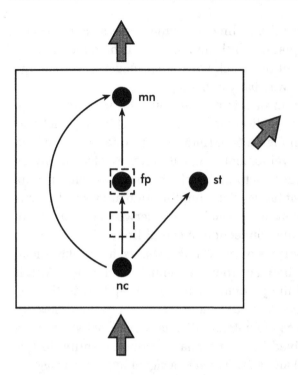

**Figure 5–6   Abstract picture of the notochord inductive system**
Solid circle, developmental 'entity'; thin arrow, causal link within
this system; thick arrow, causal link connecting this system to
others; nc, notochord; fp, floorplate; mn, motor neuron; st,
sclerotome; dashed box, 'window' down to molecular detail of
individual entity or link.

mentioned above, and to consider how various inductive subsystems
connect up to give the overall developmental programme of the organism concerned.

This extensive view is represented by the large arrows in the figure,
indicating causal influences coming into and out of the relatively
restricted subsystem (the box) upon which we initially focused. And
because we know that loops occur, it is clear that the incoming signals
need not necessarily go to the base (notochord) nor the outgoing ones
necessarily emanate from the tips (mn/st) of our subsystem. That is just
one of various possibilities. The form that the overall programme takes,
in terms of its pattern of interconnections in any particular organism, is
of fundamental importance, as is the question of how it alters in evolu-

tion (with more phenotypically complex animals probably resulting from more complex programmes, for example).

Few would doubt the fundamental importance of whole-organism developmental programmes. However, many would regard their contemplation as premature (due to the limitations of our current knowledge of development) or even unproductive in the long term (because the patterns of interconnection in overall developmental programmes will turn out to be too complex within most animals, and too variable between them, to allow any useful generalizations).

My own view, however, is less negative. I think that even now, before we have any more developmental facts at our disposal, we can distinguish between, and be confident in the general importance of, three aspects of the pattern of interconnections characterizing overall developmental programmes. These are:

1. As development proceeds, animals grow in size, their cell number increases, they regionalize, and different regions come to have different fates. *Divergences* (as described above) must underlie this, and one universal aspect of the overall developmental programme is thus a divergent hierarchy of causal links – that is, a morphogenetic tree (Arthur 1982a, 1984, 1988).

2. Despite this increasing cell number, organismic size and regionalization, developing animals remain tightly coordinated systems all the way through to adulthood. This implies that there must be one or more means of ensuring that slight departures from synchrony (or from any other 'target relationship') can be corrected (i.e. that there is a stabilizing system in place: Waddington's (1957) 'homeorhesis'). Thus a second universal aspect of the overall developmental programme must be some form of *cross-links* between processes going on in parallel in different regions of the embryo. This can be visualized as horizontal links between 'branches' of the morphogenetic tree (Arthur 1988) or between quasi-autonomous developmental modules (Raff 1996, chapter 10).

3. So far, we have concentrated on the causes of things *happening* in development. This is perfectly reasonable, of course, but it does tend to obscure the fact that repeatable developmental products are only possible if there are also effective, coordinated ways of *stopping* things happening at the correct times and in the correct places. Such 'stopping mechanisms' must involve, among other things, *feedback loops* – see, for example, Vortkamp *et al.* (1996).

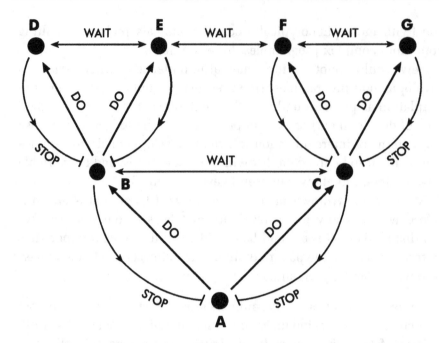

**Figure 5–7   A simple model of a developmental programme**
To see how the system works, focus initially on entity 'B'. This is
produced by A, but only becomes 'active' to induce D and E (and to
inhibit A) providing there is also a sufficient quantity of C. Activity
of B is eventually switched off by negative feedback loops from D
and E; but note that because of the 'wait' interaction between E and
F, the feeback loops that switch off B and C will both be influenced
by all of D–G. This is one of the simplest models to be capable of
incorporating all three key features of divergence, cross-links and
feedback.

(These may be formally similar to – but on a much bigger scale than
– within-cell feedback loops where accumulation of a gene product
acts to switch the gene off.) Also, programmed cell death – apoptosis – may in some cases be an important component of stopping
mechanisms, for example in the interdigital region of vertebrate
limbs (Zou and Niswander 1996) and in tooth development
(Vaahtokari, Aberg and Thesleff 1996).

The pattern of 'dots and arrows' in Figure 5–6 does not tell us anything
about the *nature* of any of the causal links. This, however, is likely to
be of fundamental importance. We have already considered (under
point 3 above) that some signals must represent 'stop' instructions.

Others that must exist include 'do' and 'wait'. And complex instructions, particularly the 'double negative' of stopping an inhibition, are probably also common. Thoughts on such a classification of the nature of causal links are still at an early stage (but see Wolpert (1990) for an explicit example of this approach). When we have a more comprehensive classification, and can assign each real developmental process to one or other category (e.g. most inductions = 'do' instructions) then different categories can be represented diagrammatically as different kinds of arrows. A first step in this direction is made in Figure 5–7, which represents a possible relationship between 'do', 'wait' and 'stop' instructions. This very simple model provides a possible explanation of how these three types of instruction may interact to produce both regionalization and stability (see legend).

Clearly, much remains to be done in this area, but it is already apparent that the pattern of interconnections in an overall developmental programme is a crucial issue. I will return briefly to this issue at the end of Chapter 6, at which stage we will be better able to visualize its genetic dimension.

# CHAPTER SIX

# Developmental Mechanisms: Genes

## 6.1   Introduction

In the previous chapter, we looked at developmental mechanisms from a starting point of the kinds of signals that cells receive: cell–cell contacts, mid-range and long-range morphogens. While genes were seen to be involved both in generating and responding to these signals, the approach adopted was not primarily a genetic one. We now embark upon the complementary, gene-centred approach. As noted in Chapter 5, these two approaches should ultimately converge to provide a comprehensive understanding of how developmental mechanisms work.

The approach, in the present chapter, is still developmental rather than evolutionary – phylogenetic comparisons will follow in Chapter 7. Thus I will concentrate here on a single organism, *Drosophila melanogaster*, which is still the best-known animal system from a developmental genetics perspective – though *Caenorhabditis*, with less than 1,000 cells, may be 'catching up'. Even with this restriction of focus, though, the relevant literature is now voluminous. One of the reasons for this is that the number of genes involved in development has turned out to be larger than was once anticipated. An early review of the genetics of *Drosophila* development was provided by Ingham (1988). More recent reviews are also available, but tend to be comparatively restricted in scope (see, for example, Blair 1995).

I will attempt to give a brief overview of *Drosophila* developmental genetics in the following section, with particular reference to the main body axes. I will then focus on two specific families of genes: (a) the homeotic (or 'selector') genes of the Antennapedia and Bithorax complexes (to which the vertebrate *Hox* genes are homologous); and (b) the *hedgehog* gene and its close relatives (whose vertebrate homologues include *Sonic hedgehog*, as we saw in the previous chapter). I will briefly revisit the issue of overall developmental programmes in Section 6.5.

The amount of information now available in this area is so vast, and the rate at which it is being added to so rapid, that it is easy for readers whose field of expertise lies elsewhere to feel swamped by it. To minimize this problem it is wise to keep sight of a small number of key questions which can be asked of any developmental gene. These include: (1) what is the gene's product, how does it function at a molecular level, and, in consequence, are its effects cell-autonomous or do they extend more widely as a result of diffusion or secretion?; (2) what other genes/factors switch it on/off and what does it in turn switch on/off?; (3) is it part of a cluster of genes? – if so, how is the cluster structured and what relevance does this have for the control of its constituent genes? (Readers may recognize the last two of these as relating respectively to the 'interactional architecture' and 'chromosomal architecture' referred to in Chapter 2). We will cover all of these questions to varying degrees in the following sections.

It has become increasingly clear over the last few years that many genes have more than one developmental role. It should thus be borne in mind that even if the role of a particular gene product in a particular process during a particular period of developmental time is reasonably well understood, it may be that at other times and in other regions that same product – or a closely-related one – may perform somewhat different developmental tasks. Associated with this, the 'intermittent' (as opposed to 'single pulse') pattern of gene expression illustrated in Figure 1–8 is now known to be common, and not just something that only happens in a few unusual cases.

## 6.2    Overview of the Genetics of *Drosophila* Body Axes

The development of the first-instar *Drosophila* larva that hatches from the egg takes about 24 hours (providing the temperature is high:

around 25–27 °C). Yet before it hatches, all the larval tissues and structures are fully formed, and the special blocks of tissue from which the adult will subsequently be made at metamorphosis – the imaginal discs and histoblast nests – are present in the various thoracic and abdominal segments. Many of the key decisions regarding the layout of anteroposterior and dorsoventral body axes, and the assigning of cells to segments (or parasegments: see below) take place in the very earliest developmental stages – in the first 2 or 3 hours after fertilization leading up to the cellular blastoderm (see Figure 6–1); and indeed some decisions – arguably all – ultimately have their origins in the oocyte within the maternal egg chamber.

Many genes are involved in these key decisions, and this is hardly the appropriate forum for a detailed discussion of all of them. Nevertheless, readers need at least to have an outline mental picture of the role of genes in establishing the major body axes, which constitute the beginnings of the body plan. This is what I aim to provide in the present section. The account that follows, then, is not intended to be complete; but it is, of course, intended to be correct as far as it goes. One way in which I have attempted to simplify the details is to deal with functional groups of genes where possible (e.g. pair-rule genes: see below), rather than with *individual* genes. I will give examples of genes within each group rather than attempt an exhaustive listing of all of them. As well as being more concise, this approach avoids the risk of rapid obsolescence run by putative exhaustive listings: new genes belonging to the various classes have been discovered in the last few years (see, for example, Baumgartner *et al.* 1994), and there may yet be others to follow.

## The Anteroposterior Axis and Segmentation

The main features of anteroposterior axis determination and patterning are now well established, even though many of the detailed mechanisms remain obscure. Five main classes of genes are involved:

### 1. Maternal-effect Genes

These are active in the nurse cells, and their RNA products are transported to the oocyte where they form localizations: for example, the concentration of *bicoid* RNA is highest at the anterior end, while that of *nanos* RNA is highest at the posterior end (Figure 6–1). These localizations of mRNA lead to gradients in the concentration of the corre-

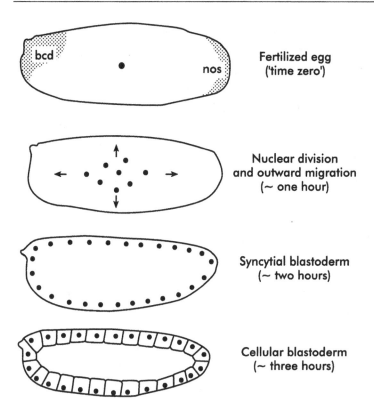

**Figure 6–1  The earliest developmental stages in *Drosophila***
The approximate spatial distribution of the mRNAs from two important maternal genes, *bicoid* and *nanos*, are shown (top); these, together with other maternal mRNAs, initiate the anteroposterior axis (anterior to the left here, as usual).

sponding proteins, following translation. Thus, the oocyte exhibits anteroposterior polarization while it is still within the egg chamber. At least another ten maternal-effect genes contribute to this polarization. (See Wang and Hazelrigg (1995) and references therein for anterior determinants; Wang and Lehmann (1991) for posterior determinants.)

### 2. Gap Genes

These are the first zygotic genes (there are at least nine of them) to respond to the maternally-established gradients. Mutations in these cause the deletion of groups of segments, and thus gaps in the normal anteroposterior pattern: hence the name. Because they respond to certain concentrations of the products of the maternal-effect genes, they form broad stripes of expression. Not all gap genes are hierarchically equivalent, though: some may act to switch on others. In particular, *hunchback* may form a protein gradient to which other gap genes respond (Kraut and Levine 1991; Struhl, Johnston and Lawrence 1992). This sort of non-equivalence characterizes the other gene groups as well; and this serves to remind us that what we are actually dealing with is a complex network, not a neat, linear series of steps.

### 3. Pair-rule Genes

So-called because mutations in them cause the deletion of alternate segments, these genes respond to patterns of gap gene expression and in turn act to control the segment polarity genes (see below). Some of the pair-rule genes are expressed in seven stripes along the anteroposterior axis, the locations of which are determined (through mechanisms that are not yet entirely clear) by the preceding and broader bands of gap gene expression. Again, there is non-equivalence within the class, some pair-rule genes (e.g. *hairy*) being described as primary; others (e.g. *fushi trarazu*) which are activated by the primary ones are referred to as secondary (Ingham 1988). Overall, there are at least nine genes in the group. Their stripes represent one of the first clear signs of segmentation (Hafen, Kuroiwa and Gehring 1984; MacDonald, Ingham and Struhl 1986; Lawrence and Johnston 1989): the anterior borders of *ftz* expression stripes mark the start of even-numbered parasegments, while the anterior borders of *even skipped* (*eve*) stripes mark the start of odd-numbered parasegments. (Parasegments are anteroposterior units that are out-of-phase with segments (see Figure 6–2) and they are important developmental units in *Drosophila*: see Martinez-Arias and Lawrence (1985).)

### 4. Segment Polarity Genes

The ten or more genes of this group are functionally downstream of the pair-rule group (Ingham, Baker and Martinez-Arias 1988). Segment polarity genes are responsible for within-parasegment patterning. Their interaction both with pair-rule genes and with each other is highly complex, but one of the key features of pattern is the formation of stripes of *engrailed* (*en*), *wingless* (*wg*) and *hedgehog* (*hh*) gene expression. While pair-rule genes initiate these stripes, they are maintained and stabilized by interactions among the products of the seg-

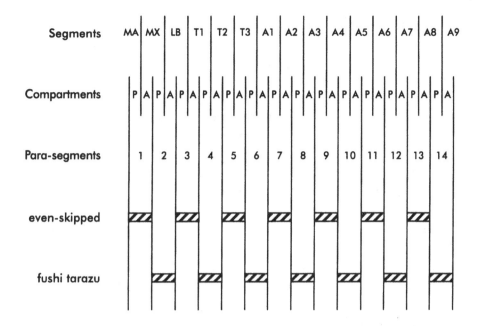

**Figure 6–2  Segments, parasegments and pair-rule gene expression patterns**

The head segments are mandibular (MA), maxillary (MX) and labial (LB). Thoracic segments are T1–T3; abdominal segments are A1–A9. Compartments are described as anterior (A) or posterior (P) according to their location in the segment (not parasegment); however, the expression patterns of pair-rule genes such as *eve* and *ftz* in the blastoderm correspond to parasegments, revealing these to be important developmental units.

ment polarity genes themselves (see Tabata and Kornberg 1994, and references therein). The position of these various stripes along the anteroposterior axis depends on the frame of reference: for example, *en* (and *hh*) stripes mark the posterior compartment of a segment but the *anterior* compartment of a parasegment.

### 5. Homeotic (Selector) Genes

Through the activity of the groups of genes described above, pattern along the anteroposterior axis is *elaborated* from its beginnings as anterior and posterior localizations of RNAs from different genes of the maternal-effect group to a point where all of the segmental units are 'visible' as stripes of expression of genes such as *engrailed* and *hedgehog*. However, while the genetic processes described so far result in the development of fourteen parasegmental units, they do not cause these units to differ from each other. This 'becoming different' is brought about by a separate process involving the homeotic (or selector) genes, which occur in two complexes, both located on chromosome 3; the Antennapedia and Bithorax complexes (ANT-C and BX-C). These genes – which are switched on approximately in parallel with (not after) the segment polarity genes – will be described in more detail in Section 6.3. Their discovery started with the observation of mutant phenotypes where a correct part was in an incorrect place (e.g. leg-for-antenna or wing-for-haltere: hence 'homeotic', a term coined by Bateson (1894)) and much of the pioneering work was done by Lewis (1963, 1978). They have also been termed selector genes (Garcia-Bellido 1975) to reflect the fact that they act to select one developmental pathway that becomes actualized (for a compartment) out of a range of possible pathways.

It is important to re-emphasize that the above account is a much simplified one, intended as an overview from which to progress to more detailed discussions of particular genes and (ultimately) to making phylogenetic comparisons. Many complexities are thus obscured. Some of these are listed below, together with appropriate references through which interested readers may follow them up.

First, there is not a simple linear cascade here (maternal → gap → pair-rule → segment polarity → homeotic) but rather, as I mentioned earlier, a complex network which also includes within-group interactions and feed-back loops (John, Smith and Jaynes 1995). Second, gene expression is regulated by a mixture of positive and negative effects (activations and repressions). Some authors have stressed the importance of a mixture of the two (e.g. Riddihough and Ish-Horowicz 1991), while others have particularly emphasized the importance of negative regulation (Manoukian and Krause 1993). Third, there are signs of redundancy of mechanisms, which may represent a way of ensuring repeatability, as discussed in Chapter 4 (Rivera-Pomar *et al.* 1995). Fourth – and perhaps an example of redundancy – some genes function both maternally and zygotically (Schulz and Tautz 1995). Fifth, the early classes of zygotic genes (gap and pair-rule) all encode transcription factors which can move to varying degrees within the syncytium, whereas segment polarity genes function after cellularization of the blastoderm and are much more heterogeneous, encoding cytoskeletal, secreted, and other types of protein (tabulated by Perrimon 1994).

Sixth, although I have concentrated on *Drosophila*, the genetic mechanisms outlined above apply in other dipterans (Sommer and Tautz 1991), and some of them apply across a wider range of taxa (see Chapter 7). Finally, the dorsoventral patterning mechanisms discussed below interact with anteroposterior patterning – the two systems are not completely independent (González-Reyes, Elliott and St Johnston 1995).

## The Dorsoventral Axis

As in the case of the anteroposterior axis, discussed above, the primary determination of the dorsoventral axis requires maternally-acting genes, whose products establish gradients to which zygotic genes respond. Dorsoventral bands of expression of these latter genes then enable the dorsoventral pattern to be elaborated.

There are at least twelve genes in the maternal group. Their interactions are complex (see review by St Johnston and Nüsslein-Volhard 1992), and I will not consider them here, as I wish to concentrate on what happens further downstream. Suffice it to say that one important outcome of the concerted action of several maternal-effect genes is the establishment of a dorsoventral gradient in the distribution of the product of the zygotic *dorsal* gene (so called because it is a ventralizing agent, and hence mutations in it can have dorsalizing effects).

The *dorsal* product is a transcription factor, and the dorsoventral gradient that is established is not in the overall concentration of this protein, but rather in its incorporation into the nucleus: ventral nuclei contain more Dorsal protein than dorsal nuclei (Rushlow *et al.* 1989; Steward 1989; Roth, Stein and Nüsslein-Volhard 1989). This gradient is caused by the phosphorylation, and consequent splitting off, of a protein called Cactus, which otherwise binds to Dorsal, in the ventral region (Whalen and Steward 1993). Unbound molecules of Dorsal protein can be translocated into the nucleus, while bound ones cannot. (This means that *cactus* is a dorsalizing gene, and mutational inactivation of it is ventralizing, so in this sense it is the opposite of *dorsal*.)

A group of some eight or so zygotic genes respond to the dorsoventral gradient of Dorsal protein in the nuclei of the cellular blastoderm: see Figure 6–3. I will briefly discuss three of these: *decapentaplegic* (*dpp*), *snail* (*sna*), and *short gastrulation* (*sog*).

The *dpp* gene (Irish and Gelbart 1987) encodes a protein that is a member of the transforming growth factor beta (TGF$\beta$) family. While

**133**

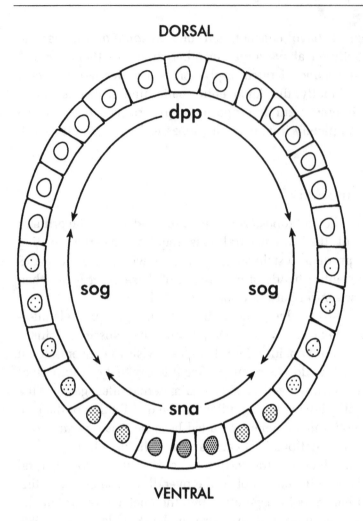

**DORSAL**

**VENTRAL**

**Figure 6–3 Transverse section of cellular blastoderm, showing expression domains of selected dorsoventral patterning genes**
Density of stippling indicates extent of nuclear localization of the Dorsal protein. The approximate dorsoventral bands of expression of three downstream genes are shown: *decapentaplegic* (*dpp*), *short gastrulation* (*sog*) and *snail* (*sna*).

the sequence of steps by which *dpp* is activated is not yet clear, what *is* clear is that its expression is inhibited by the transcription factor Dorsal. So, in those cells where Dorsal has been translocated into the nucleus to a high degree – that is, in the ventral and lateral cells – *dpp* is switched off. This leaves a dorsal and dorsolateral span of some 40 percent of the embryo's transverse-section periphery over which *dpp* expression is observed (Figure 6–3).

The *snail* gene exhibits a pattern of expression which is the inverse of *dpp* – that is, it is restricted to a ventral band of cells (Ray *et al.* 1991). Its product (a zinc finger protein) is clearly involved in initiating the process of gastrulation (wherein ventral cells buckle inwards), as *snail* embryos are characterized by a failure to gastrulate.

In between dorsal and ventral bands of *dpp* and *snail* expression is a broad ventrolateral band of expression of the *short gastrulation* (*sog*) gene (François *et al.* 1994). This gene is known to have a ventralizing effect, and it is possible that this is mediated in two ways: diffusion of *sog*'s product (or fragments thereof) (a) dorsally, where it antagonizes the effects of *dpp* and (b) ventrally, where it acts to regulate as-yet unknown combinations of downstream genes. Downstream targets of *dpp* are better known, and include the genes *tinman* (*tin*) and *bagpipe* (*bap*: Azpiazu and Frasch 1993).

Different dorsoventral bands of tissue give rise to different parts of the embryo, and ultimately to different structures and organs. It thus comes as no surprise that particular downstream targets of dorsoventral gradient genes have been implicated in the formation of particular organs (for example, *tinman* has a role in heart formation). We are thus at a stage where we can see – albeit still patchily – a network of genetic and cellular events leading from maternal initiation of the dorsoventral heterogeneity in the oocyte to the correct relative dorsoventral positioning of larval tissues and organs.

## The Left–Right Axis

Comparatively little is known about this axis in *Drosophila* or any other organism, and there are many issues here that need to be addressed – such as the asymmetric placing of certain organs (including the human heart). Population biologists have paid considerable attention recently to small, quantitative departures from perfect left–right symmetry (see Markow (1995) for a review), but have been concerned more with consequences (for fitness) than with causes. The genetic control of chirality in gastropods has been understood for some time (see Sturtevant 1923; Freeman and Lundelius 1982), but the molecular details remain largely unknown. Some recent molecular studies of asymmetry implicate TGF$\beta$: see Levin *et al.* (1995), Meno *et al.* (1996), Collignon, Varlet and Robertson (1996) and Lowe *et al.* (1996).

## 6.3 The Antennapedia and Bithorax Complexes

We now return to the elaboration of pattern in the anteroposterior axis, and focus on the role of these two gene complexes (abbreviated to ANT-C and BX-C) in causing the developmental fates of different segments to diverge. We will concentrate on the three key questions identified in the introduction to this chapter – relating to the nature of gene products, interaction with upstream/downstream genes, and chromosomal architecture.

Non-drosophilists may wish to be reminded, at this point, of the general chromosomal context. A female *D. melanogaster*'s chromosomes are in four pairs: X, 2, 3, 4. Autosomes 2 and 3 are 'long' (considerably longer than the X-chromosome), while autosome 4 is the very short 'dot' chromosome. The bands visible in preparations of polytene chromosomes from larval salivary glands are used as a frame of reference for the locations of genes. Groups of bands are recognized, the number of which broadly reflects relative chromosomal length. Groups 1–20 are on the X-chromosome; 21–60 are on chromosome 2; 61–100 are on chromosome 3; 101 and 102 are on chromosome 4. Each of these groups is divided into subgroups A–F; and each such subgroup contains a small number of bands (variable, but often in single figures). The bands represent more tightly condensed chromatin than the paler interband regions.

Chromosome 3 – the home of both ANT-C and BX-C – is approximately metacentric, with bands 61–80 on the 'left' arm and 81–100 on the 'right'. Both gene complexes are on the right chromosomal arm: ANT-C in band subgroups 84A and B; BX-C in band subgroup 89E. Descending to the molecular level, both ANT-C (Kaufman, Seeger and Olsen 1990) and BX-C (Duncan 1987) are around 300 kb in length. Clearly, the distance between them – from 84B to 89E – is vastly greater than their combined lengths, and represents a stretch of many thousands of kilobases. (In other insects, and indeed in other phyla, equivalent genes are found in a *single* chromosomal cluster, which appears to have been split in the dipteran lineage: see Chapter 7.)

Early views on the number of genes in the Bithorax complex (Lewis 1963, 1978) are now known to have been overestimates (see review by Duncan 1987), because some mutations in regulatory regions (e.g. the postbithorax mutation) were misinterpreted as separate loci. It is now clear that there are only three genes (i.e. transcription units) in the complex: *Ultrabithorax, abdominal-A* and *Abdominal-B*; see Figure 6–4. The Antennapedia complex contains some eighteen genes (Kaufman *et al.* 1990; Denell 1994; Figure 6–4), but only five of these are homeotic in the strict sense: *labial, proboscipedia, Deformed, Sex combs reduced* and *Antennapedia*. All eight homeotic genes spread through both complexes contain Antp-type homeoboxes, and their products are transcription factors which are known to bind to varying combinations of their own promoter, the promoters of some of their seven counterparts, and the regulatory regions of downstream genes, many of which remain to be identified. (See Table 6–1 for working definitions of homeotic, homeobox and related terms.)

The key role of the homeotic genes in determining the developmental fate of parasegments is illustrated by the severity of the phenotypic effects observed when these genes are mutated or deleted: for example, deletion of the BX-C results in all larval segments posterior to T2 developing as repeats of T2 (Struhl 1981), which makes sense in the light of the segmental patterns of gene expression (see below). Equally, their key role is also illustrated by their high degree of conservation over large taxonomic distances. Yet these genes *have* evolved, albeit slowly: we will return to this story in Chapter 7.

The patterns of expression of the homeotic genes within ANT-C/BX-C have attracted considerable interest, largely because of the phenomenon of colinearity, wherein the linear order of the genes along the chromosome, and the anteroposterior order of segments within whose cells they are expressed, are the same. (Strictly, this is spatial colinearity: in some systems exhibiting this phenomenon, there is also a temporal dimension involving the order in which the genes are switched on: see Duboule 1994.)

The spatial colinearity pattern of ANT-C and BX-C is shown in Figure 6–5. It must be stressed that this concerns gene expression in the embryonic epidermis. These genes are expressed in other tissues, including the imaginal discs, and the expression patterns vary somewhat both between tissues and between developmental stages, but a colinear pattern is found in all cases.

As will be apparent from the figure, however, the colinearity is imperfect: *proboscipedia* is out-of-order. The reason for this is not clear. This is hardly surprising since the reason for the colinearity itself is not clear – even though it is 'pregnant with meaning'. One thing that is apparent, however, is that colinearity is not a chance phenomenon – the odds against that are too great. With a few simplifying assumptions,

(a)

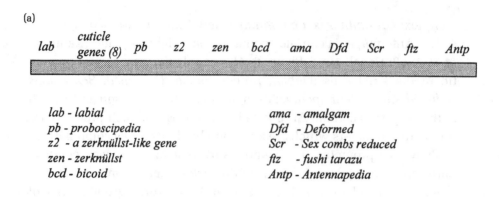

lab - labial
pb - proboscipedia
z2 - a zerknüllst-like gene
zen - zerknüllst
bcd - bicoid

ama - amalgam
Dfd - Deformed
Scr - Sex combs reduced
ftz - fushi tarazu
Antp - Antennapedia

(b)

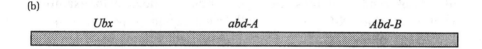

Ubx - Ultrabithorax
abd-A - Abdominal-A
Abd-B - Abdominal-B

**Figure 6–4 The genes of the Antennapedia (a) and Bithorax (b) complexes**
Information sources: (a) Kaufman *et al.* (1990); (b) Peifer *et al.* (1987).

the probability of getting perfect colinearity by chance in a group of eight genes is 2/8!, which works out at 1 in 20,160; and even the probability of getting 'one-out' imperfect colinearity is very low: 1 in 2,520.

How is the anteroposterior expression pattern of homeotic genes achieved, and what does this pattern then go on to achieve? Here, we return to the issue of 'interactional architecture' and the identification of upstream and downstream genes. While large parts of the picture are still obscure, enough has been discovered for it to be clear that the gene-switching patterns are extremely complex. Several genes (and gene groups) act to regulate the homeotic genes; they in turn regulate many other 'direct target' genes; and since some of the targets are other homeobox genes (i.e. code for transcription factors), they then go on to

**Table 6–1. Working definitions of 'hom-terms'**

The table should enable readers to appreciate the various non-correspondences that exist in this field. For example, not all homeobox-containing genes are *Hox* genes.

| 'Hom-term' | Working definition |
|---|---|
| Homeotic transformation | The transformation of a structure or region into the form appropriate for a different structure or region (e.g. leg forming where antenna should have been). Can be caused by mutation or by environmental agents. |
| Homeotic mutation | Mutation of a homeotic (=selector) gene that causes a homeotic transformation. |
| Homeotic gene | A gene which can mutate in such a way as to cause a homeotic transformation. |
| Homeotic complex (HOM-C) | A chromosomal cluster of homeotic genes (e.g. the Bithorax complex in *Drosophila*). |
| Homeobox | A 180-bp stretch of DNA which codes for a protein homeodomain. Found in all homeotic genes, but also in many others. |
| Homeodomain | A 60-aa stretch of protein which binds to DNA and identifies the protein concerned as a transcription factor. Classified into families etc. (e.g. Antp family) by degree of sequence similarity. |
| *Hox* gene | Originally, *Hox* (and HOX) were used for vertebrate homologues of the homeotic genes in the *Drosophila* Antennapedia and Bithorax complexes. *Hox* is now used for genes belonging to these complexes in *all* phyla, though specific gene names beginning with *Hox* (e.g. *Hox-A2*) are still restricted to vertebrates. |

regulate yet further downstream genes, and so on. Add to this the fact that the homeotic genes have regulatory effects on each other, and the full scale of the complexity begins to be apparent.

A simplified picture of the interactional architecture of the ANT-C and BX-C genes is given in Figure 6–6. This identifies some of the upstream and downstream genes, but does not do justice to the full complexity of the interactions. There is not space here for all the molecular detail; but let us take a more in-depth look at one homeotic gene, just to get a glimpse of some of the intricacies of regulation that

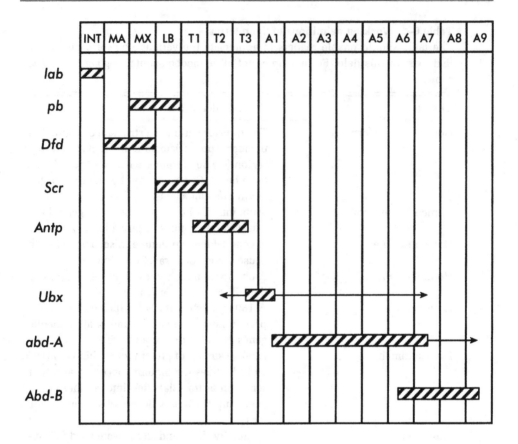

**Figure 6–5    Spatial colinearity in the Antennapedia and Bithorax complexes**

The genes are listed in their linear order along the chromosome from *labial* (the most centromere-proximal) to *Abdominal-B* (the most distal). It can be seen that the two variables 'gene order' and 'anteroposterior expression zone' are almost perfectly correlated. Expression zones given are for the larval epidermis; arrows indicate areas of weaker expression. INT = intercalary segment; other abbreviations as in Figures 6–2 and 6–4.

we meet at the molecular level. I have chosen *Ubx*, because mutations of this gene and its regulatory regions produce the famous four-winged fly, whose photograph is now commonplace (see, for example, Plate 5.1 in Lawrence 1992).

The structure of the *Ubx* region is shown in Figure 6–7. As can be seen, the gene consists of four exons and three introns, with the homeobox located in exon 1. Regulatory regions are found both within

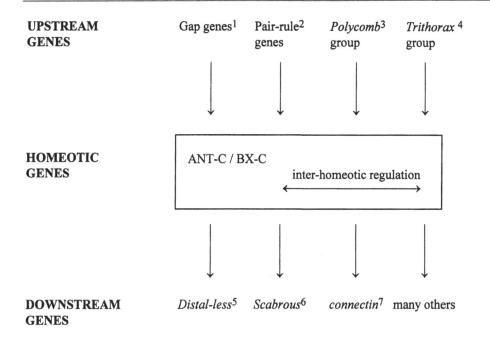

**UPSTREAM GENES** — Gap genes[1] — Pair-rule[2] genes — *Polycomb*[3] group — *Trithorax*[4] group

**HOMEOTIC GENES** — ANT-C / BX-C — inter-homeotic regulation

**DOWNSTREAM GENES** — *Distal-less*[5] — *Scabrous*[6] — *connectin*[7] — many others

**Figure 6–6  Genes that are functionally upstream and downstream of those in ANT-C/BX-C**
References, in order of superscripts: 1. Casares and Sanchez-Herrero 1995; 2. Ingham and Martinez-Arias 1986; 3. Simon *et al.* 1992; Chiang *et al.* 1995; 4. Breen and Harte 1991; Sedkov *et al.* 1994; 5. Vachon *et al.* 1992; 6. Graba *et al.* 1992; 7. Gould and White 1992.

the large intron and upstream of the gene. *Ubx* expression is controlled by a variety of gene products (including its own) binding to different regulatory regions, and having both positive and negative effects. For example, the *hunchback* product binds to the postbithorax regulatory region and acts as a repressor (Zhang *et al.* 1991); it also binds to the anterobithorax and bithorax regulatory regions, again in a repressive manner (Qian, Capovilla and Pirrotta 1991). *Ubx* is capable of both positive and negative autoregulation (Irvine *et al.* 1993), and some of this occurs indirectly via *dpp* (Thuringer, Cohen and Bienz 1993).

There is not one but at least six *Ubx* products – isoforms of the protein produced by different RNA-splicing patterns (Subramanian, Bomze and Lopez 1994). These appear to have both tissue and stage specificity, and this specificity is evolutionarily conserved across other species of *Drosophila* (Bomze and Lopez 1994). The DNA-binding

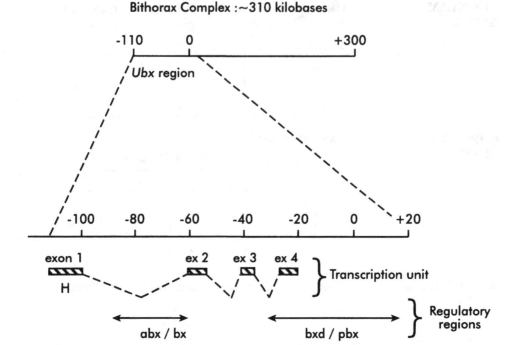

**Figure 6–7 Structure of the *Ubx* region of the Bithorax complex**
The regulatory regions, previously thought to be separate genes, are named after the mutant phenotypes to which mutations in them gave rise: abx = anterobithorax; bx = bithorax; bxd = bithoraxoid; pbx = postbithorax. The zero point is where a rearrangement breakpoint allowed the original isolation of BX-C DNA (Bender *et al.* 1983). H = homeobox.

characteristics of *Ubx* products are partly determined by their homeodomain sequence (Ekker *et al.* 1992), but can also be altered by the protein produced by *extradenticle* (Chan *et al.* 1994; van Dijk and Murre, 1994; Johnson, Parker and Krasnow 1995).

The repressive effect of *Ubx* proteins on *Antp* (Appel and Sakonju 1993) can be seen as part of (a) inter-homeotic regulating effects in general, and (b) more specifically the frequent repressive effects of more 'posterior' homeotic genes on more anterior ones. It would appear that the bands of expression of particular BX-C and ANT-C genes in particular parasegments are initiated by gap gene products such as *hunchback* and sharpened up and maintained by: (i) regulatory influences from the pair-rule genes; (ii) inter-homeotic product repression and 'competition' for DNA-binding; (iii) autoregulatory effects; and (iv) the binding of Polycomb protein (Chiang *et al.* 1995) and Trithorax protein, which may operate by modifying chromatin structure (Farkas *et al.* 1994).

Returning now to a more general perspective of ANT-C and BX-C: the extent to which our three key questions can be answered is clear. First, with regard to how the genes work: homeotic gene products are transcription factors, and they achieve their effects by regulating downstream genes. Second, with regard to upstream and downstream interactions: some of the homeotic regulators and homeotic targets have been identified, but some remain to be discovered; and many details of molecular mechanisms of interaction still need to be elucidated. Third, with regard to chromosomal architecture: the genes are clustered in a (split) linear sequence that matches the anteroposterior sequence of gene expression. The reasons for this colinearity are not yet apparent, but the probability of it being due to chance is sufficiently low that it can be regarded as negligible. Given the antiquity of the homeotic gene complex(es), and the consequent opportunity for 'reshuffling', there is clearly a selective reason for the maintenance of colinearity, which may relate to *cis*-regulatory mechanisms. (Despite this, it has been shown that at least some aspects of homeotic gene function are insensitive to chromosomal rearrangements: Randazzo *et al.* (1991), Hendrickson and Sakonju (1995).)

## 6.4    The *hedgehog* Gene and Limb Development

Throughout the last two sections, I have concentrated on the egg-to-larva developmental system rather than the disc-to-adult system (see below), even though many of the genes concerned are active in both. In the present section, however, I will go some way to redressing the balance by examining the genetic basis of appendage formation. Since dipteran larvae are legless, we will be concerned here with the role of genes in the formation of adult structures – both legs and wings.

*Drosophila* legs and wings have their origins in imaginal discs located in the thoracic segments of the larva (see review by Cohen 1993). Specifically, there are paired T1, T2 and T3 leg discs and also paired discs for the wings (T2) and halteres (T3). These discs – which give rise to both appendage and adjoining section of body wall – are composed of epithelial cells which proliferate and become committed in various ways as the larva in which they are situated grows. Eventually, during the pupal period, they give rise to the appropriate parts of the adult.

Many of the genes that have a developmental role in the embryo/ larva also have such a role in the discs – so we will encounter, below, several genes with which we are already familiar from the account of axis formation given in Section 6.2. While the ways in which these genes operate in the discs differ in detail from the ways in which they operate in the embryo, there are also broad – and in some cases very striking – similarities.

The gene interactional architecture of leg formation in *Drosophila* is illustrated in Figure 6–8a, and the spatial layout of some of the gene expression patterns involved is shown in Figure 6–8b. I will now devote a little space to 'fleshing out' these two diagrams.

As in the embryo, *engrailed* expression is restricted to posterior compartment cells, and this gene acts as a 'posterior determinant' which activates *hedgehog*. However, an additional gene, *invected*, has recently been discovered to play a role in the determination of posterior compartments (Simmonds *et al.* 1995). The relationship between *engrailed* and *invected* is not yet entirely clear, though one possibility is that the former is involved in anteroposterior compartment boundary determination, while the latter acts in a more generalized way to specify posterior cell fate.

The product of the *hedgehog* gene is a secreted protein which is present in all posterior compartment cells and in a band of anterior compartment cells which abuts the anteroposterior boundary. In this band, Hedgehog causes the activation of either *dpp* or *wingless*. Which of these is expressed depends on whether the cell concerned is in the anterodorsal or anteroventral compartment. That is, 'heritable' gene switching associated with cell lineage history appears to determine how anterior compartment cells respond to receipt of the Hedgehog protein secreted by their neighbours. (In the wing, the situation is different: here all cells in the band concerned express *dpp* rather than *wg*; recent studies of downstream effects of *dpp* in the wing have been conducted by Lecuit *et al.* (1996) and de Celis, Barrio and Kafatos (1996). Also, an inhibitory effect of the Dpp protein on *wg* expression has been shown by Penton and Hoffmann (1996).)

The activating effect of *hh* on *dpp* and *wg* may operate through interference with repression of these target genes by *patched* (Ingham, Taylor and Nakano 1991; Taylor *et al.* 1993). It appears that

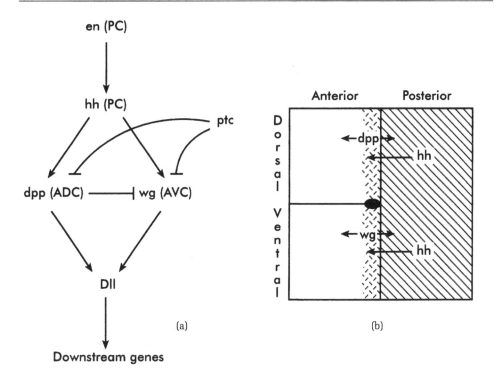

(a)                              (b)

**Figure 6–8   Genes involved in leg formation in *Drosophila***
(a) Interactional architecture: *en, engrailed, hh, hedgehog, wg, wingless, dpp, decapentaplegic, Dll, Distalless, ptc, patched*; PC, posterior compartment; ADC, anterodorsal compartment; AVC, anteroventral compartment. (Note: for recent evidence that the trans-membrane protein Patched is the receptor for Hedgehog, see Stone *et al.* 1996 and Marigo *et al.* 1996.) (b) Spatial layout of expression patterns: Hedgehog protein is secreted by posterior compartment cells, and thus the band of high Hedgehog concentration in the anterior compartment is found adjacent to the A/P boundary. This causes ADC cells to secrete Dpp and AVC cells to secrete Wg. The central dark patch is *Dll* expression. Reprinted with permission from *Nature* (Basler and Struhl, 1994. **368**, 208–214). Copyright (1994) Macmillan Magazines Ltd.

these interactions involve cAMP-dependent protein kinase A, since knocking out the gene (*pka*) which encodes this protein has a similar developmental effect to either supplying Hedgehog protein or inactivating *patched* (Lepage *et al.* 1995).

The three-way contact-point for anterodorsal, anteroventral and posterior compartment cells acts as a focal point for determination of the leg's proximodistal axis. It seems that the close proximity of *dpp-*

and *wg*-expressing cells at that point (and nowhere else in the leg disc) acts to switch on the gene *Distal-less* (*Dll*; Diaz-Benjumea, Cohen and Cohen 1994). Exactly how *Dll* establishes the proximodistal axis is not clear, but the mechanism must involve switching effects on genes further downstream – *Dll* contains a homeobox, and thus encodes a transcription factor.

It is clear, then, that adult limb formation is no different from embryonic body axis formation in that both involve a complicated network of gene interactions, many of which (particularly downstream ones) remain to be discovered. Both transcription factors and secreted proteins are involved; and gene regulation operates through the interaction of activating and repressing influences. Many of the gene switching patterns (e.g. *en* → *hh*) are similar in the two systems, but differ in their precise spatial layout and in their downstream consequences.

## 6.5 Developmental Programmes and an Evolutionary Message

Towards the end of Chapter 5, I said that we would return at this point to overall developmental programmes – the rationale being that we should now be better able to see their genetic dimension. Well, we have certainly examined the genetic details of several aspects of *Drosophila* development: what conclusions, if any, have emerged regarding the structure of developmental programmes?

As in the previous chapter, I would like to focus on the pattern of interconnection of causal links. We saw several aspects of this previously: divergences, convergences, loops, cross-links and autoregulatory feedback – all superimposed on the basic idea of a cause-and-effect cascade. It is now clear that *all* of these phenomena can be observed at an explicitly genetic level. For example, the interaction between *dpp* and *wg* to activate *Distal-less* represents a convergence, while we saw autoregulatory feedback behaviour in *Ubx*.

In Chapter 5, we noted that as well as the *pattern* of links (expressible diagrammatically as dots and arrows), the *nature* of each link was also important. Specifically, we distinguished, following Wolpert (1990), between 'do', 'wait' and 'stop' instructions. Once again, all of these can now be seen in explicitly genetic form. Positive control, such as the switching on of *hedgehog* by *engrailed*, represents a 'do' signal; while repression (e.g. of *Antp* by *Ubx*) represents a 'stop' – or 'don't', depending on the time sequence. 'Wait' signals in genetic guise are harder to spot. However, in any case of target gene activation that requires cooperation between two activators to take effect, whichever arrives first (assuming that one or other does so by some time interval, however small) can be interpreted as a 'wait' signal.

So, our previous conclusions on the nature of developmental programmes have been confirmed, rather than modified, by taking a gene-centred view. It is abundantly clear that no one simple model (e.g. a linear stepwise cascade) is appropriate either for the *overall* developmental programme of the whole organism or for any appreciably sized chunk of it, such as the programmes for segmentation, dorsoventral polarity or limb formation. Without question, all of these are highly complex, involving an interplay between multiple types of signal in a network that includes a diversity of feedback loops.

It is prudent to ask, at this stage, what effect this conclusion has on (a) the elusive 'theory of development' discussed in Chapter 4 and (b) evolutionary theory – specifically that part of it dealing with the evolution of developmental systems and of the resultant morphology. I will deal with these two effects in turn.

It seems hard at present to resist Weiss's (1940) pessimistic conclusion (see quotation in Section 4.4) that there is no 'theory of development', but only a plethora of diverse mechanisms. However, much developmental detail remains to be discovered – particularly regarding downstream genes, as noted earlier. It is still possible that when our fragments of knowledge join up to produce a more complete picture, we will see regularities of pattern that are currently hidden. Perhaps we will even come full circle from the simplicity of early theories such as Child's (1941) gradients, through the 'messy complexity' we can now see, to some future unifying principle that will coexist with, and in some sense 'explain' all the genetic and cellular details. But at present we can only speculate on this possibility.

Turning now to evolutionary theory, it is ironic that the sheer complexity of interaction between genes that underlies development carries a very simple and clear evolutionary message – concerning the importance of what Whyte (1965) calls 'internal selection'. This message will recur intermittently, and will be seen from various angles, in subsequent chapters, with an in-depth discussion in Chapter 9. It will end up contributing to one of the book's main conclusions. This is a good point at which to begin to broaden out from the brief initial statement of this message given in Section 5.3.

In discussions of morphological characters, the distinction between external adaptation and internal coadaptation is a familiar one. The divergence of tetrapod limbs into avian wings, mammalian legs, pinniped fins etc. is caused by selection for adaptation to particular environments and to ways of moving about within them. The retention of certain bone/joint arrangements – such as the articulation between humerus and radius/ulna *despite* adaptive divergence is caused by the 'need' for internal coadaptation. This 'need' also operates through selection, but this time *internal* selection (and in this instance stabilizing, though there is not a general correspondence between external/directional and internal/stabilizing – rather all four combinations are possible).

Now consider this same distinction, but focusing on developmental genes rather than the adult structures that they will ultimately produce. The immediate 'environment' of any such gene (or its product) is an extensive array of interacting genes (or their products): upstream, cross-linked or downstream in developmental terms. Selection pressures on any one gene in the all-pervasive network or 'web' of the overall developmental programme will have much more to do with evolutionary changes in those other genes than with changes in environmental variables such as temperature or food supply. Just as genes for digestive or detoxifying enzymes may be selected upon because of changes in *external* molecules with which they interact (see Day, Hillier and Clarke (1974) for a potential example involving alcohol dehydrogenase), so may developmental genes be selected upon because of changes in the *internal* molecules (primarily the products of other such genes) with which *they* interact. Mainstream evolutionary theory has so far paid insufficient attention to this internal selection, particularly in relation to the evolution of development where it is likely to be most important. I hope that, among other things, this book persuades many evolutionary biologists that we need to make good this deficiency.

# CHAPTER SEVEN

# Comparative Developmental Genetics

## 7.1 From Development to Evolution

This is an important watershed in the structure of the book, for we now turn back again from development itself to its evolution. Thus while the previous chapter was largely restricted to the developmental genetics of *Drosophila*, the present one will range widely across the animal kingdom, to examine the pattern of similarities and differences in the genetic control of development among phyla. I will concentrate, at some points, on arthropods and chordates; but nematodes, platyhelminthes, annelids and others will feature from time to time.

Although the taxonomic breadth is now greater, the approach will still be selective, and still for the same reason: to allow a reasonably in-depth treatment of the examples chosen. The particular genetic systems that I focus on in Sections 7.2 to 7.5 all relate to important ongoing research themes and debates; and all have been introduced, to varying degrees, through consideration of their mechanics in Chapters 5 and/or 6. As we proceed through these selected systems, I will attempt to draw at least one important general message on the evolution of development from each; and these messages will then be brought together in the final section.

It is appropriate to recall, at this stage, that one of the main aims of evolutionary developmental biology – as listed in Section 4.2 – is to map developmental genetic control systems onto the phylogeny of descriptive embryologies that we already have for the animal kingdom. The present chapter will allow us to examine the results of many studies conducted over the last couple of decades on the form this mapping takes. Three robust messages deriving from this accumulating body of work are: (i) confirmation of animal monophyly and the strongly conserved nature of developmental genes (see Slack *et al.* 1993); (ii) realization of just how important gene duplication events were in the early evolutionary elaboration of developmental control systems – a point made in some detail by Ohno (1970) long before most of the studies reviewed here had been conducted; and (iii) appreciation of the ubiquitous evolutionary importance of changes in interactional architecture and spatiotemporal expression patterns.

Of the four case studies discussed below, one – evolution of *Hox* genes – will be considered in greater detail than the others. The reason for this in-depth coverage is not just that more details are available for *Hox* genes than others, but that the story of their evolution now constitutes the 'classic' case study in comparative developmental genetics (see Carroll 1995). In time, no doubt, the evolution of other such families will be equally well understood, including families of genes that are developmentally downstream of the *Hox* genes, which are very numerous (Mastick *et al.* 1995) and as yet relatively poorly characterized, particularly in phylogenetic terms.

While phylogenetic comparisons of developmental genes are readily informative about the importance of certain types of mutational change in evolution (a subject which will be discussed in Chapter 8), they are much less directly informative about the mechanisms of spread of variant ontogenies at the population level. This perhaps accounts for the fact that despite the nascent 'meeting of minds' between developmental biologists and palaeontologists (see, for example, Akam *et al.* 1994b), the resultant combined endeavour remains largely detached from population genetics theory. Perhaps, at the population level, no special mechanisms are required, and 'ordinary' directional selection will suffice; but this rather dull conclusion may be premature. I will defer discussion of this issue until Chapter 9.

## 7.2 Phylogeny of *Hox* Genes

### What is a *Hox* Gene?

Answering this question will involve expanding on the working definition given in Table 6–1. Now that several hundred homeoboxes have been sequenced, it is possible to produce a sort of 'taxonomy' of homeoboxes, in the form of a nested hierarchy. This has been done by Bürglin (1994), who recognizes two superclasses, each divided into many classes which are in turn subdivided into families. The exact arrangement of taxonomic units suggested by Bürglin will no doubt change as further data accumulate, but since 346 sequences were known at the time, many of the main conclusions about homeobox relatedness are likely to be broadly correct.

All of the *Drosophila* homeotic genes (of both ANT-C and BX-C), together with all their vertebrate homologues, belong to the *Antennapedia* class, or to the closely related *labial* and *Abdominal-B* classes which have about 60–70 percent sequence similarity (measured at the amino acid/homeodomain level) with *Antp*. (Within-class percentage amino acid similarities are higher: values in excess of 90 percent are common.) The three main criteria that can be used to delimit the category of genes known as *Hox* genes are: (a) having a high degree of homeobox sequence similarity with *Antp*, and so being members of the *Antp* or closely related classes; (b) being located in a chromosomal cluster of such genes; and (c) being known to be capable of homeotic mutation or having a spatiotemporal expression pattern – usually with a pronounced anteroposterior component – that would suggest such mutations should be possible.

There is no perfect way to combine these criteria to obtain a clean separation of *Hox* genes from non-*Hox* genes, especially in a pan-metazoan phylogenetic context. I will work on the basis that *Hox* genes must satisfy criterion (a) but I will be flexible with regard to criteria (b) and (c); from an evolutionary perspective, this seems the safest way to proceed, as both cluster membership and expression pattern might be subject to sudden changes. Since we are dealing, then, with a group of genes that have *Antp*-like homeoboxes, and whose products – all of them transcription factors – have Antp-like homeodomains, we should examine the structure of the Antp homeodomain, and consider its function – in particular its specificity for binding to particular base sequences in downstream (or 'target') genes.

Figure 7–1 shows the primary structure of the Antp homeodomain, and indicates that, at the level of secondary structure, there are four alpha-helices (two of them joined together), as well as four short non-helical regions. Amino acids known to be involved in the recognition of particular DNA sequences are arrowed (based on information in Bürglin 1994 and Hunter and Kenyon 1995). Amazingly, the 'core' sequence recognized by homeodomains may be as short as four bases, although other bases must also be involved (Ekker *et al.* 1994; Gehring, Affolter and Bürglin 1994). This is important in evolutionary terms: the shorter the target sequence, the easier it will be for mutation to alter the range of downstream genes whose expression is controlled by any particular *Hox* gene. (Studies on *Ubx* in *Drosophila* suggest that the number of downstream targets for that single *Hox* gene may be as great as 170: Mastick *et al.* 1995.)

## Phylogenetic Extent of *Hox* Genes

The broad category of homeobox-containing genes extends beyond the animal kingdom into the other main eukaryotic groupings (plants and fungi: see Long *et al.* 1996 for an example in *Arabidopsis*). However, *Hox* genes, as a subset within this broader category, appear to be restricted to, and distributed throughout, the Animalia (Slack *et al.* 1993). They may, therefore, be regarded as an animal synapomorphy.

Although it appears that there are no *Hox* genes in protozoans (searches have so far proved negative: Degnan *et al.* 1995), the most 'primitive' animal phyla such as Porifera and Cnidaria *do* contain *Hox* genes, as do all other animal phyla that have so far been investigated – albeit we still have no relevant information for many of the thirty-five or so existing phyla. Table 7–1 represents an attempt to summarize the known taxonomic spread of *Hox* genes across phyla within the animal kingdom.

We can attempt to establish the precise number of *Hox* genes in each phylum (or other higher taxon). However, any such counts may be inaccurate for two reasons. First, the number of *Hox* genes may vary from species to species within a higher taxon. Second, within any species, the number of genes found may be an underestimate of the number residing in the genome: after initial studies on any species it is best to regard the claimed number of *Hox* genes as a minimum; only after more exhaustive study we can be more confident of accuracy.

| 1 | 2 | 3 | 4 | 5 | 6 | 7 | 8 | 9 | 10 | 11 | 12 | 13 | 14 | 15 | 16 | 17 | 18 | 19 | 20 |
|---|---|---|---|---|---|---|---|---|----|----|----|----|----|----|----|----|----|----|----|
| R | K | R | G | R | Q | T | Y | T | R | Y | Q | T | L | E | L | E | K | E | F |
| Arg | Lys | Arg | Gly | Arg | Gln | Thr | Tyr | Thr | Arg | Tyr | Gln | Thr | Leu | Glu | Leu | Glu | Lys | Glu | Phe |

| 21 | 22 | 23 | 24 | 25 | 26 | 27 | 28 | 29 | 30 | 31 | 32 | 33 | 34 | 35 | 36 | 37 | 38 | 39 | 40 |
|----|----|----|----|----|----|----|----|----|----|----|----|----|----|----|----|----|----|----|----|
| H | F | N | R | Y | L | T | R | R | R | R | I | E | I | A | H | A | L | C | L |
| His | Phe | Asn | Arg | Tyr | Leu | Thr | Arg | Arg | Arg | Arg | Ile | Glu | Ile | Ala | His | Ala | Leu | Cys | Leu |

| 41 | 42 | 43 | 44 | 45 | 46 | 47 | 48 | 49 | 50 | 51 | 52 | 53 | 54 | 55 | 56 | 57 | 58 | 59 | 60 |
|----|----|----|----|----|----|----|----|----|----|----|----|----|----|----|----|----|----|----|----|
| T | E | R | Q | I | K | I | W | F | Q | N | R | R | M | K | W | K | K | E | N |
| Thr | Glu | Arg | Gln | Ile | Lys | Ile | Trp | Phe | Gln | Asn | Arg | Arg | Met | Lys | Trp | Lys | Lys | Glu | Asn |

**Figure 7–1  Structure of the Antennapedia homeodomain**

The amino acid sequence (primary structure) is shown in both single-letter and three-letter codes. The positions of alpha-helical regions (secondary structure) are shown with bars. Amino acid positions known to be important in determining the target DNA sequence are indicated with vertical arrows. It is likely that additional sites important in target specificity remain to be discovered. Modified from Bürglin (1994).

**Table 7-1. Taxonomic distribution of *Hox* genes**

| Higher taxon | Representative species | *Hox* genes | Reference |
|---|---|---|---|
| Protozoa | Various | 0 | Degnan *et al.* 1995 |
| Porifera | *Tethya aurantia* | s | Degnan *et al.* 1995 |
| Cnidaria | *Hydra vulgaris* | m* | Shenk *et al.* 1993 |
| Platyhelminthes | *Phagocata wood-worthii* (flatworm) | m | Bartels *et al.* 1993 |
| | *Echinostoma trivolvis* (fluke) | m | Bartels *et al.* 1993 |
| | *Polycelis* spp. (flatworm) | m | Balavoine and Telford 1995 |
| Nematoda | *Caenorhabditis elegans* | m | Bürglin *et al.* 1991 |
| Mollusca | *Haliotis rufescens* (gastropod) | m | Degnan and Morse 1993 |
| Annelida | *Ctenodrilus serratus* (polychaete) | m | Dick and Buss 1994 |
| Crustacea | *Artemia fransiscana* (branchiopod) | m | Averof and Akam 1993 |
| Chelicerata | *Limulus polyphemus* | m | Cartwright *et al.* 1993 |
| Insecta | *Drosophila melanogaster* | m | Lewis 1978 |
| Echinodermata | *Strongylocentrotus purpuratus* (urchin) | m | Ruddle *et al.* 1994b |
| Hemichordata | *Saccoglossus kowalevskii* (acorn worm) | m | Pendleton *et al.* 1993 |
| Acraniata | *Branchiostoma floridae* (amphioxus) | m | Garcia-Fernàndez and Holland 1994 |
| Vertebrata | *Homo sapiens* | m | Boncinelli *et al.* 1988 |

s = single; m = multiple; * = includes unpublished material.
Note that the table is intended to be comprehensive at the Phylum level (in terms of studies conducted at the time of writing) but only illustrative at the species level.

Another aspect of the number of 'different' *Hox* genes within a genome concerns the distinction between replication of genes within a cluster and replication of clusters – (see Figure 7–2). It can be seen that the thirty-eight vertebrate genes fall into four distinct clusters: *Hox*

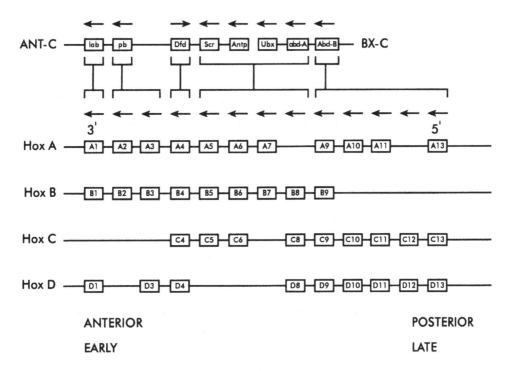

**Figure 7–2  The four vertebrate *Hox* clusters (A–D), together with their probable homologies with insect *Hox* genes of the Antennapedia complex (ANT-C) and Bithorax complex (BX-C)** Arrows indicate the direction of transcription. Modified from De Robertis (1994).

*A*, *Hox B*, *Hox C* and *Hox D*. (The terminology now used, which is a considerable advance on earlier usage, is due to Scott 1992.) These clusters are unlinked – in mice they are located on chromosomes 6, 11, 15 and 2, respectively (see review by McGinnis and Krumlauf 1992). It appears that an original cluster of thirteen genes replicated to produce four such clusters, and that different secondary losses in different clusters gave rise to the thirty-eight genes found in today's higher vertebrates. Before this cluster replication there must have been repeated duplication/replication of an initial *Hox* gene, to produce the first thirteen-member cluster.

The terms orthology and paralogy (Fitch 1970) are sometimes used in connection with *Hox* gene clusters. Orthology indicates cross-species homology, while paralogy indicates cross-locus homology within a species. However, in the context of the vertebrate *Hox* clusters this two-way distinction is not ideal as there are *three* kinds of comparison that we may wish to make: between $A_1$ and $A_1$* (within-cluster, within-locus, between species: hence, orthologous); between $A_1$ and $A_2$ (within-cluster, between-locus, within species: hence, paralogous); and between $A_1$ and $B_1$ (between cluster, within locus, within species: also paralogous). I will thus generally avoid using 'orthology' and 'paralogy' as far as possible.

Figure 7–3 shows the approximate number of *Hox* genes in those phyla for which we have information, arranged as a cladogram consistent with that shown earlier (Figure 3–5). Although I have no doubt that some of the numbers shown will change a little as further studies come in, several hypotheses suggest themselves, as follows.

The advent of animal multicellularity was associated with the appearance of the first *Hox* gene. The whole of the animal kingdom, *including* Porifera, is monophyletic. The origin of body plans was associated with *Hox* gene replication (or repeated duplication) and the consequent appearance of a *Hox* cluster. After that, the relationship between complexity of *Hox* genes and of body plans becomes rather messy. The condition of possessing two or more *Hox* clusters has arisen at least once (vertebrates) and possibly twice or more (data on four claimed paralogy groups in the chelicerate *Limulus* provided by Cartwright, Dick and Buss (1993) being as yet unconfirmed). The modification of phylum-level body plans (as opposed to their origins) may have more to do with altered spatiotemporal expression patterns than with altered numbers of genes, as suggested by Carroll (1995) and others. Maintenance of *Hox* genes in clusters, given the time available in which these could have disintegrated and the genes in them dispersed, must be governed by selection, probably related to *cis*-acting coordinate control mechanisms, as noted in Chapter 6. However, this selection appears to be relatively weak in drosophilids, possibly related to their non-sequential segmentation during embryogenesis (see Von Allmen *et al.* 1996).

While the truth or otherwise of these hypotheses will be of considerable interest, variation in *Hox* gene numbers, clustering and chromosomal locations is only part of the story. We also need to enquire about DNA sequences, patterns of expression, and 'interactional architecture'. These topics are dealt with in general terms in the following subsection, specific chordate and arthropod examples then follow.

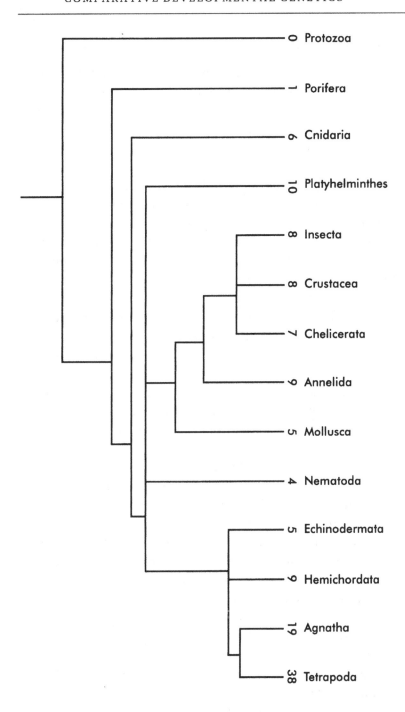

Figure 7–3  Approximate number of *Hox* genes in a selection of animal phyla

## Kinds of *Hox* Gene Evolution

What are the important variables involved in the evolution of a developmental gene (or gene group) that is regulated by upstream genes and whose product(s) in turn regulate(s) the expression of downstream genes? In the form given, the question is a general one, related to a much broader category of genes than the *Hox* genes, which are merely representative examples about which we happen to have a reasonable amount of data. Posing the question in its most general form is deliberate: we need to (a) develop a general conceptual framework within which we can recognize different categories of evolutionary change in developmental genes; and (b) be able to interrelate these categories with examples of evolutionary change in particular developmental genes in particular lineages.

Figure 7–4 is an attempt to answer the above key question. As indicated in the figure, at least five variables characterizing the group of genes in focus (*Hox*, in this case) are important from an evolutionary perspective. These should not be regarded as independent, of course; rather, they are all interconnected. For example: altered chromosomal location may result in an altered combination of *cis*-controlling elements; altered base sequence may affect the binding of a transcription factor produced by an upstream gene, thus producing an altered pattern of expression; and so on. Also, all of the five variables listed for *Hox* genes (or equivalent) in Figure 7–4 also apply to the upstream and downstream genes; and variation in the numbers and identities of these implies variation in the overall 'interactional architecture'.

If we wish to take a more holistic view of development, then it is necessary to recognize that the interactions illustrated in Figure 7–4 represent only a small part of the overall cascade/network of interactions that extends throughout ontogeny from particular distributions of maternal gene products in the egg to, ultimately, the precise spatial arrangement of structures that represents adult morphology. A change in the pattern of expression of a developmental gene at any stage in this sequence is likely to have knock-on effects on the expression patterns of genes further downstream, and hence, on organismic structure. Figure 7–5 – which represents an entirely *hypothetical* animal – illustrates two distinct ways in which expression patterns could be altered (either in individual animals by mutation or in lineages/clades by mutation and selection).

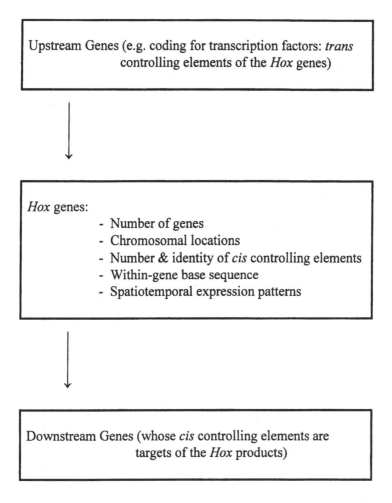

**Figure 7–4  Key variables involved in the evolution of a group of developmental genes**
All five variables identified for *Hox* genes apply also to upstream and downstream genes.

In Figure 7–5B, a mutation causing an altered base sequence in a maternal gene results in altered properties of the gene product (mRNA or protein) which causes it to be less tightly clustered to the poles of the egg. This altered pattern causes all the zygotic developmental genes to have altered patterns of expression, and a modified morphology is the end result of this cascade. Clearly, a single base change (i.e. a 'point' mutation in the relevant maternal gene) could be sufficient to bring about this whole altered ontogeny/morphology.

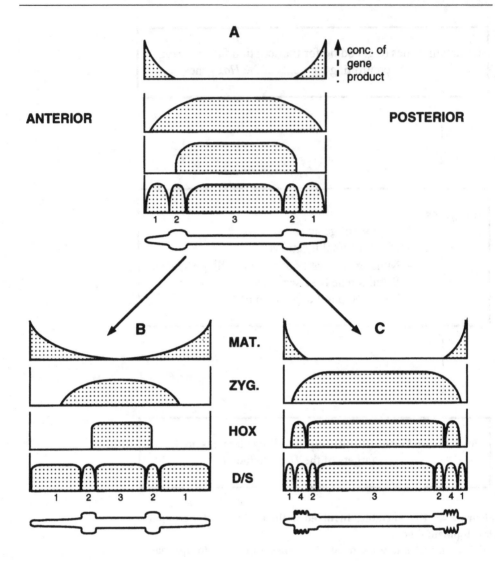

**Figure 7–5  Evolutionary changes in the spatial expression patterns of developmental genes in a hypothetical vermiform animal**

MAT = maternal gene; ZYG = early zygotic developmental gene; HOX = *Hox* gene; D/S = downstream developmental genes. *Switching patterns in A*: ZYG is switched on by mid/low concentration of MAT product, and *Hox* by high concentration of ZYG product. D/S 1 gene is activated by lack of *Hox* product, D/S 2 by low/medium *Hox* product and D/S 3 by high concentration of *Hox* product. (See text for explanation of changes occurring in B and C.)

In Figure 7–5C, an altered morphology is produced by a quite distinct mechanism. Here, the *Hox* gene duplicates, and the duplicate copy has somewhat different properties to the original. (Some of these could arise immediately, because of the duplicate's necessarily non-identical chromosomal location; they would also be enhanced, over evolutionary time, by sequence divergence in the duplicated gene.) The difference illustrated involves the duplicate being switched on only by low/moderate concentrations of the product of the upstream gene, in contrast to the original copy, which is activated by high concentrations. If (after sequence divergence) the transcription factor made by the new gene has altered target specificity, then it may activate downstream genes (labelled 4 in Figure 7–5) which were already present in the genome but were not previously involved in this particular cascade. This activation, in turn, helps to produce an altered morphology.

Of course, all of this is very simplistic – especially the relationship between the downstream genes and morphology. Also, there are many further possible types of evolutionary change that are not illustrated in Figure 7–5. But at least we have begun to visualize some of the different ways that developmental genes can mutate and evolve, with consequent effects on morphology. Anyone who was unconvinced by my assertion (in Chapter 1) that the supposed distinction between micro- and macromutation was an unhelpful oversimplification will by now, I hope, have been persuaded; and I will develop this point further in Chapter 8. But now we should return from the hypothetical to the actual, and examine some case studies of *Hox* gene evolution.

## Some Aspects of *Hox* Gene Evolution in Arthropods and Chordates

There have been many recent studies in this area, and I will not attempt to cover all of them. There are already several good reviews covering arthropods, chordates or both: see Krumlauf 1992; Duboule 1992; Tabin 1992; Krumlauf 1993; Ruddle *et al.* 1994a, b; Carroll 1995; Akam 1995; Holland and Garcia-Fernàndez 1996.

In general, chordate workers (e.g. Garcia-Fernàndez and Holland 1994) have given comparatively more emphasis to the importance of gene duplication and divergence, while arthropod workers (e.g. Averof and Akam 1995) have tended to stress the importance of interactional architecture and patterns of expression. This difference in emphasis will be reflected in the examples below. However, whether it reflects an actual difference between the groups concerned remains to be seen. Although advanced chordates exhibit the most elaborate *Hox* complexes, I suspect that in a general sense the apparent taxon/mechanism link is false, and that inasmuch as gene duplication and changes in expression pattern can be separated (and Figure 7–5 urges caution in attempting such a separation), both are important in the evolution of *all* major clades. Possibly, as we proceed from *Bauplan* to *Unterbauplan* origins, gene duplications become less important, and non-duplication-related changes in expression patterns more so. This may be true regardless of whether the body plans concerned are arthropodan, chordate, molluscan or whatever – although the transition may occur earlier in some groups than others.

Figure 7–6 shows the cladistic relationships and *Hox* gene numbers for a range of chordate genera, and for a hemichordate outgroup – the acorn worm *Saccoglossus*. The morphologies of the more primitive of these animals (acorn worm, amphioxus and lamprey) are illustrated in Figure 7–7, as these are generally less well known.

An obvious question to ask at this stage is: is there a relationship between number of *Hox* genes and morphological complexity? Arriving at a clear answer is made difficult by the fact that there is no single universally applicable definition of complexity. We could, of course, take a subjective view of complexity, but there is a danger that such a view might be influenced inadvertently by the *Hox* gene number, if this were known in advance of ranking morphological complexities – which is now unavoidable.

Definitions of the complexity of an organism (or automaton or ecosystem) include the number of components that it is made up of (von Neumann 1966), the number of different *types* of component (Saunders and Ho 1976), and the information content (Margalef 1968; Yockey 1992). When these are all highly correlated, there is no problem, but if they vary in opposite directions then ranking complexities for the species concerned becomes problematic. For example, if species A has more segments but species B has a greater diversity of segment types, which is more complex? This situation is, of course, frequently encountered.

Since *Hox* genes are responsible for determining segmental identities in the anteroposterior axis, perhaps the number of different *types* of bodily component along that axis is the best measure of complexity for our present purposes. Taking this view, the repetition of rather

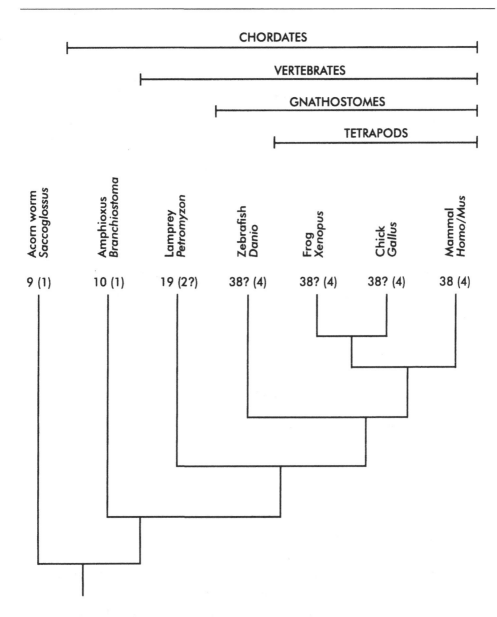

**Figure 7–6 Cladistic relationships and *Hox* gene numbers in a range of chordate genera (and a hemichordate outgroup)**
Numbers in brackets are numbers of *Hox* complexes. Tunicates are not included as their number of *Hox* genes remains to be resolved: see Holland *et al.* 1994b; Ruddle *et al.* 1994a, b; Di Gregorio *et al.* 1995. Question marks indicate uncertainty over some of the estimates. Sources of data: Pendleton *et al.* 1993; Garcia-Fernàndez & Holland 1994; Boncinelli *et al.* 1988. *Note added in proof*: it now appears that tetrapods generally have 39 *Hox* genes; while some fish may have only about 30 – see Aparicio et al. (1997).

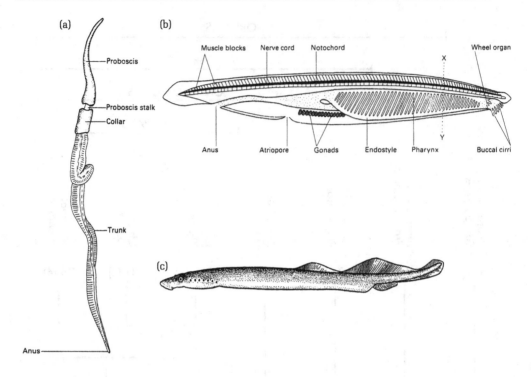

**Figure 7–7  Morphology of an acorn worm (a), an amphioxus (b) and an agnathan (c)**

Note the lower segmental diversity than in 'higher' chordates, particularly the tetrapods. From Young 1962 and Barnes *et al.* 1988. *The Invertebrates: A New Synthesis.* Blackwell, Oxford, with permission.

similar segments that characterizes the acorn worm and amphioxus can be regarded as morphologically simple, by comparison with the other groups shown in Figure 7–6. There is then an increase in complexity from amphioxus through lamprey and zebrafish to the tetrapod clade. Within that clade, however, complexity is broadly similar. For example, it seems more accurate to consider a frog and a mouse as representing different morphologies at approximately the *same* level of complexity rather than to consider a mouse as being the more complex of the two. (We must at all costs avoid *scala naturae* thinking (as Panchen 1992 has warned) wherein we see a ladder of complexity with ourselves at the top.)

If the ranking of complexity just put forward is correct, then there is clearly a partial correspondence with number of *Hox* genes: the simplest chordate morphologies have the fewest *Hox* genes, and the most complex morphologies have the most. However, the patterns of increase in the two variables do not map exactly. In particular, it appears that the shift from fish to tetrapod body plan involved relatively little further elaboration of *Hox* gene number, albeit we still only have extensive data for a relatively small number of species.

Actually, this is not surprising, and we should hardly expect an exact correspondence. Once a reasonable number of *Hox* genes have appeared through duplications, there is considerable scope for divergence between existing genes to result in altered patterns of spatial expression and altered interactional architectures, as noted above – and in some cases such alterations will result in more complex morphologies (see Sordino, van der Hoeren and Duboule 1995 for a possible example involving the fin-to-limb transition). The *ad hoc* nature of evolution must always be borne in mind. There is no single route to any particular state such as 'increased complexity'. Rather, any viable route to that state will occur with a frequency that derives, ultimately, from its ease of production through mutation (see Chapters 8 and 10). In the present case, a lineage's journey along one route – proliferation of *Hox* genes – eventually decreases the importance of that route by opening up others.

We will now examine a selection of studies on *Hox* gene evolution in arthropods. Here, insects are of course the best known and thus constitute a sensible starting point. It is less clear, however, which other higher taxa to use in a comparative study. As we saw in Chapter 3, the existence of a uniramian grouping (insects + myriapods) has been questioned by some recent molecular studies (Boore *et al.* 1995; Friedrich and Tautz 1995; Popadíc *et al.* 1996) which suggest that crustaceans, rather than myriapods, are the sister group of the Insecta; and this issue remains to be resolved. Given this controversy, and the fact that it makes sense to look at *Hox* gene evolution against an agreed cladistic background, I will focus, here, on relationships between Annelida, Crustacea and Insecta. There can be no reasonable doubt that the appropriate cladistic pattern for these three groups is (A + (C + I)).

Insects, as we saw in Chapter 6, have eight *Hox* genes – although this number is affected by exactly how the category is delimited. (Akam *et al.* (1994a) include *zerknüllt* and *fushi tarazu* as *Hox* genes because they emphasize the importance of cluster membership.) Although the cluster is split in *Drosophila* into ANT-C and BX-C, this appears to be unique to the dipteran clade. Other insects (for example, beetles: see Beeman *et al.* 1993) have a unitary *Hox* cluster.

Both annelids and crustaceans have at least five, and probably more like eight or nine, *Hox* genes (Dick and Buss 1994; Akam *et al.* 1994a). This is of particular interest because, in terms of morphological complexity, the 'proto-annelid'-to-insect (or to crustacean) transition has some similarities to the protochordate-to-tetrapod one: both involve an increase in intersegmental diversity. Yet while this type of transition was associated with an elaboration of *Hox* genes in chordates, it is not associated with a comparable elaboration in the annelid/arthropod clade. (Note that cladistic uncertainty in the protostomes makes it less certain that arthropods evolved from proto-annelids than that tetrapods evolved from protochordates.) Perhaps the move from a simpler, non-segmented protostome body plan to a proto-annelid was accompanied by *Hox* gene proliferation. This possibility is supported by the limited *Hox* cluster of the nematode *C. elegans* (four genes: Bürglin and Ruvkun 1993; Hunter and Kenyon 1995). However, (a) this cluster may be secondarily simplified; and (b) the apparent existence of up to ten *Hox* genes in flatworms (Figure 7–3) is hard to explain under this view.

The diversification of arthropod body plans, then, was not primarily based on changes in *Hox* gene number, but rather on changes in the spatial expression patterns and interactional architecture of the *Hox* genes that were already present in the appropriate stem group. This is true both of the divergence of arthropod classes (Averof and Akam 1995) and of the divergence of *Unterbaupläne* within them (Warren *et al.* 1994; Carroll *et al.* 1995). We will now examine these studies in a little more detail.

Averof and Akam (1995) investigated the expression patterns of four *Hox* genes: *Antp*, *Ubx*, *abd-A* and *Abd-B* in the branchiopod crustacean *Artemia fransiscana*. The crustacean body plan exhibits much variation in the number of segments and appendages (see Barnes *et al.* 1988, chapter 8), but there are often multiple thoracic segments, all with paired biramous limbs, and *Artemia*'s body plan is of that general kind. This contrasts with the typical insect arrangement of well-defined head, thorax and abdomen, with a standard three thoracic segments each with a single pair of (uniramous) legs.

Averof and Akam's findings for *Artemia*, and comparative information on insects, are summarized in Figure 7–8. It should be noted in particular that there has been a major divergence in the anteroposterior

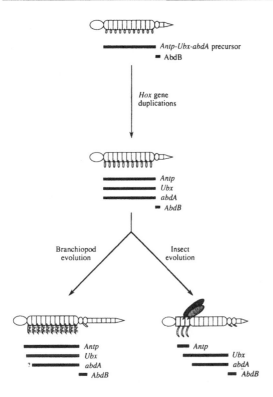

**Figure 7–8   Model of the evolution of insect/crustacean body plans, and associated changes in the expression patterns of four _Hox_ genes, proposed by Averof and Akam (1995)**

Reprinted with permission from _Nature_ (Averof and Akam, 1995. **376**, 420–423). Copyright (1995) Macmillan Magazines Ltd.

expression domain of _Antp_, and that in both cases these domains correspond approximately to the sequence of leg-bearing, 'thoracic' segments. However, Averof and Akam propose a homology between the _Artemia_ 'thorax' and the whole insect thorax + abdomen (up to the genital region). They also propose a homology between the insect and crustacean genital regions, both of which are characterized by expression of _Abd-B_. Further studies, particularly on other crustacean _Unterbaupläne_, may help to ascertain whether the proposed homologies are correct. But regardless of this, and of the fact that there is variation in expression patterns between tissues (not shown in Figure 7–8), the link between _Hox_ gene expression pattern and body plan divergence is clear.

A similar message emerges from the studies by Warren *et al.* (1994) and Carroll *et al.* (1995) on the divergence of limb arrangements in different groups of insects – but these workers give particular emphasis to the *negative* role of the *Hox* genes. One example of a negative controlling influence is that larval abdominal limbs in Lepidoptera develop in regions which represent 'holes' in *Ubx* and *abd-A* expression. (There are no such holes in *Drosophila*; and dipteran larvae correspondingly lack abdominal legs.) However, a combination of positive and negative effects seems more likely since *Antp* is activated in the *Ubx / abd-A* 'holes'. Such a combination of effects would also be compatible with Averof and Akam's (1995) results, as depicted in Figure 7–8.

## Conservation of Colinearity

Colinearity of chromosomal sequence of genes and anteroposterior expression domains was noted in *Drosophila* in Chapter 6. There is also a temporal (activation sequence) dimension to colinearity in vertebrates – and probably also in short germ band insects. (The directionality of the correspondence is: 3'/anterior/early versus 5'/posterior/late.) It is now clear that colinearity (sometimes called 'collinearity') applies very generally across both insect and vertebrate *Hox* complexes (see, for example, Krumlauf 1992; Dekker *et al.* 1993; Duboule 1994) and probably those of many other phyla too. Colinearity is maintained in transgenic mouse embryos containing regulatory elements imported from *Drosophila*. One of the few exceptions to colinearity occurs in regenerating limbs of axolotls (Gardiner *et al.* 1995); but the aberrant expression patterns found there (which are coincident, rather than sequential) do not characterize embryonic limb development in the same species, where colinearity again applies.

The conservation of colinearity across many taxa must occur due to functional/selective reasons, and not because of a lack of variability in *Hox* clusters. Although some of the literature tends to engender a mental image of '*the* insect cluster' or '*the* four vertebrate clusters', the apparent uniformity implied is inaccurate. Even between congeneric species the exact configuration of *Hox* genes can alter, as Randazzo, Cribbs and Kaufman (1991) and Von Allmen *et al.* (1996) demonstrated in their studies of different *Drosophila* species. And indeed there are within-*species* polymorphisms of *Hox* genes, as shown recently by Gibson and Hogness (1996).

In a general sense, it is probably the case that colinearity is phylogenetically widespread because variants which depart from it tend to damage a fundamental developmental control system which is required for anteroposterior patterning. We need, of course, to be more specific than this, and further studies of *Hox* genes will no doubt eventually enable such specificity. Colinearity still has much to teach us, both about development itself and about its evolution.

## 7.3   Dorsoventral Polarity in Arthropods and Chordates

We now turn from anteroposterior to dorsoventral patterns, thus following the same sequence as in Section 6.2, but now, of course, taking a comparative perspective. We will focus on a recent study by Holley *et al.* (1995). This appears to support a relationship between dorsoventral polarity in arthropods and chordates proposed by Geoffroy Saint-Hilaire (1822), namely that each is an 'upside-down' version of the other. This proposal has its origin in the observation that certain structures which are dorsal in arthropods (e.g. the heart) tend to be ventral in chordates, while other structures (notably the main nerve cord) show the opposite pattern.

In the previous chapter, we examined some of the genes responsible for determining dorsoventral pattern in *Drosophila*. These included *decapentaplegic* (*dpp*), which is expressed in a broad dorsal band, and *short gastrulation* (*sog*), whose broad ventrolateral band of expression abuts that of *dpp* (Figure 6–3). Like the *Hox* genes, these participate in a gene-switching cascade that ultimately produces the correct spatial arrangement of morphological structures. (However, *dpp* and *sog* are unlike the *Hox* genes not only in the body axis along which they operate but also in terms of how they produce their effects: the Dpp protein, for example, belongs to the TGF$\beta$ family, and so acts in an extracellular, rather than intranuclear, capacity.) As we saw in Chapter 6, one of the genes downstream of *dpp* – *tinman* – is involved in heart formation. Presumably other such downstream genes control the formation of other dorsal structures, while 'ventral downstream genes' (including those downstream of *sog*), have a similar structural role in the ventral region.

Let us now examine an explicitly evolutionary version of Geoffroy Saint-Hilaire's theory. (Recall that 1822, when the original version was proposed, was pre-Darwinian; so the original theory was essentially one of 'natural pattern', much like proposals originating from the trans-formed (or 'pattern') school of cladistics over the last couple of dec-ades.) Translated into evolutionary language, Geoffroy's proposal can be put as follows: "there was a reversal in the dorsal-ventral axis after the divergence of the common ancestor of insects and vertebrates" (Holley *et al.* 1995). The homogeneity of dorsoventral polarity within each group suggests that its reversal in one of them occurred during, or at least not *long* after, their divergence.

If this view is correct, what should we expect to find with regard to the developmental–genetic control mechanisms involved in determin-ing dorsoventral pattern in chordates? We should find at least three things: first, homologues of *dpp* and *sog* (but with a reasonable degree of sequence divergence, given the lengthy period of elapsed time since lineage separation); second, an opposite pattern of expression (with the *dpp* homologue expressed ventrally); and third, opposite effects (the *dpp* homologue ventralizing and the *sog* homologue dorsalizing, through their effects on appropriate groups of downstream genes, which may themselves have been conserved since lineage divergence). There is now evidence for all three of these things, and we will ex-amine them in turn.

The product of the *chordin* gene from the toad *Xenopus* shares nearly 30 percent of its amino acid sequence with the product of *sog* in *Drosophila* (François and Bier 1995). The *Xenopus* gene *bmp-4* (named after its bone morphogenetic protein product) is functionally homologous to *dpp* (Holley *et al.* 1995), though at the time of writing the exact percentage sequence similarity between them remains to be established. The patterns of expression are indeed opposites, as pre-dicted from the evolutionary version of Geoffroy's theory. That is, *bmp-4* is expressed ventrally in *Xenopus* (Fainsod, Steinbesser and de Robertis 1994), while *chordin* is expressed dorsally (Sasai *et al.* 1994).

That *bmp-4* has a ventralizing, and *chordin* a dorsalizing effect in vertebrate development is hardly surprising, given their expression patterns; and these effects have indeed been confirmed – see Jones *et al.* (1992) and Sasai *et al.* (1994), respectively. Holley *et al.* (1995) build on these results and confirm the functional equivalence of the verte-brate and insect systems, by conducting a series of experiments invol-ving interspecific (*Drosophila/Xenopus*) injections of the appropriate mRNAs.

Of the four possible such injections, Holley *et al.* (1995) conducted three: *dpp* and *sog* mRNAs from *Drosophila* injected into *Xenopus* embryos; and *chordin* mRNA from *Xenopus* injected into *Drosophila* embryos. Both *Drosophila* gene products had effects in *Xenopus* that mimicked the effects of their *Xenopus* homologues rather than their own effects in their native *Drosophila*. That is, *dpp* ventralized and *sog* dorsalized. *chordin* mRNA at first appeared to have no effect in *Drosophila*, but when attached to the N-terminal region of the *sog* mRNA, to facilitate its processing in *Drosophila*, it did then have the expected ventralizing effect – again in line with the developmental mechanics of the host embryo. (The *sog* mRNA N-terminal region, when injected alone, had no effect.)

While the case for homology and for axis inversion in one of the two lineages is not conclusive (see Lacalli 1995 and Peterson 1995 for doubts), it is certainly very persuasive (de Robertis and Sasai 1996). Also, an alternative hypothesis involving the independent origins of complex and yet very similar dorsoventral patterning mechanisms seems rather implausible (but see next section). So it looks as if Geoffroy Saint-Hilaire was right after all, even if it took the rest of us almost two centuries to catch up with him.

Taken together, this comparative case study and that involving the *Hox* genes tell a story of developmental mechanisms that are highly conserved over great taxonomic distances. Thus while the plethora of different mechanisms of development operating within any one species does indeed convey a rather heterogeneous picture, as discussed in Chapters 4–6, this picture is counterbalanced somewhat by tantalizing glimpses of similarity of at least some mechanisms across many taxa. Perhaps we should begin to think in terms of a matrix in which developmental mechanisms form the columns and species the rows. At present, among-column variance looks as if it is much greater than among-row variance. But of course, such speculations are premature, as evolutionary developmental biology has so far opened only a small minority of the 'boxes'.

## 7.4    Limb Formation, *hedgehog*, and the Nature of Homology

We have touched on the concept of homology at various stages in the book, from its intrinsic link with the idea of body plans (Chapter 2) to specific proposals of homology (last two sections). These multiple points of contact are hardly surprising, as homology is one of the key concepts in evolutionary biology – despite its non-evolutionary origin (Owen 1847, 1848: see Table 4–1). The whole of the living world can be pictured in terms of an array of nested homologies, with those of more restricted taxonomic scope being grouped within others of broader scope. The genetic code is homologous across the whole of the living world; *Hox* genes are homologous throughout the animal kingdom; the notochord is a panchordate homology; feathers constitute a homology restricted to birds. In general, the characters involved in the broadest homologies tend to appear earliest in individual ontogeny, as we saw in Chapter 2. This relationship holds for the four examples given, and in many other cases also. There are, however, exceptions: it is a statistical relationship, not an absolute one. Whatever else can be said of evolution, it rarely produces patterns that are neat and tidy.

Despite the widespread acceptance and usage of the concept of homology, its familiarity to evolutionary biologists tends to mask some considerable problems that are associated with it. The nature of these depends on the perspective taken, which in turn depends on one's disciplinary background. Some cladists wish to exclude evolution from the *definition* of homology, so that an *explanation* of it in evolutionary terms is not 'circular reasoning' (Brady 1985). In that case, homology is hard to define. (This is also true, of course, of Owen's original non-evolutionary concept.) My own view is that it is perfectly acceptable to define homology in evolutionary terms, and indeed that to avoid doing so in present-day studies seems almost perverse. Homologous characters are then those whose similarity, in different taxa, derives from their common origin in a shared ancestral lineage. This view is expanded upon in Figure 7–9. Note that the 'characters' involved can be molecular, developmental, behavioural or whatever: they need not necessarily be anatomical, although it was of course in the anatomical sphere that the concept first arose.

When this view is taken, the definition of homology becomes relatively straightforward. What is difficult, in many cases, is being certain that a particular pair or group of characters, from different taxa, are or are not homologous. (The alternative, as noted in Figure 7–9, is that they are 'homoplastic' – a term originally introduced by Lankester (1870) – and that the observed similarity derives from convergent evolution which can effectively mask their independent origins.) Of course, many cases are clear: the homologous nature of tetrapod legs

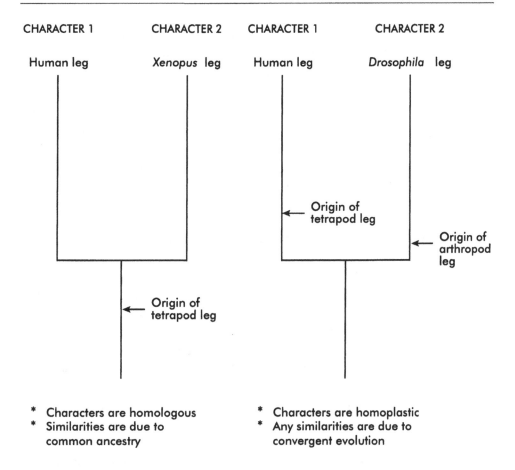

**Figure 7–9  Phylogenetic 'definition' of homology and homoplasy**
The kind of homology considered here is taxic homology. Where
two or more structures within the same organism are homologous –
such as the human arm and leg – this constitutes serial homology.

is not in doubt; nor, until recently, was the homoplastic relationship
between these and arthropod legs (see below). But many cases are less
clear cut, and it is thus sensible to examine what criteria are best
employed to make the correct decision on the nature of the relationship – homologous or homoplastic – in any particular case.

   Some of the criteria that have been employed are somewhat character specific. For example, using the 'principle of connections' (due to
Geoffroy Saint-Hilaire: see Panchen 1994), homologous relationships
among bones can be determined by looking at the other structures –
including other bones – with which they interconnect. Thus vertebrate

radius and ulna bones are generally found running between the humerus and the carpals. While this approach can be generalized to morphological characters other than bones, it remains an intrinsically structural/anatomical approach, and is difficult to apply to many other sorts of character. However, it might be possible to look at the position of the product of a developmental gene within a cascade or network of interactions in a similar way. I am not aware of anyone having taken this approach, but this seems an interesting possibility for the future. Probably 'connections' within cascades (i.e. the interactional architecture) will be more evolutionarily variable than anatomical connections. If so, then while similar such 'connections' may be an argument in favour of developmental homology, different ones are not a strong argument against it.

The best approach to get us away from character specificity of criteria for homology is to look at the overall *complexity* of the character that is similar between two or more taxa. The greater the shared complexity, the lower the probability that different lineages will have hit upon it independently, and thus the higher the probability that the relationship is one of homology. Paired outgrowths of unspecified structure from the bodies of two or more animals under consideration are not too impressive in this respect, if no further information is available. But if the paired outgrowths in both/all cases have similar detailed structure (including similar connections of constituent parts – so that criterion is subsumed here), a similar developmental origin, and so on, then the case for homology is much stronger. One cautionary comment, though: any complex character whose various details may all have arisen as pleiotropic effects of a single gene would be relatively easily hit upon by mutation – as in the case of polydactyly – and so would provide weaker evidence of homology than an equally complex character with a multifactorial basis.

Let us now focus on insect and vertebrate limbs. As noted above, these are generally regarded as homoplastic. That is, when the protostome and deuterostome branches of the bilaterian radiation split more than 500 *my* ago, their last common ancestor was limbless. Limbs were then 'invented' independently by arthropods and chordates. The tetrapod leg originated from a fin (see review by Vorobyeva and Hinchliffe 1996), and fins in turn originated from relatively insignificant outgrowths from the trunks of amphioxus-like protochordates. The arthropod limb probably arose during the early history of the arthropod clade, so that (for example) insect and crustacean limbs may be homologous (see Averof and Cohen 1997), just as are mammal and reptile limbs; but the two groups are not homologous with each other.

Given this proposed homoplastic relationship between insect and vertebrate limbs, we would not expect their 'detailed complexities' to be similar; and mostly they are not. Their gross structures are sufficiently different that it is not even meaningful to look for similarity of 'connections'. (The use of vertebrate terms like 'femur' for parts of insect legs is, of course, only an irritating distraction here.) Their embryological origins are also very different. In general, we have a classic homoplastic relationship: similarity of function (i.e. analogy) but, because of independent origin, very different structural details.

Given this situation, it is perhaps odd that some important aspects of the developmental–genetic control of limb development are rather similar between insects and vertebrates. In particular, the *hedgehog* gene (which we encountered in Chapter 6) is expressed in the posterior compartments of wing and leg imaginal discs in *Drosophila* (Tabata, Eaton and Kornberg 1992). Its close homologue, *Sonic hedgehog* (Chapter 5) has a similar pattern of expression in the limb buds of mice and chicks (Riddle *et al.* 1993) and in the fin buds of zebrafish (i.e. again a restriction to the posterior regions: Figure 7–10 top; see Fietz *et al.* 1994).

If the genetic control of limb formation was entirely distinct from the control of development in the trunk, then these expression patterns of *hedgehog* family genes would present us with strong evidence that the prevalent view of metazoan phylogeny was wrong and that arthropod and chordate limbs were homologous. However, as we saw in Chapters 5 and 6, *hedgehog* family genes are also involved in developmental processes in the trunk. This provides a possible explanation for developmental–genetic similarity despite homoplasy.

The scenario goes something like this. The trunks of all bilaterians are homologous, and indeed the zootype concept (Slack *et al.* 1993) suggests that many aspects of the developmental–genetic control of trunk patterning are common to all bilaterians. When, in *any* lineage, limbs begin to be formed, the cells involved represent novel proliferations whose origin (in cell lineage terms) is the trunk, even though their new-found developmental fate is ultimately to grow outwards from it. They will carry out with them particular arrays of 'on' and 'off' developmental genes, as a result of their clonal history. This is shown in very simplified form in Figure 7–10 (bottom). In reality, the pattern of cellular movements and differentiation involved in the for-

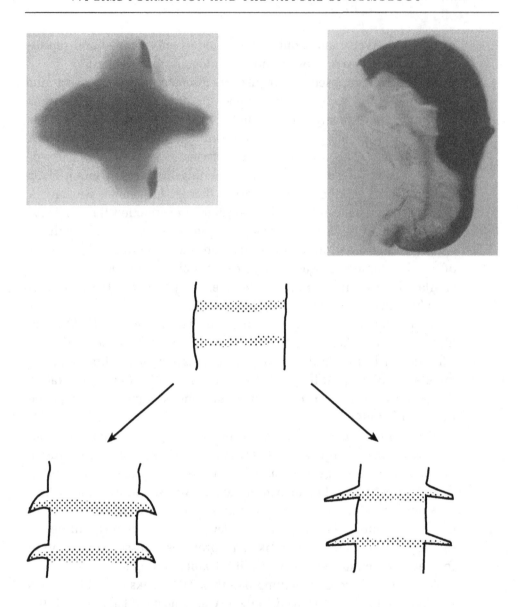

**Figure 7–10   Expression of *hedgehog* genes in posterior limb primordia, and implications for limb homology/homoplasy**
Top: *Sonic hedgehog* expression in posterior mesenchyme of mouse limb bud (left); and expression of *hedgehog* in posterior wing disc of *Drosophila* (right). From Fietz *et al.* 1994, in *The Evolution of Developmental Mechanisms*. The Company of Biologists, Cambridge.
Bottom: A way in which homoplastic limbs could exhibit developmental–genetic similarities due to outgrowth from a homologous trunk and utilization of trunk patterning genes.

mation of a limb primordium is very complex: see Bate and Martinez-Arias (1991) for a detailed description of the origin of limb imaginal discs in the *Drosophila* embryo.

Clearly, this sort of thing could happen twice or many times in phylogenetic history, with the result that homoplastic limbs have similar expression patterns of developmental genes because of their origins from homologous trunks. Perhaps, then, there is no need to revise our picture of metazoan limb phylogeny after all. (Whether a broadly similar form of argument can be used to defend the previously agreed homoplasy of arthropod and chordate segmentation against recent suggestions of homology based on developmental-genetic similarity (Holland *et al.* 1997; De Robertis 1997) remains to be seen.)

## 7.5    Phylogeny of Cadherin Genes

So far, all of the genes we have examined in this chapter are involved primarily in the process of pattern formation, rather than cell differentiation *per se*. Often, different segments will contain a very similar array of cell types, and the main difference – for example between wing and haltere in a dipteran – is in the overall pattern, rather than in the individual components out of which it is built.

Cadherin genes, in contrast, have direct links with both cell differentiation and pattern formation. Often, differentiation involves, among many other things, the expression of particular cadherin genes and the consequent 'peppering' of the cell membrane with a particular type of cadherin protein: N-cadherin in certain neural cells, for example. But because the main function of cadherins is usually homophilic binding (as we saw in Chapter 5), they help to determine the shape of blocks of tissue, and thus pattern formation. They are even involved (albeit in a negative way) in morphogenetic movements, as there is a rapid loss of cadherin gene expression in cells which are about to become migratory (Levi *et al.* 1991).

A typical mature cadherin protein has seven linearly arranged domains: extracellular domains EC1 to EC5 (the first four of which are closely related to each other), followed by a transmembrane domain and, finally, a single cytoplasmic domain (see Pouliot 1992). The EC1 domain, which is at the amino-terminal end of the molecule, appears to be the most highly conserved. It consists of about 110 amino acids, and it contains the HAV motif (histidine–alanine–valine) that plays an important role in mediating intercellular adhesion.

Very high amino acid sequence similarities are observed among 'core' members of the cadherin family. For example, K-cadherin from rats has around 97 percent sequence similarity with human cadherin-6 (Xiang *et al.* 1994; Shimoyama *et al.* 1995). However, as well as the 'core' cadherins, there are now recognized to be other, more divergent, members of a cadherin 'superfamily'. These include protocadherins (Sano *et al.* 1993), the product of the *Drosophila Fat* locus (Mahoney *et al.* 1991), the product of the human *ret* oncogene (Kuma, Iwabe and Miyata 1993; Rimm and Morrow 1994), fibroblast growth factor receptors and influenza strain A haemagglutinins (Byers *et al.* 1992).

Pouliot (1992) produced a phylogeny of vertebrate cadherin genes, based on sequences of the conserved EC1 domain which were analysed using PAUP. Part of this phylogeny is shown in Figure 7–11. In general, there appears to be a tendency for interspecies comparisons of an equivalent cadherin type to yield tighter clustering than intertype comparisons within species. This result links well with the 'matrix' approach suggested at the end of Section 7.3. There, we considered a species × developmental mechanism matrix, and noted the possibility of a higher variance between mechanisms (within species) than between species (within mechanisms). We now observe a similar pattern in a species × developmental gene matrix.

Pouliot (1992) makes clear his view that gene duplication underlies the production of the diversity of sequences covered by his phylogeny. He states (p.747): "As with many other gene families, the [cadherin] superfamily appears to have been generated by successive rounds of gene duplication." Also, the pattern of phylogenetic clustering allows us to make inferences about the temporal order of these successive duplication events and subsequent mutational divergence. In particular, where many species cluster tightly for a particular type and there is a relative 'gulf' between that and another cadherin type, it suggests that the various types were produced by gene duplication in an early common ancestor. These types will hence have had more time for divergence than the different specific homologues of a particular cadherin (e.g. rat K, human 6), whose shorter divergence period corresponds with the time since separation of the lineages concerned.

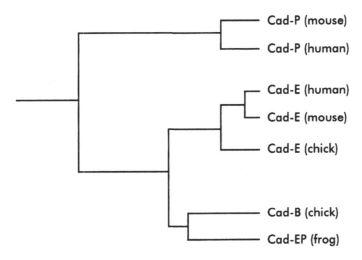

**Figure 7–11 Phylogenetic relationships among some vertebrate cadherins**
Note that two of the apparent clusters are based on cadherin types, while none are based on species. So (for example) cadherins P and E diverged before the specific versions of each. P = placental; E = epithelial; EP = E and P-like; B = brain. Data from Pouliot (1992).

## 7.6 Emergent Evolutionary Messages

At the end of Chapter 6, having considered the developmental genetics of only a single organism, one evolutionary message was already apparent. The complexity of the network of interacting genes and gene products that underlies development indicates that many of the selective pressures acting upon a developmental gene are likely to be internal (i.e. related to its interactions with other such genes) rather than external (i.e. directly related to population density, food resources, temperature or other environmental variables). This 'internal selection' has been generally underestimated, even ignored in many instances, in neo-Darwinian theory – as pointed out by Whyte (1965). I will discuss this topic in detail in Chapter 9.

Several additional messages emerge from the current chapter's comparative examination of developmental genetics, and one of them – the importance of *co*evolution – connects directly with the concept of internal selection. In conventional evolutionary biology – that is, focusing on *external* selective agents – one important distinction between different kinds of evolutionary change is that between situations in which the selective agent itself can evolve and those in which it cannot. In the former case, causality is two-way and we have a coevolutionary process. This has been much studied in relation to various kinds of interspecific interaction, including competition (Arthur 1982b), predation and parasitism (Fain 1994). The equivalent of such 'external coevolution' which now comes into focus is the 'internal coevolution' of developmental genes whose products interact. An evolutionary change initially affecting any particular point in a network of such interactions may have selective consequences which ramify out from the point initially affected, and any resultant changes elsewhere may feed back selective pressure on that 'initial point'. Study of such internal coevolutionary processes has barely begun – but it is likely to be of considerable importance in the future. Key questions can already be anticipated, including that of whether there is a predominant directionality of the evolutionary changes involved: perhaps more often from upstream to downstream genes than vice versa?

Some of the other 'emergent evolutionary messages' have already been discussed quite extensively in previous sections, and all that is required here is a brief recap. With regard to *pattern*: the animal kingdom is indeed monophyletic and characterized by nested homologies; and many of our morphologically based phylogenies have now been confirmed by comparative developmental studies. With regard to *mechanisms*: changes in interactional architecture and spatiotemporal expression pattern of developmental genes underly *all* evolutionary divergence in ontogeny/morphology; how readily such changes occur depends on many things – among them the length of DNA sequence recognized by a controlling agent such as a transcription factor; when mutations are thought of in a developmental context, the false micro/macro 'dichotomy' disappears; gene duplication is of much importance, particularly in the origin of the metazoa and the initial stages of diversification of body plans.

Let us now return to the 'matrix' picture encountered twice in this chapter: a matrix whose columns are either developmental mechanisms or developmental genes, and whose rows are species. If it is indeed true that there is typically less variation going up or down a column than there is in traversing a row, then this argues for the establishment of different types of developmental gene (and mechanism) at a very early stage in metazoan evolution. This is then followed by lineage divergences involving morphological changes that are based on modification of the exact ways in which the already available developmental genes and mechanisms work, rather than on the evolutionary 'invention' of entirely novel mechanisms. This view has many connections throughout the nascent (and still very heterogeneous) discipline of evolutionary developmental biology: from Slack *et al.*'s (1993) genetics-based zootype to the palaeontology-based Vendian/Cambrian 'explosion' (see Valentine 1994; but also Fortey *et al.* 1996 and Wray *et al.* 1996) to Kauffman's (1995) complexity theory-based picture of evolution being "dramatic at first, then dwindling to twiddling with details later", which I discuss further in Chapters 10 and 11.

We will now enter the territory of conventional evolutionary biology, armed with sufficient developmental insight (a) to view afresh the interplay between mutation (Chapter 8) and selection (Chapter 9); and (b) to seek explanations of large-scale evolutionary patterns in the interaction between developmental and population-level processes. This endeavour will occupy the remaining five chapters.

# CHAPTER EIGHT

# Gene Duplication and Mutation

## 8.1   Introduction

Several of the cladistic and phylogenetic studies examined in Chapter 3 suggested that the whole of the animal kingdom – including such outlying groups as Porifera – represents a monophyletic clade, which originated from a unicellular eukaryote. The comparative developmental–genetic studies discussed in Chapter 7 confirmed this proposed monophyly; and the idea of a general genetic mechanism underlying the development of all animals is neatly captured in the 'zootype' concept (Slack *et al.* 1993).

If this scenario is broadly correct, then *all* the genes of *all* animals may have arisen from the gene pool of a single population of a single (unknown) pre-Cambrian species, through the processes of gene duplication and mutation. (The possibility of subsequent 'injection' of some other genes into the animal clade through horizontal transfer via transposable elements should also be borne in mind.) The present-day unicellular eukaryote *Saccharomyces cerevisiae* (baker's yeast) has approximately 6,000 genes (see Chothia 1994). While this is clearly not a candidate for being a very close relative of the ancestor of all animals, it is probable that that ancestor had a comparable gene number – somewhere in the 5,000–8,000 range. So, our own complement of perhaps some 70,000 genes has arisen through an approximately tenfold increase in the number of genes, coupled with mutationally driven divergence.

The group of molecular events which we refer to as 'mutations' can be delimited in various ways, thus producing definitions of varying breadth – though all are based ultimately on the idea of heritable changes in DNA base sequence. At the narrow end of this spectrum, 'mutation' (*sensu stricto*) is sometimes considered to include only small-scale changes such as base substitution, addition or deletion, which are then *contrasted* with bigger-scale microscopically visible changes in karyotype, which are referred to as chromosomal rearrangements. At the opposite, broad end of the spectrum, *all* heritable DNA sequence alterations, including those resulting from chromosomal inversions, translocations, etc. are included under the general heading of 'mutation' (*sensu lato*). I will be adopting the broader view of mutation herein, because all heritable DNA changes are potentially important for evolution. However, for the purposes of the present discussion, I have chosen to deal separately with processes through which new genes originate (Section 8.2) and processes through which existing genes are modified.

The entry of 'gene' into the discussion above is potentially misleading, as many heritable changes in DNA are not in genes at all. Animal genomes are complex and heterogeneous, and over the last couple of decades we have come to recognize introns, upstream/downstream non-transcribed/non-translated regions, pseudogenes, intergenic 'spacers', integrated pieces of mobile DNA, and repetitive regions of various kinds, including extensive repeats of very short DNA sequences in special regions such as telomeres. (For general accounts of such genomic features, see Lewin 1994 and Strachan and Read 1996.) I will be concentrating, herein, on mutations that affect developmental processes. At present, it seems likely that mutations in some kinds of non-genic DNA will have important developmental consequences, while mutations in other kinds will not. For example, small changes in the base sequence of non-transcribed controlling regions may have major effects; equivalent changes in some other regions (e.g. telomeres) are likely to be less important.

From an evolutionary perspective, it is conventional to distinguish between somatic and germline mutations. The effects of the former are restricted to the cellular descendants (in terms of clonal proliferation) of the cell in which the mutation occurred; while the effects of the latter are transmissible to *organismic* descendants and hence may contribute to evolutionary change. This distinction derives from Weismann's (1904) view of the 'continuity of the germ-plasm', which is still generally accepted. However, Buss (1987) has pointed out that the physically discrete individuals and early soma/germline separation seen in (for example) dipterans and mammals is itself an evolutionarily derived state, which we should not assume was necessarily present in the earliest metazoans. If it was not, then early evolutionary processes may have been somewhat different from later ones. In particular, Buss argues that intra-organismic 'selection' among cell lineages may have been important. This possibility remains to be fully explored by evolutionary biologists.

Overall, mutation is a huge and complex subject (see Auerbach 1976; Cooper and Krawczak 1993) and it will be necessary, as usual, to be selective. I intend to concentrate on those aspects of mutation that are relevant to the evolution of development, and, as part of that, to the origin of body plans. In particular, I will focus on the role of mutation in (a) altering the spatiotemporal expression pattern of a gene, (b) altering the structure of the interactional architecture within which the gene and its product are embedded, and (c) the relationship between these two. The overall aim will be as anticipated in the introductory chapter: to begin to correct the 'lopsidedness' of current neo-Darwinism, in which, at a theoretical level, our concepts of selection are far better developed, and more sophisticated, than our concept of the evolutionary role of mutation.

This developmental/evolutionary approach to mutation will occupy Sections 8.4 and 8.5. In order to set the scene, the origin of genes (which subsequently diverge mutationally) will be considered in Section 8.2; while essentially non-developmental approaches to mutation and its molecular causes will be examined in Section 8.3. As will be seen, these broadly 'genetic' approaches are a two-edged sword. They have been enormously helpful in elucidating the nature of mutation in general. Yet they also, in a sense, form a dominant conceptual mould from which it is necessary to break away if developmental aspects of mutation are to be more thoroughly considered.

## 8.2    The Creation of New Genes

The evidence that many – perhaps even most – genes arose through duplication events is now overwhelming (Chothia 1994; Brenner *et al.* 1995; Labedan and Riley 1995). Early examples included the haemoglobin genes (Ingram 1961) and the genes coding for trypsin and other proteolytic enzymes (Hartley *et al.* 1965). Reviews of more recent studies emphasizing the evolutionary importance of gene duplication can be found in Errington (1991; transcription factors), Gupta (1995; chaperonins) and King and Millar (1995; peptide hormones). Also, we have already encountered, in previous chapters, the importance of gene duplication in the elaboration of *Hox* complexes (Garcia-Fernàndez and Holland 1994).

There are several ways in which gene duplications may occur. Some of these have been apparent since the early days of genetics, while others have only come into view more recently, following particular discoveries; for example, of replicative transposition by mobile genetic elements. It can be argued that, from an evolution-of-development perspective, the precise molecular causes of gene duplications are of less importance than their consequences, following mutational divergence, on expression patterns and interactional architecture. However, it seems prudent to give a brief overview of possible causes, if only to indicate their diversity.

Ohno (1970), reviewing the mechanisms of gene duplication known at the time, devoted three chapters to the subject. He distinguished between whole genome mechanisms and those specific to a particular chromosome (and indeed to a particular region of the chromosome). In the former category were the processes of auto- and allopolyploidy; while in the latter were unequal exchange of material between sister chromatids and between homologous chromosomes. The last of these processes, occurring during meiosis in the germline, seems likely to be of particular evolutionary importance (see Figure 8–1). Polyploidy, while undoubtedly of evolutionary significance, appears to be much less widespread among animals than plants; and in general where clusters of genes are involved – be they *Hox* or globins – it seems inevitable that 'localized' processes restricted to particular chromosomal regions are responsible.

In some cases, genes thought to have arisen through duplication are dispersed rather than clustered. While this could arise from tandem duplication followed by translocation, it could also arise from the movement of transposable elements. Some such elements contain functional genes, and on each occasion that these transpose in replicative manner, a duplication necessarily results. The relative importance of this process compared to the more localized process of unequal exchange remains to be clarified. However, a large proportion of eukaryotic DNA is thought to be attributable to transposable elements, and their importance as agents of gene duplication and mutation is likely to be considerable.

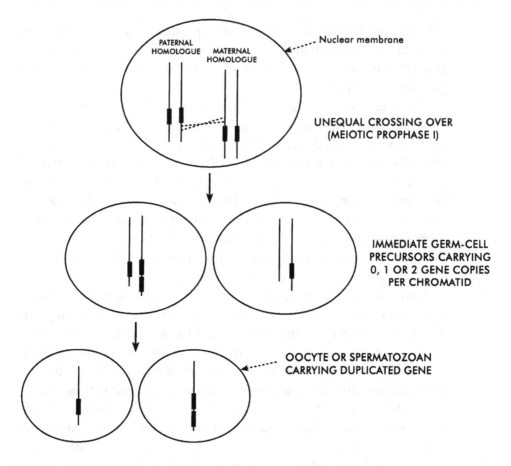

**Figure 8–1  Gene duplication through unequal crossing over**
Following a fertilization involving the germ cell carrying the duplicated gene, a variety of other events are possible, including further duplications that result, ultimately, in the formation of a gene cluster. (The *Hox* complexes may have arisen through this kind of process.)

In addition to transposable elements, the main complication that has arisen since Ohno's (1970) work concerns the discovery of the multicomponent nature of eukaryotic genes, and the further elaboration of the multicomponent nature of the corresponding proteins. As we saw in Chapter 6, genes are split into (a) expressed regions and (b) intervening regions that are transcribed but not translated (hence *exons* and *introns*: Gilbert 1978). Proteins consist of structural and/or functional *domains* (see, for example, Campbell and Baron 1991). In some cases, exons and domains coincide (Blake 1978; Südhof *et al.* 1985; Durkin, Wewer and Chung 1995) while in others it appears that there is no simple correspondence between the two (e.g. Kretsinger and Nakayama 1993).

It has been suggested that genes may evolve through a process known as 'exon shuffling' (see reviews by Patthy 1991 and Dorit and Gilbert 1991), whereby individual exons are moved between transcription units. As in the case of gene duplication, there are several ways in which this could occur. One of these involves transposable elements (see Figure 8–2). The fact that most introns are non-coding (but see Tycowski, Shu and Steitz 1996), and that they are usually much longer than exons, means that there are considerable stretches of DNA in which interexon breaks can occur, without causing obvious damage to gene function – providing there are no frameshifts. This opens up enormous possibilities for the creation of novel transcription units.

Whether this process of exon shuffling could have been important in the origin of animal body plans depends on when introns first arose. There have until recently been two alternative theories of the origins of introns: (a) the 'introns-early' theory (Doolittle 1978; Darnell 1978), which sees split genes as ancestral, in which case modern prokaryotes have lost them; and (b) the 'introns-late' theory (Hickey and Benkel 1986 and references therein), in which the typical uninterrupted prokaryote gene is seen as the ancestral condition, into which introns intruded later as an evolutionary innovation. If the latter theory is correct (which it probably is; see below), then the question arises of when introns originated, and whether they appeared all-at-once or over a protracted period.

The prevalence of introns within eukaryotic genes (unicells, fungi, plants and animals) and their general absence in prokaryotes makes the 'introns-late' theory the more parsimonious of the two. Also, recent studies on a gene previously thought to support the 'introns-early' theory (Gilbert and Glynias 1993) – triose-phosphate isomerase (TPI) – are highly persuasive in rejecting this view (Logsdon *et al.* 1995). Moreover, these authors showed that the taxonomic distribution of introns was variable, with some being widespread, others much more restricted. While the issue remains open to debate, it seems likely that the first introns arose in eukaryotic unicells; and that others followed before, during and after the proliferation of animal body plans.

When a new gene arises either through duplication of a complete 'parental' gene or through the creation of a hybrid transcription unit by

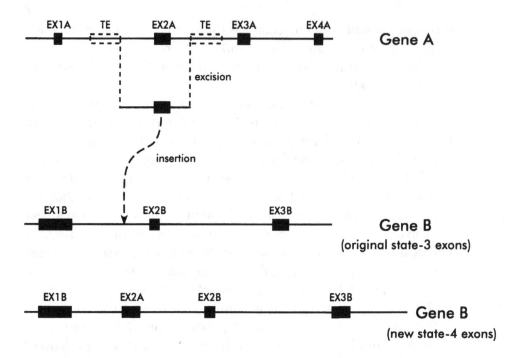

**Figure 8–2  Exon shuffling through faulty transposable element excision mechanisms**

Instead of the opposite ends of the same element being the points of excision, the right-hand end of the first element and the left-hand end of the second one are used. The result is an excised exon flanked by two partial introns. This 'unit' is then inserted into another gene, thus creating a hybrid transcription unit. EX, exon; TE, transposable element.

exon shuffling, it will necessarily be subject to ongoing change through mutation. We will now consider the kinds of mutation that may occur.

## 8.3  Mutation: the Classical Approach

It is axiomatic that the 'classical' approach to the study of mutation is a genetic one; from its outset, the study of inheritance – transmission genetics – was based on observing numerical patterns in the transmission of mutant characteristics from parents to offspring. These characteristics were often the result of the mutation of a single gene – giving rise to short pea plants, black moths, insecticide resistance, polydactyly and so on. But as the focus of genetics altered, through the present century, from externally visible phenotypes to proteins and DNA sequences, so too did the way in which mutations were viewed. The result of this ongoing shift of focus is a confusing variety of ways of classifying (and hence of thinking about) mutations.

Some frequently encountered classifications are presented in Table 8–1. This is not intended to be an exhaustive list, either of criteria or of categories within criteria, but it is sufficiently extensive to help us to visualize the nature of the problems at hand, which are two-fold. First, in many cases there is no simple connection between a particular category within one criterion and the categories listed for another. For example, a chromosomal rearrangement ('genetic scale') could belong to any of the 'fitness' categories from lethal to beneficial. Second, and more important for our present purposes, most categories under most criteria have no simple link with development. For example, a point mutation involving a base substitution could have major, moderate, minor or no developmental consequences depending on the nature of the gene, the location within it of the mutation, and exactly what base pairs are involved.

A notable exception to this lack of categories that connect with development is the micro/macro dichotomy; but as we have seen (in Chapters 1 and 7) this dichotomy is unsatisfactory and needs to be replaced by a more complex and developmentally explicit classification. Before attempting such a classification it is worth briefly considering each of the ten criteria listed in Table 8–1 to draw out some points of interest in relation to development, evolution or both.

1. *Origin*  It is probably the case that all categories listed contribute (with unknown and indeed variable relative frequency) to evolutionary changes in developmental processes. The importance of transposable element insertions in causing mutations in the Bithorax complex in *Drosophila* was revealed by Bender *et al.* (1983). For their more general mutagenic importance, see Lambert, McDonald and Weinstein (1988).

2. *Scale*  Again, all categories are important in the evolution of development. Changes of intermediate scale – for example biased gene conversion, unequal recombination and slippage – affecting the evolution of gene families may lead to fixations at the population level independently of conventional selection and drift, through 'molecular drive' (Dover 1982, 1986). Some authors have

**Table 8-1. Ways of classifying mutations**

| Criterion | Categories |
| --- | --- |
| 1. Origin/causation | Spontaneous; caused by chemical mutagens; caused by irradiation; caused by insertion of transposable elements |
| 2. Scale of genetic change | Point mutation; intermediate scale; chromosomal rearrangement |
| 3. Nature of genetic change | *In point mutations*: substitution; addition; deletion. *In chromosomal rearrangements*: inversion; translocation; deletion (etc.) |
| 4. Phenotype of heterozygote | Dominant; recessive; neither; partially dominant; codominant |
| 5. Effect on triplet code | Silent; one amino acid affected; frameshift |
| 6. Effect on gene product | Altered; truncated; absent ('null') |
| 7. Cellular location | Germline; soma |
| 8. Effect on fitness | Lethal; sublethal; detrimental; nearly neutral; neutral; beneficial |
| 9. Relation to adaptation | Random; directed |
| 10. Effect on morphology | Micro; macro (this is an inadequate classification: see text) |

argued for a major role for chromosomal rearrangements in speciation (White 1978; King 1993).

3. *Nature*   Base substitution in developmental genes is likely to be of considerable importance in the divergence of initially identical duplicate copies. Additions and deletions cause frameshifts and are hence likely to be detrimental. Among chromosomal rearrangements, inversions are probably the commonest type to be polymorphic (often stably so) within populations – for examples see Dobzhansky (1961) on *Drosophila pseudoobscura* and Butlin *et al.* (1982) on the seaweed fly *Coelopa frigida*. Inversions can cause position effects; they sometimes affect morphology, but often only in a minor, quantitative manner (Butlin and Day 1984).

4. *Heterozygote*   Dominance relationships can be modified by selection (Fisher 1930). Parallel polymorphisms in related species can have opposite dominance relationships – for example chirality in snails, where sinistrality is dominant in *Partula* (Murray and

Clarke 1966) but recessive in *Lymnaea* (Freeman and Lundelius 1982). The dominance or otherwise of a new mutation affects its 'selectability', as it will initially only be present in heterozygous condition.

5. *Triplet code*  In coding regions, 'silent' mutations in the third (redundant) base of triplets are likely to be in the neutral/nearly neutral category (but see Clarke 1970), and if the gene concerned produces a protein morphogen of some kind, such mutations are unlikely to affect developmental processes directly. Frameshifts, as noted earlier, will tend to be detrimental – even lethal – and so will be removed by selection, except in special cases such as (non-functional) pseudogenes. Non-silent substitutions, causing different amino acids to be incorporated at translation, may be of particular developmental importance where the gene concerned makes a transcription factor, intercellular morphogen (such as *Sonic hedgehog*) or other proteinaceous signalling agent.

6. *Gene product*  In general, 'altered' products are likely to be the most important in the evolution of development. Complete absence of product will often be detrimental or lethal; likewise for truncated products, except that minor/moderate truncations may contribute to mutational divergence following duplication events.

7. *Cellular location*  Although Buss's (1987) point about the non-universality of a germ/soma split needs to be noted (see Section 8.1), in evolutionary studies we are generally dealing with germline mutations. Those occurring early in the germline (e.g. oogonia, spermatogonia) have the potential to give rise to 'clusters' of phenotypically mutant progeny a generation earlier than in the case of those occurring late (oocyte, spermatocyte). For a dominant mutation, such a cluster may occur in the $F_1$ (as opposed to $F_2$) generation if it occurs in an early germline cell. For a recessive mutation, the equivalent shift is to $F_2$ from $F_3$.

8. *Fitness*  The main problem besetting Goldschmidt's (1940, 1952) macromutational theory of the evolution of development was the severely detrimental effect of macromutational changes in key developmental genes. However, most known smaller-effect mutations of these genes that have been well studied are also detrimental. (The neutral theory of molecular evolution, beautifully described by Kimura (1983), is not directly relevant to the evolution of development/morphology, as Kimura himself pointed out –

but for a suggestion of *possible* neutral evolution of a developmental process see Wray and Bely (1994).)

9. *Adaptation*   Underlying the Darwin/Mendel/Weismann 'axis' of evolutionary theory is the assumption that mutations occur at random with regard to what is 'desired' in adaptive terms. For example, it is assumed that black moths appeared at the same frequency before and after the industrial revolution, despite its effects causing them to be newly advantageous. (There are, of course, no reliable data with which to test this assumption.) However, Cairns, Overbaugh and Miller (1988) provided what appears to be evidence that in some cases environmental conditions can induce specifically adaptive mutations in bacteria; and their results appear to be repeatable and extendible to some eukaryotic unicells (Hall 1992; Rosenberg, Harris and Torkelson 1995). However, directed mutation remains a controversial subject (see Chapter 9); and at any rate there is no evidence for its occurrence in any *animal* species.

10. *Morphology*   As noted previously, there has been an unfortunate tendency for some neo-Darwinians to think of mutations as 'big' and 'small' − macro and micro − in terms of their phenotypic effects. The former are thought of as evolutionarily irrelevant 'monsters', while the latter are considered to provide the variation on which positive directional selection is able to act, in appropriate circumstances, to produce evolutionary change. This false dichotomy can be traced back to Darwin himself, who used the famous phrase *Natura non facit saltum* (and was criticized by T.H. Huxley for doing so: see Darwin 1887). It then featured in the debate between mutationists and biometricians at the start of the present century (see Chapter 10); was reflected in the distinction between major genes and 'polygenes' in the early days of quantitative genetics; and has been propagated by the recent polemic writings of Dawkins (1986), as we saw in Chapter 1. In general, advocates of a major role for 'macromutations' in evolution (notably de Vries (1910) and Goldschmidt (1940, 1952)) have been considered to be heretics, and have been excluded (in most cases correctly) from mainstream evolutionary theory.

What is wrong with the micro/macro dichotomy? What leads me so confidently to label it as 'false'? There are three problems, as noted in Chapter 1. First, there is a continuous distribution of magnitudes of mutational effects on the phenotype, not two distinct categories. Second, 'micro/macro' usually refers to the effect on adult morphology, but it is clearly preferable to take a four-dimensional, ontogenic view of a mutation's effects. Even those that have very large effects on the adult are likely to start off, in early development, by producing rather minor changes, perhaps initially affecting only one or a few cells out of many. Third, it is crucial that we distinguish not just sizes and timings of effect, but also *types* of effect. We now know enough about developmental genetics that we can visualize the types of possible mutational change much more clearly than hitherto; and given that, arguments (such as those of Fisher 1930) based on a very abstract notion of generalized 'magnitude' of effect are really no longer adequate.

The history of *rejection* of a micro/macro dichotomy is as long as the history of the dichotomy itself; and in some cases involves the same individuals. Thus, while Darwin (1859) did not think that evolution *ever* made 'jumps', he also made clear his view that 'monstrosities' graded into 'variations'. And Fisher (1930), whose views were instrumental in the rejection of macromutational inputs to evolution, explicitly recognized, and indeed presented graphs of, the truly *continuous* distribution of mutational effects (recall Figure 1–5). Darwin even brought in an ontogenetic perspective (though Fisher did not); and the following passage (1859, chapter 3) makes clear some of his views in this area:

> It is commonly assumed, perhaps from monstrosities often affecting the embryo at a very early period, that slight variations necessarily appear at an equally early period. But we have little evidence on this head – indeed the evidence rather points the other way; for it is notorious that breeders of cattle, horses, and various fancy animals, cannot positively tell, until some time after the animal has been born, what its merits or form will ultimately turn out.

What neither Darwin nor Fisher nor any of their respective contemporaries could have been expected to do was to produce a scheme of *types* of mutational effect on development and, hence, on morphology. The requisite information was simply lacking. And while some of it is still lacking, enough is now known of the developmental genetics of a wide range of organisms that we should at least be able to formulate the beginnings of such a scheme. I attempt this in the following two sections.

## 8.4    Mutation: a Developmental Approach

Mutations can affect developmental systems in many different ways. This is a very complex issue to address, and there is little in the way of established conceptual framework to use as a scaffolding: rather, we have to devise this framework as we go. It may help to consider first the cellular and genetic context in which mutations occur; and to proceed to a consideration of mutations themselves only after this context has been described in some detail. This strategy is adopted below.

### The Cellular and Genetic Context

For our present purposes, genes can be divided into three groups. First, there are those that are expressed in all or virtually all functioning, nucleated cells, and whose products are concerned with essential cellular activities. These are the so-called housekeeping genes; examples are the genes for glucose-6-phosphate dehydrogenase (involved in glycolysis) and for cytochrome c (involved in the electron transport chain). Second, there are genes which are only expressed in particular cell types or tissues and whose products (a) contribute much to the nature of the tissues concerned but (b) do not have a primarily controlling role. Examples in mammals are the genes for actin and myosin (muscle), haemoglobin (red blood cell precursors) and keratin (skin). These can be regarded as the terminal targets for decision-making networks of upstream genes. Finally, there are the developmental genes themselves: those that have both upstream controllers and downstream targets. Many examples were covered in Chapters 5–7, including genes encoding transcription factors (e.g. *Hox*) and growth factors (e.g. *dpp*).

What follows concerns only this last category of 'developmental genes'. However, the boundaries between the three categories are not entirely clear cut: for example, cadherin genes may in some cases be developmental, in others terminal targets. Also, accumulated mutations can ultimately move a gene from one category to another. These caveats should be constantly borne in mind. ('Gene' should be interpreted here and below as referring not only to a coding region but also to its adjacent controlling regions. In general, mutations in controlling regions will alter upstream interactions while those in coding regions will alter downstream interactions: more on this later.)

One of the most important characteristics of any developmental gene is its spatiotemporal pattern of expression; and the context for this is the four-dimensional cellular array that we call the developing organism. Because three dimensions are easier to think about, I will simplify the approach adopted here by using, again, the concept of developing organisms as inverted cones first introduced in Figure 1–7 (Section 1.7). At any point in ontogeny, the organism is pictured as a disc growing by cell proliferation until it reaches a predetermined adult size, at which point growth ceases. The general principle illustrated in Figure 1–7 was that, other things being equal, *the earlier a developmental gene was first switched on, the greater would be its ontogenetic (and adult phenotypic) effect.* However, there are several complications to this picture, which now need to be discussed (these are illustrated in Figure 8–3).

### Clonal Proliferation Rates May Vary

Not only can different cell proliferation rates be induced experimentally (using the *Minute* technique: see, for example, Garcia-Bellido, Ripoll and Morata 1973) but variation in such rates also occurs naturally in most – probably all – developing animals. At any given time, some clones will be proliferating faster than others. As development proceeds, these relative proliferation rates may themselves alter. One extreme possibility is the programmed cell death (apoptosis) of all the descendants of a clone that was proliferating rapidly at an earlier stage.

### Expression Zones and Effect Zones May Not Coincide

Both the region of an embryo affected by a particular developmental gene at a certain time and the span of ontogenetic time over which the gene exerts effects will tend to be greater (i.e. more extensive and longer respectively) than the region and timespan within which it is expressed (i.e. being transcribed). Spatial realms of effect will be broadened in the case of genes making mobile morphogens; and temporal realms of (indirect) effect will be extended in the downstream direction in the many cases where target gene expression persists after the transcription of the gene in focus has been switched off. Thus mutational effects may be more extensive (in space and time) than would be predicted by examining expression patterns alone.

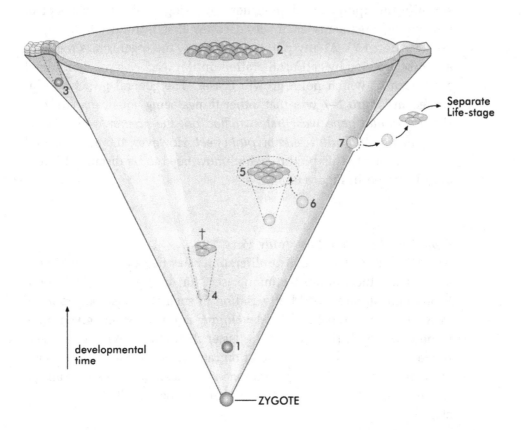

**Figure 8–3  A more complex view of the 'developing organism as inverted cone' model**
1. Cell in which 'gene A' is first switched on. The descendants of this cell (2) form the centre of the disc. 3. Cell in which 'gene A' is reused to mark the proximal end of the appendage. 4. Cell (in which 'gene B' is active) that proliferates slowly and whose descendants die, resulting in a negligible effect of 'gene B' on the adult disc, despite its earliness of activity. 5. Clone of cells in which 'gene C' (making a mobile morphogen) is active. This gene's region of effect is greater than its region of expression. 6. Cell that migrates and comes under the influence of the product of 'gene C'. 7. Cell leading to another life stage, when the disc is considered to be a larval stage of a complex life cycle.

### Cells May Migrate and Come Under the Influence of Different Morphogens

Some developmental processes – notably gastrulation – involve extensive movement, over considerable distances, not just of individual cells but of large groups of cells. In any such migration event, the cells concerned may move to an area pervaded by different intercellular morphogenetic signals from those pervading their area of origin. The expression patterns of the developmental genes carried by those cells may be altered both by the cell migration itself, and by their coming under the influence of different upstream controlling influences.

### Expression May be Intermittent in Time and/or Patchy in Space

In Figure 1–7, a developmental gene was pictured as being initially switched on in a particular cell at a particular developmental stage, and expressed only in the clone of cells to which that original cell gave rise. However, this is an over-simplification. Bilateral symmetry will often lead to physically separate, mirror-image duplicate regions of expression. Genes expressed in limb buds will have at least as many patches of expression as there are limbs. And since many genes are involved in limb *and* trunk patterning – as we saw for *hedgehog* and *Sonic hedgehog* in previous chapters – the number of expression regions will be greater again (unless they fuse together). In dipterans, many patterning genes are used both in the embryogenesis of the larva and in the imaginal discs. Their expression in these two contexts generally differs in time as well as space. It is important to note that the interactional architecture of a developmental gene may be different in different regions and at different ontogenetic stages.

### Complex Life Cycles are the Rule Rather Than the Exception

It is tempting, from our human perspective, to picture simple, single phase life cycles such as the mammalian one as the norm; but this is not the case now, and it has probably not been the case through most of metazoan history. Cnidarians, platyhelminths, insects, crustaceans, tunicates, amphibians and others exhibit a great variety of complex life cycles. In these, one or more cells in a particular life stage are diverted into a different developmental pathway, and give rise to a distinct life stage. There are many different ways in which this can happen, but in all of them the important point – from a developmental genetic perspective – is that genes may be reused in different stages (as in the dipteran larva/disc example above). The complete spatiotemporal expression pattern of a developmental gene is that characterizing the *whole life cycle*. (There is a link here with the question of whether larvae violate von Baer's laws: this issue will be discussed in Chapter 11.)

Given these five complications to the initial simplified picture presented in Figure 1–7, should we still believe the 'general principle' suggested above: namely that the earlier a developmental gene is first switched on, the greater will be its ontogenetic/phenotypic effect? In absolute terms, the answer has to be 'no'. An early gene may be active in a clone that proliferates very slowly or even dies, and has few knock-on effects on others. A late gene may code for a growth hormone which permeates the whole body and exerts an all-pervasive influence on its development. However, in probabilistic terms I would argue that the answer is 'yes' – *in general*, earlier acting developmental genes *do* have more major effects.

There are at least three reasons for making this assertion. First, nature is reasonably economical: large numbers of early developmental genes are not 'wasted' in governing only the fate of limited/doomed clones. Second, late effect growth hormones may alter such things as body size, shape and the timing of reproductive maturity, but they usually do not have major effects on the body plan. Third, and most importantly, *the fundamental asymmetry of causal relationships in development is likely to have an overriding effect*. Most 'early' genes have ramifying effects that cascade through a network of downstream genes, even after the early gene concerned has been switched off (see above). Mutations in early genes will thus tend, probabilistically, to have greater morphological effects than those in later ones; and this statistical trend has profound evolutionary consequences, as discussed in Chapter 11. This is an appropriate point to move on from the cellular/genetic context to the mutations themselves.

## The Developmental Effects of Mutation

I should briefly re-emphasize, here, three restrictions in my approach. First, the account below deals only with developmental genes: neither housekeeping nor 'terminal target' categories are considered. Second, mutations in developmental genes that do not have developmental *effects* are also excluded from consideration (although these must occur and they may be important in the future in combination with subsequent mutations). Third, the molecular causes of mutation are not considered. My underlying assumption here is that different causes do not connect cleanly with different developmental effects. (If that turns out to be wrong in some cases, it would not invalidate what follows – rather it would add another dimension.)

The key question is this: what are the important functional features of a developmental gene that may be altered by mutation? In fact, there are just two such features: interactional architecture and the spatiotemporal distribution of the gene's product. I will deal with them in that order, and will conclude with a brief discussion of the interplay between the two.

### 1. Interactional Architecture

I previously drew attention to the distinction between interactional and chromosomal architecture (Chapters 1 and 6). While the two sometimes overlap extensively in that clustered developmental genes tend to have important interactions with their chromosomal neighbours, in general there is limited overlap, and a gene will typically interact with several others from various chromosomal locations. It is this functional, or interactional architecture that is vital in the control of development. Mutations which alter it are likely to cause an altered morphology – unless this is prevented by redundancy of control systems and/or by canalizing mechanisms.

The main components of interactional architecture are shown in Figure 8–4; all of these can be altered by mutations in the gene concerned. In general, incoming upstream signals (both positive and negative) will be received by the controlling region(s), while downstream, outgoing signals (again both positive and negative) will emanate via the gene product (RNA or protein) – although in some cases the downstream effect may be indirect, as in the case of an enzyme that catalyses the production of a morphogen from an inert precursor. Autoregulatory effects (positive or negative feedback – with only the latter being strictly regulatory) may take place through a flow of information directly or indirectly from the product to the controlling region – as we saw in the *Ultrabithorax* gene in Chapter 6. Cross-links are probably variable in nature, and are as yet less well understood.

If we focus in on a particular developmental gene that is embedded in a network of interactions (like that shown in Figure 8–4) then mutation can have four kinds of effect on any individual interaction: it can

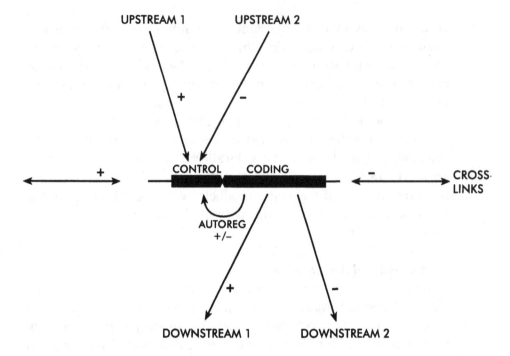

**Figure 8–4  Components of interactional architecture**
All of these can be altered by mutations in the gene concerned.
Most can also be altered by mutations in the various interacting
genes. (The possible exception here is autoregulation – but even a
gene's ability to autoregulate might be altered by, for example,
mutation in an upstream gene encoding a transcription factor.)

be deleted, brought into existence, qualitatively modified (e.g. from
positive to negative signal) or quantitatively modified (e.g. altered affinity of a control region for a transcription factor, causing a different
dose/response pattern). Given eight possible types of link (upstream,
downstream, cross and auto × positive, negative), there are thirty-two
distinct types of possible mutational change.

However, it seems likely that interactions are often altered in
groups rather than individually. For example, a mutational change in
the DNA sequence of a controlling region may simultaneously alter the
affinity of that region for incoming upstream, cross-link and autoregulatory signals. Also, if we 'focus out' from an individual gene to a more
extensive view, then other possibilities emerge. For example, a control
cascade A → B → C could be 'collapsed' to A → C. The larger the

network of interacting genes we consider, the more such 'compound' mutational changes in interactional architecture become possible.

### 2. Spatiotemporal Product Distribution

There are two aspects of this: the spatiotemporal expression pattern of the gene concerned, and the degree of mobility and durability of its product. Both of these can be altered by mutation. A gene may become expressed in regions of the embryo in which it was previously inactive due to a mutation in a controlling element that causes it to be switched on by a morphogen pervading that region but not previously able to activate the gene concerned there. Alternatively, the spatiotemporal distribution of a gene's product may be altered because a sequence change in the coding region causes a previously secreted protein to be 'confined' or a previously labile protein to be less readily degradable.

As in the case of interactional architecture, it is useful to have an overview of the possible types of change that mutation can produce. The possible types of change in spatiotemporal product distribution are illustrated in Figure 8–5 for a gene whose product is initially (i.e. premutation) present as a single anteroposterior band of a particular width during a certain developmental period.

Let us examine spatial changes first. A single region of product can increase or decrease in size, shift in position, alter in shape, or exhibit some combination of these. If we bring in the possibility of multiple product regions, then these can increase or decrease in number as well as some or all of them undergoing changes in size, shape or position. That is an exhaustive list: all possibilities are subsumed within one or more of the above categories. For example, the development of 'holes' in a previously continuous distribution of product is a change in shape. (Recall the 'holes' in *Antp* expression corresponding to the position of larval legs in Lepidoptera discussed in Chapter 7.)

The above considerations relate to the between-cell component of the spatial distribution of a gene product. However, its within-cell distribution is also important. For example, we saw in Chapter 6 that the degree to which the product of the *dorsal* gene was transported into nuclei was of considerable developmental significance. Clearly, mutations affecting its 'transportability' might have major developmental consequences.

The equivalents of the above spatial changes for *temporal* product distribution are: extended period, diminished period or shifted period

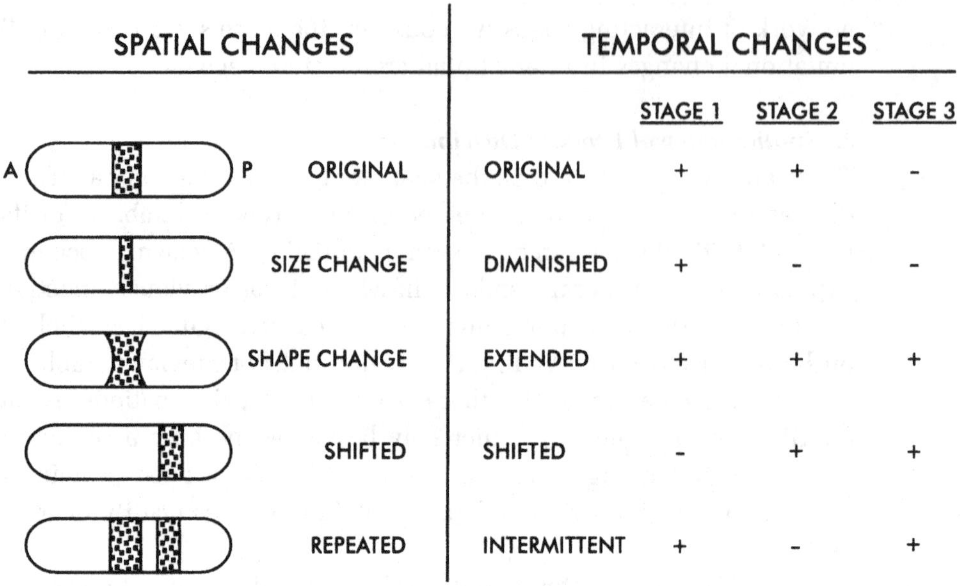

| SPATIAL CHANGES | | TEMPORAL CHANGES | | | |
|---|---|---|---|---|---|
| | | | STAGE 1 | STAGE 2 | STAGE 3 |
| A ⬭ P | ORIGINAL | ORIGINAL | + | + | - |
| ⬭ | SIZE CHANGE | DIMINISHED | + | - | - |
| ⬭ | SHAPE CHANGE | EXTENDED | + | + | + |
| ⬭ | SHIFTED | SHIFTED | - | + | + |
| ⬭ | REPEATED | INTERMITTENT | + | - | + |

Figure 8–5  **Possible types of mutational change in the spatiotemporal distribution of a gene's product**
To relate the two parts of the Figure, assume that the spatial layouts (left) are all 'stage 1', and that a plus in the chart of temporal changes (right) represents a band of product identical to that shown in the topmost embryo (left), while a minus indicates an absence of product. (Complex combinations of spatial and temporal changes are also possible, but are not illustrated here.)

of gene product; coupled with the possibility, in the case of intermittent gene activity, of adding or deleting separate periods of expression.

In most cases, the changes pictured in Figure 8–5 could be produced by altered expression pattern and/or gene product characteristics. There are exceptions, however: a distinct second period of product presence could only be produced by a new wave of gene expression if all early period product had been previously degraded.

Finally, the simplified sort of diagram shown distinguishes only between areas (or stages) in which gene product is present and those from which it is absent. In reality, variation in the *concentration* of product within those areas in which it is found adds another dimension to the picture. One interesting possibility is that a change in the concentration of a hormone or other widespread mobile morphogenetic agent might explain Thompson's (1917) famous 'transformations', wherein body shape in one species is approximated by a systematic distortion of body shape in another. If cell proliferation responds in some way to a threshold concentration, and if the spatial fall-off of the hormone to subthreshold levels is steeper in some regions, shallower in others, then an increase or decrease in overall concentration will give rise to more spatially extensive changes in some regions, and hence an altered shape.

### 3. *The Interplay Between the Two*

How are interactional architecture and spatiotemporal expression patterns related? To what extent does a change in one imply a change in the other? These are very difficult questions. It seems likely that changes in upstream interactions (due to mutation in the gene in focus and/or in upstream genes) will give rise to different expression patterns, and that these in turn may alter downstream interactions, especially in regions or stages to which expression has been extended. This interplay between interactional architecture and expression patterns may cascade all the way from the point in developmental time at which the mutated gene is first expressed to modification of adult body form. It cannot, of course, cascade 'backwards'. Thus we end where we began, both at the beginning of this section and at the beginning of the book: in general, early acting developmental genes will have more major effects than their later counterparts. This probabilistic trend cannot fail to have evolutionary consequences. But do those consequences merely take the form of selective elimination of all early effect mutations, because of their disruptive effects, and their consequent preclusion from contributing to evolutionary change? Or do some such mutations somehow get through the 'selective screen' and lead the way into uncharted morphospace? The latter seems more likely, given the very simple but important observation that even the most highly conserved and earliest acting developmental genes are indeed different in different taxa. The final section of the present chapter, together with Chapters 9 to 11, will explore this fascinating issue.

## 8.5    Mutation and the Evolution of Development

In Section 8.3, we briefly examined each of ten 'conventional' ways of classifying mutations, and noted that none were particularly helpful in elucidating the *developmental* changes that mutations can produce. In

Section 8.4, those developmental changes provided the focus of attention, and we saw that mutation can alter interactional architecture and spatiotemporal product distribution in many ways (at least thirty-two in the former case; a harder-to-determine number, but certainly in double figures, in the latter).

Having a conceptual framework of this kind is useful in relation to understanding the developmental effects of mutations on individual organisms. However, since evolutionary change takes place in *populations*, we now need to relate our developmental categories of mutation to *fitness* categories (see Table 8–1). What follows are several interconnected hypotheses about the form this relationship may take.

First of all, let us consider how the probability of a single mutation in a developmental gene being advantageous varies according to the time of the gene's onset of activity and, in broad terms, the type of effect on development the mutation has. I will start with a simple conceptual model and make it more complex – and hopefully more realistic – in stages. This is similar to the approach of the population biologist who starts with simple mathematical models (e.g. of population growth) and gradually makes them more complex. The aim in both cases is to arrive at the model of minimum complexity that adequately describes reality.

The simplest possible relationship between a gene's time of onset of activity ($t$) and the probability of a mutation in it being selectively advantageous ($a$) is a smooth curve, as shown in Figure 8–6A. However, as I have previously suggested (Arthur 1984, 1988), there is likely to be a slight increase in $a$ for mutations in the very earliest acting developmental genes (Figure 8–6B). This is because they can potentially alter *all* downstream events in a coordinated way, thus avoiding the problem of mismatch between altered and unaltered ontogenetic processes. The best examples of this are provided by switches in chirality in gastropods. These are known (a) to have occurred in evolution (see, for example, Clarke and Murray 1969 and Van Batenburg and Gittenberger 1996) and (b) to be underlain by mutation in a gene (or possibly two genes) affecting the nature of cell divisions in early cleavage (Freeman and Lundelius 1982). However, a symmetry-reversal is a rather special kind of developmental change. It remains to be seen whether other kinds also contribute to the 'blip' in $a$ at $t \sim 0$.

There are likely to be other developmental stages at which problems of 'mismatch', or decreased internal coordination, are somewhat reduced. One possible example is the origin of any quasidiscrete devel-

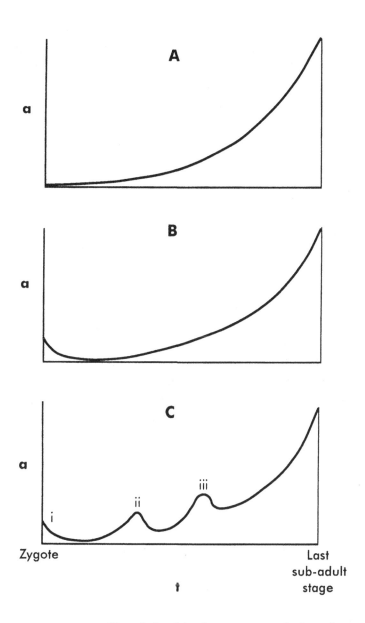

**Figure 8–6  Possible relationships between a gene's time of onset of activity (*t*) and the probability of a mutation in it being selectively advantageous (*a*)**

(A), (B), (C) Different assumptions: see text. In (B) and (C), the subsidiary peak(s) are all much smaller than the main (right) one – much more so than shown, assuming a linear probability scale. In (C), the relative sizes of i, ii and iii shown are arbitrary.

opmental 'module', such as limb buds. Here, the rationale for an increase in $a$ for a gene affecting initiation of a limb bud is the same as that used in considering the initiation-of-cleavage genes, except for two things. First, within-module coordination may be maintained, but the relationship with the rest of the developing body may be problematic. Second, if the gene concerned is an earlier gene being reused, then the probability of a mutation being selectively advantageous may be reduced – unless there are separate controlling regions for the different gene usages and the mutation concerned is in one of these.

This line of argument leads to the prediction of other minor peaks in $a$ at certain developmental stages (Figure 8–6C). These almost certainly exist, but their numbers and relative sizes (a) are very difficult even to hypothesize at present, and (b) should be expected to vary considerably across different taxa. The pattern shown in Figure 8–6C is an arbitrary choice.

Let us now turn to the relative probabilities of mutations of different developmental *kinds* (as outlined in Section 8.4) being advantageous. I would like to advance a hypothesis here, namely that mutations that add links (to any given architecture) are more likely to be advantageous than those which delete links. If this is so, it may help to explain the evolutionary increase in mean organismic complexity (see Chapter 11).

So much for single mutations in single genes. Major evolutionary changes, including the origins of body plans, involve multiple mutations in many genes. Of particular interest here is the *coevolution* of developmental genes that interact with each other (see Dover 1992, 1993; Fryxell 1996). Let us consider a particular scenario: a mutation in an early acting developmental gene has spread through a population (perhaps because of a rare overall *net* increase in fitness: see Chapter 9) but hidden within that net positive effect are negative effects on internal coordination related to the way in which the mutant early gene interacts with its downstream targets. Selection may then favour mutations in these downstream genes that would previously have been detrimental. Because later genes are generally less constrained (Figure 8–6), this coevolution will generally flow from upstream to downstream, so that coevolution tends to cascade, in evolutionary time, in the same direction as developmental causality cascades in ontogenetic time. If so, then we can begin to see an internal, developmental–genetic aspect to the origin of body plans and the radiation of *Unterbaupläne*, to complement the external, ecological side – for example, in the form of newly-available niches.

There is nothing in the argument so far to make the origin of a new body plan more or less likely as evolution proceeds. To put it another way, there is nothing that could be used to 'predict' the concentration of such origins in the early stages of animal evolution, as appears to have been the case (see Chapters 1 and 3). But a relatively simple extension of the above reasoning can render such 'prediction' possible.

Let us return again to the abstract picture of a developing organism as an inverted cone. In considering mutational changes in single developmental genes and coevolutionary changes in interacting genes, we have tacitly assumed that the overall context – the general shape and size of the cone and the general structure of the interactional architecture that generates it – remains fixed. This is hardly realistic. If the hypothesis advanced earlier is correct – that mutations adding interactions are more often advantageous than those deleting them – then both genetic and cellular complexity will tend to increase over evolutionary time. This has clearly happened in the case of real animals: despite much 'noise', a sequence from unicell to ball-of-cells to primitive bilaterian to complex arthropod or chordate (for example) can be discerned. The bulk of this increase in complexity probably occurred between about 600 and 400 *my* ago. Both during that period and subsequently, selection for buffered developmental systems is likely to have increased the number of cross-linkages – and indeed, the complexity of interactional architecture generally. So even those recent animals (such as flatworms) that *appear* no more complex than their distant ancestors may actually be more complex in terms of the structure of their developmental programme.

The more complex the organism and/or its developmental system, the harder it is likely to be for individual mutational changes to produce a coordinated overall effect that stands a chance of being beneficial. Thus we can 'predict' that the probability of new animal body plans originating should decline over time from a high point occurring not too long, in evolutionary terms, after the origin of animal multicellularity.

The approach taken in this section is clearly highly speculative. But if it is right, it provides a possible explanation of 'directionality' in evolution in the sense of big 'experiments' giving way to progressively more restricted modifications, as we saw at the end of Chapter 7. This idea of directionality – which is absent from conventional neo-Darwinian theory – is important in Evolutionary Developmental

Biology; and it can be found, in various forms, in other recent books on the subject, notably that of Thomson (1988). But we must avoid the 'Goldschmidt trap' of failing to combine developmental excitement with population-level rigour. Hence the next chapter.

# CHAPTER NINE

# The Spread of Variant Ontogenies in Populations

## 9.1    Introduction

One of the pillars of developmental genetics has been description of the phenotypic consequences of mutations in particular developmental genes (see, for example, Nüsslein-Volhard and Wieschaus (1980)). Many such studies have been conducted over the last couple of decades on a small but increasing range of organisms (notably *Drosophila, Caenorhabditis, Danio, Xenopus* and *Mus*). As a result, there is now a considerable body of information at our disposal on the ways in which mutations can alter individual ontogenies. Some of these alterations are trivial, while others – such as the famous four-winged *Drosophila* – are quite dramatic. (We looked at many examples of developmental mutations in Chapter 6, and considered a conceptual framework in Chapter 8.)

Have these various developmental mutations contributed to evolutionary change? There are two components to this question: 1. Do the mutations concerned occur spontaneously in natural populations as well as through the use of mutagenizing procedures/agents in the laboratory? 2. If they do, then are they able to spread through one or more populations, and to displace their alternative alleles?

Spontaneous occurrence in natural populations is hardly in doubt. We cannot rule out the possibility that some synthetic mutagens which an organism never encounters in the wild may produce some specific types of mutation that would not otherwise occur at all. However, *in general*, the difference between field and laboratory is merely one of frequency, with mutagenizing procedures simply causing particular mutations to occur more often than they would naturally, in order to facilitate their study.

Ability to spread through populations, on the other hand, most certainly *is* in doubt for many mutations of developmental genes, and this issue will provide the focus for the present chapter. No theory of evolutionary mechanism which fails to address this issue is ever likely to gain widespread acceptance. Goldschmidt's theory is the classic example here. He proposed that specific homeotic mutants had been involved in key evolutionary innovations (Goldschmidt 1952) despite the fact that the mutations concerned were known to be seriously detrimental or even lethal.

Clearly, the probability of a developmental mutation spreading through a population is highly variable, depending on the magnitude, timing and type of ontogenetic change, and its compatibility with the prevailing environmental conditions. We began to address this 'probability variation' in Section 8.5, and the present chapter will allow further examination of this issue. It seems likely that the probability of population-level spread to fixation is zero for many developmental mutations, regardless of the precise environmental conditions, because of their detrimental effect on internal coordination. However, I will argue that it is mistaken to assume (a) that this is true for *all* individually recognizable mutations, and (b) that all developmental change in evolution is underlain by the individually 'invisible' mutations that contribute to the familiar normal distribution curves of 'quantitative' phenotypic characters. So, as in previous chapters, I will be urging a 'middle course' between the macromutational 'heretics' and the more extreme of the neo-Darwinian gradualists. (Some loci are involved both in key early developmental decisions and in late-stage quantitative variation: see Mackay and Langley 1990 and Chapter 11.)

This is an appropriate point at which to consider the possible mechanisms of population-level spread for mutations generally. There are at least three such mechanisms: directional selection, genetic drift, and various forms of non-Mendelian 'drive'. These may have been supplemented by others in *early* animal evolution, particularly given the less pronounced germ/soma split in primitive animals (Buss

1987) and the possibility of 'directed mutation' in unicellular eukaryotes (Hall 1992; but see also Lenski 1989).

The view that I will be taking herein is that the evolution of animal development is largely brought about by an interplay between mutation and selection. Consequently, having dealt with mutation elsewhere (primarily in Section 7.2 and Chapter 8), I will devote most of the current chapter to selection. But this is not to say that the other mechanisms do not occur – clearly some of them do. Indeed, all populations are subject to stochastic variation, and hence drift, all of the time. This and other non-selective mechanisms of spread will receive brief coverage in Section 9.6.

Since the interplay between mutation and selection is already central in mainstream evolutionary theory, one possible interpretation of the above paragraph is that the perspective taken herein will be an entirely conventional one. That would, however, be incorrect, largely because I believe that the kinds of selection normally considered in neo-Darwinian theory are not in themselves sufficient to give a complete picture of how developmental systems evolve. I will, consequently, be arguing for a shift in emphasis, and in particular for a recognition of the importance of internal selection (Whyte 1965), co-evolution of developmentally-interacting genes (Dover 1992, 1993; Fryxell 1996) and a process known as $n$-selection involved in the invasion of novel niches (Arthur 1984, 1988).

A caveat is in order here about the relationship between *frequency* and *importance* of evolutionary events (especially as Gould (1977a) has argued that debates in evolutionary biology are normally about relative frequency). These two attributes are sometimes wrongly assumed to be positively correlated in a straightforward manner, so that, in general, evolutionary events with the highest probabilities will make the greatest impact on the overall evolutionary process – for example, allopatric speciation is much commoner than sympatric, therefore it is more important. However, while such a positive relationship may be true in some individual cases (like the speciation example) it is seriously misleading as a *general* view. Clearly, the relative frequency of different kinds of mutation does not map neatly to their evolutionary impact. Indeed, since the majority of mutations are disadvantageous, and since these do not generally contribute to evolutionary change, we could say that there is a strongly *negative* relationship between frequency and importance in this respect.

Less obvious, perhaps, is the possibility that a negative relationship between frequency and importance may also prevail within the sub-group of mutations that *do* contribute to evolution. The more pronounced the effects of a particular type of mutation, the lower will be its probability of being selectively advantageous (the 'Fisher principle': see Sections 1.5 and 8.5), and so the less frequently will it spread through populations. But those same major effects imply that on the rare occasions when such a mutation does spread, its impact will be considerable, in terms of the invasion of novel morphospace – albeit a doubtless imperfect 'invasion' which may later be both improved and diversified through selection on 'modifier' genes in the lineage concerned.

Evolutionary events with low probabilities of happening are particularly important (a) in the long term and (b) in relation to infrequent occurrences such as the origin of body plans. For example, a particular developmental mutation with a probability of $10^{-6}$ of occurring, in any generation, in a population of a given size, will occur, on average, once every $10^6$ generations in a population fluctuating around that mean size. Such events can be 'written off' in a geographically localized microevolutionary study spanning (say) ten generations; but over the entire species/lineage lifespan and geographical range they may be very important. Also, since body-plan origins are by their nature very rare evolutionary events (but see Valentine 1994), any theory of their origin that is based on a highly probable evolutionary process is immediately suspect: if it is so probable, why are there so few major body plans?

There are twin dangers lurking here. One is a readiness to latch on to any improbable mechanism as an explanation for the origin of body plans. The other is a tendency to avoid explicit consideration of body-plan origins altogether because of the difficulties of testability inherent in any low-frequency process. I am attempting, in this book, to steer a course between these twin extremes of mysticism and pragmatism.

## 9.2 Population Genetic Models of Directional Selection

Within population genetics itself, there has always been much interest in the maintenance of variation within populations, where the emphasis is on various forms of stabilizing or balancing selection. (See Ohta (1996) and Kreitman (1996) for views on the current state of the selectionist/neutralist controversy.) However, for our present purpose – to understand the evolutionary diversification of animal developmental systems – the emphasis is on evolutionary *changes* and, accordingly, on *directional* selection. The principle behind this form of selection is simplicity itself; and some models of directional selection reflect this. However, a great many complications can be superimposed on the basic picture; and the models which take these into account are correspondingly complex.

I will consider first a very simple model of potentially general application; and will then examine some of the possible complexities. Readers wishing to pursue models of selection more widely should consult the early 'classics' of population genetics theory (Fisher 1930; Wright 1931; Haldane 1932) and/or more recent texts such as those of Cook (1971), Hartl and Clark (1989), Maynard Smith (1989), Ridley (1993) and Charlesworth (1994). (Ridley's book represents a good starting point for those without a population genetics background; Charlesworth's book is the most advanced.)

The biological context for our simple model of directional selection is as follows. We consider a local panmictic population of an (unspecified) animal species that is geographically isolated to the extent that there is no (or negligible) migration to/from other populations. We assume that this animal is diploid, and reproduces sexually with mate choice being random. Generations are discrete and non-overlapping, so that the population effectively has no age structure; that is, the animal has a perfectly semelparous life cycle. We further assume that the population is free from stochastic effects.

Now consider a locus A for which the population is initially monomorphic, with all individuals being homozygous for a single allele, $A_1$. (At this stage, of course, no selection can act at this locus, as there is no variation.) Now suppose that a spontaneous mutation occurs in the germline of one individual so that one copy of allele $A_1$ becomes $A_2$. (The exact changes in DNA are irrelevant to the formal population model; the mutation could involve a single base substitution or a much bigger-scale change such as the 'shuffling-in' of a new exon.) We will assume that at the phenotypic level the new allele exhibits complete dominance over the 'old' one.

Initially, there may be one, a few or many individuals carrying the mutant allele, depending on (a) where in the germline the mutation occurred and (b) the reproductive success of the parent concerned. Such individuals will be heterozygotes in the first generation – but they will be phenotypically mutant because of the dominance relationship between the two alleles. In the second generation and beyond, $A_2A_2$ homozygotes may appear, as a result (initially) of $A_1A_2 \times A_1A_2$ matings.

We now have a situation in which selection can act. Suppose that the phenotypic changes caused by possession of $A_2$ result in an increased fitness in the standard sense of an increased net probability of surviving and reproducing. Of the various components of this overall fitness that are spread through the life cycle, let us assume that the only component affected here is zygote-to-adult viability. The overall context has now been defined sufficiently clearly by a series of assumptions that modelling becomes possible.

Let us choose a generation at random anywhere between the appearance of the first mutant individuals and (a considerable time later) the disappearance of the last 'wild type' ones. If, at the gamete stage in this arbitrary generation, the frequencies of $A_1$ and $A_2$ are $p$ and $q$, respectively (where $p + q = 1$, because we assume no further mutations producing $A_3$ etc.), then according to the Hardy–Weinberg law the frequencies of the *zygote* genotypes should be $p^2$, $2pq$ and $q^2$ (for $A_{11}$, $A_{12}$ and $A_{22}$, respectively).

It is conventional to denote the fitness of the fittest genotype as 1, with the other fitnesses ($w$) being fractional (or zero if the mutation is lethal). Also, the strength of selection against a less fit genotype is measured by the selective coefficient $s$, where $s = 1 - w$. Thus the fitnesses of the genotypes can be written in either of two ways:

| genotype | $A_1A_1$ | $A_1A_2$ | $A_2A_2$ | (9.1) |
|----------|----------|----------|----------|-------|
| fitness | $w$ | 1 | 1 | |
| | $1 - s$ | 1 | 1 | |

Because of these fitness differences in zygote-to-adult viability, the genotypic ratio among the adults will be:

$$wp^2 : 2pq : q^2 \tag{9.2}$$

Since $A_1A_2$ individuals produce $A_1$ and $A_2$ germ cells in equal numbers (assuming no 'driving' processes distorting normal Mendelian segregation), then the ratio of these two types of germ cell will be

$$(wp^2 + {}^1/_2 2pq) : (q^2 + {}^1/_2 2pq) \tag{9.3}$$

which simplifies to

$$p\,(wp + q) : q\,(q + p) \tag{9.4}$$

To reconvert these gametic ratios to relative frequencies, we need to divide each by the gametic total (i.e. the sum of the two components of ratio 9.4). This sum is sometimes referred to as the mean population fitness, $\bar{\omega}$. The gametic *frequencies*, then, are

$$p\,(wp + q)\,/\,\bar{\omega} : q\,(q + p)\,/\,\bar{\omega} \tag{9.5}$$

Our model has now covered one complete generation (from gamete to gamete), so the change in gene frequency per generation can now be quantified as

$$\Delta p = [p\,(wp + q)\,/\,\bar{\omega}] - p \tag{9.6}$$

for the $A_1$ allele; or alternatively, for its counterpart:

$$\Delta q = [q\,(q + p)\,/\,\bar{\omega}] - q \tag{9.7}$$

Clearly, because of the assumptions made, $\Delta p$ is negative and $\Delta q$ is equal but positive.

Since this model applies throughout *all* generations, the increases in the frequency of the $A_2$ allele will be cumulative, and will result, eventually, in its fixation in the population, when all copies of $A_1$ have been eliminated. Figure 9–1 shows population trajectories for this replacement process, starting from an arbitrary point at which the frequency of the new allele is 0.1, and assuming various strengths of selection. These simulation results in a sense merely quantify the obvious: namely (a) fixation is ultimately inevitable (though some help is needed from stochastic events: see below); and (b) the speed at which it occurs depends on the value of *s*.

It is now necessary to discuss the various assumptions underlying the above model — both some that have already been made explicit and some that have not. There is insufficient space to give a full account of all, so I will be selective.

## Stochastic Effects

A very clear exposition of the massive impact of stochastic effects was given by Haldane (1932), who showed that in reality the vast majority even of selectively *advantageous* mutations must be lost through genetic drift before they reach appreciable frequency. So a simplistic mental picture of all detrimental mutations being removed by

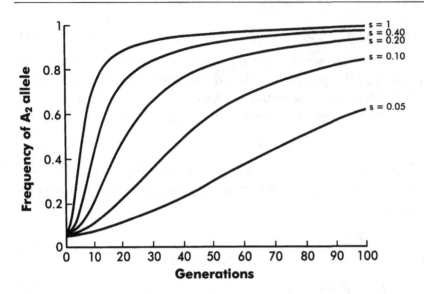

**Figure 9–1  Selection for a new dominant mutation**
Each curve represents a different selective coefficient, as indicated, in favour of a new dominant allele, $A_2$, at a locus A. Even when the value of s is very high, many generations are required to reach the starting value shown here (0.1) from the much lower value that applies when the mutation first occurs. Modified from Hartl and Clark 1989. *Principles of Population Genetics.* Sinauer, Sunderland MA, with permission.

'negative' (or 'purifying') directional selection and all advantageous ones being swept through the population by positive directional selection should be avoided. Rather, detrimental ones are lost by a mixture of selection and drift; the few advantageous ones that survive early stochasticity will finally become fixed in the population also by a mixture of selection and drift (the latter now constituting a threat to the 'old' allele whose frequency has become very low). The fate of mutations that are selectively neutral is entirely dependent on drift (see Kimura 1983).

## Dominance Relationships

If the new, advantageous mutant allele is recessive (which is more commonly the case), then it may be *dependent* on drift to raise its initial frequency to a level at which homozygotes – and hence altered phenotypes – begin to appear, because prior to such appearance selection is unable to act. Also, in addition to complete dominance in either direction, other possibilities include heterozygote intermediacy, partial dominance and codominance. The population trajectories shown in Figure 9–1 only apply to the spread of a new mutation that is completely dominant. In other cases, the trajectories are quantitatively altered but qualitatively the same in that fixation still occurs. (The only exception to this is where 'overdominance' – in the sense of the heterozygote being outside the range of the two homozygotes – applies not just to a morphological character but also to fitness: in one direction this takes the system into the realm of heterozygote advantage and hence of balancing selection.)

## Other Genetic Complications

The simple picture presented in the above model and in Figure 9–1 may be completely altered by a variety of genetic phenomena. These include: the simultaneous segregation of multiple alleles in the population; linkage effects, where the fate of alleles at one locus can be influenced by selection on a neighbouring one, because of the low frequency of recombination – so-called 'hitchhiking'; and epistatic interactions with loci *anywhere* in the genome. In this last case, where the fitness relationships between genotypes at one locus depend on the genotype at another, complex patterns of genetic change can occur that are entirely unpredictable using simple models that assume a lack of epistasis (see Lewontin 1974a, chapter 6). I will return to this point from a more developmental angle shortly.

## Population Structure

Most populations do not have perfect semelparous life cycles and a lack of overlap between generations. Rather, there is a complete spectrum from this extreme to the pronounced iteroparity and complex age structure of, for example, most populations of large mammals. Further, the sex ratio may alter somewhat from time to time; and the population density is likely to fluctuate markedly between generations – often through two or more orders of magnitude in insect populations. Selection may vary between sexes and age groups. It may also be density dependent.

### Other Ecological Complications

There is no such thing as a 'homogeneous environment'. Rather, all populations inhabit environments with complex patterns of variation at several intergrading scales in both spatial and temporal dimensions. Selection pressures may vary across environmental patches, and how this affects populations will depend on the mobility of the animals concerned. Equally, climatic and other changes occurring over time – from day to day, seasonally or in the longer term – may affect the strength and even direction of selection.

These various complications do not, of course, invalidate population genetic modelling; but their existence means that realistic models need to be correspondingly complex. However, I do not wish to devote space here to the consideration of complex models; there are several good sources for such material (e.g. Crow and Kimura 1970; Charlesworth 1994). Rather, I wish to consider a key feature of directional selection generally: the nature of the agent generating the selective coefficient. This is the subject for the next section.

## 9.3    Internal Selection

As we have seen, the selective coefficient is a measure of the strength of selection acting on polymorphic variation. In cases of continuous phenotypic variation, the strength of selection can be measured by the selective differential (see Falconer 1989 for a general account). In either case, the entity *causing* selection is referred to as the selective agent.

In the vast majority of mainstream evolutionary writings, from *The Origin of Species* itself to recent neo-Darwinian works, there is an all-pervasive emphasis on selective agents that are external to the organism. There is nothing in Darwinism or neo-Darwinism that explicitly denies the existence of internal selective agents. But their importance is *effectively* denied by the scale of their neglect. (I have borrowed this last phrase from Horder (1994) who in discussing the modern synthesis refers more generally to "the scale of its neglect for embryology".)

A good example of the external bias of mainstream evolutionary theory is provided by Berry (1982), who tabulates (p.39) a series of well-known examples of selection, listing for each a selective agent (usually in the singular; but in a few cases multiple). In virtually all cases the selective agent is external. The agents listed include: competition for food, predation, parasitism/disease, pollution, temperature, wind, climatic stress more generally, and interactions with conspecific individuals, including mate choice and breeding success. Not a single agent refers explicitly to internal developmental coordination.

Berry's book is entitled *Neo-Darwinism*, and the external bias that it exhibits is an accurate representation of the bias pervading that tradition generally. This in turn is a faithful continuation of the bias that pervades *The Origin of Species*, with its emphasis on the 'struggle for existence' against external threats, both biotic and abiotic.

There are at least two reasons for this bias. First, from a methodological viewpoint, mainstream evolutionary theory is derived ultimately from Darwin himself, and in Darwin's day natural history was a well-developed pursuit, while causal developmental biology was in its infancy. Second, from a philosophical viewpoint, many of today's neo-Darwinians – who can be less readily excused for neglecting internal processes – remain fascinated, as Darwin was, with *adaptation*, where external selective agents are particularly important. Internal *coadaptation* gets a mention from time to time, but has generally not been a comparable focus of attention. It is of course in relation to such coadaptation that internal selective agents are likely to be most important.

It is apparent, then, that the division of selective agents into predators, competitors, climate, etc. is a second-order one. The primary split is between internal and external, as shown in Figure 9–2.

This point is not new. It was made by Whyte (1965) in a work of enviable clarity and brevity. This was reviewed in *Nature* by Thorpe (1966) who described it as "profound"; yet it made little impact. Nor was the idea of internal selection new in 1965. Whyte published two earlier short papers on the subject (1960a, 1964) and he also (1965, chapter 5) reviewed scattered suggestions made by various other biologists on the possibility of internal factors operating in evolution. Because of its importance, and its puzzling neglect up to now, I will discuss Whyte's book in some detail.

First, Whyte was clear about how he regarded the modern synthesis: correct as far as it goes, but seriously incomplete. He describes the synthesis (p.12) as "a necessary but not sufficient theory of the directive factors in evolution". Second, he was clear about what was missing (p.32): "The statistical theory of populations must be

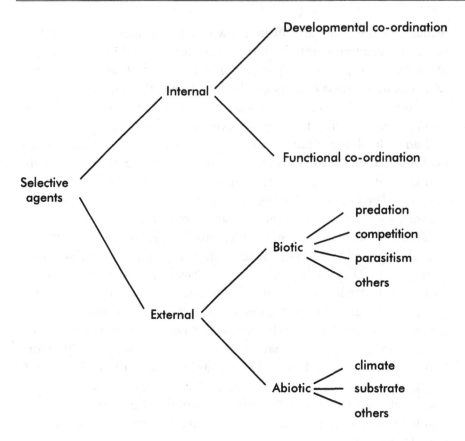

**Figure 9–2  A classification of selective agents, in which the primary division is into internal and external**

complemented by a structural theory of individual ontogenesis and of its influence on phylogeny." Third, he suggested that internal selection might be particularly important in relation to the origin of body plans (p.65): "Internal selection may prove to have been most important, and perhaps solely operative, during some or all of the major steps in evolution, which are not yet understood."

Whyte was in no doubt that his proposal of a key evolutionary role for internal selection would get a rough ride. In his chapter 6, he lists a series of responses he had received in correspondence with various leading evolutionary biologists of the time. These were generally both antagonistic towards him and (unwittingly) contradictory towards each other. They include the views (a) that there is no internal selection, (b) that there is no distinction between internal and external selection, (c) that the idea of internal selection is premature, and (d) that the idea is commonplace. A variant on the last of these views was provided by Lewontin and Caspari (1960) in their response to Whyte's (1960a) first paper on internal selection (which at that stage he called 'developmental selection'). They asserted that the idea of *negative* (or 'purifying') internal selection (wherein mutants with significantly decreased internal coordination were removed from a population) was commonplace, but they were strangely silent on whether they thought that *positive* internal selection was capable of increasing the frequency of new variants with improved coordination, as noted in Whyte's (1960b) reply. Yet despite all this, Whyte remained convinced that internal selection was hugely important. He states (p.30): "No future survey of evolutionary theory can be regarded as competent which fails to discuss internal selection." Most surveys published between then and now have 'failed' in this sense; which in turn means that Whyte largely failed to persuade the biological community of the importance of his ideas.

The only one of the four above antagonistic views with which I agree is that in 1965 the idea of internal selection may indeed have been premature. The recent burgeoning of data on comparative developmental genetics, which we saw in Chapter 7, may provide a background against which the idea of internal selection is more likely to capture the imagination, and hence to take hold. In an attempt to ensure that it does so – and at the very least that the idea is vigorously debated rather than overlooked – I give below a series of eight specific proposals which relate to the evolutionary importance of internal selection. The first of them effectively constitutes a definition, and this differs somewhat from Whyte's: he regarded internal selection as 'all-or-nothing', while I regard it as probabilistic, like its external counterpart.

1. Internal selection occurs when individuals with different genotypes at a given locus differ in fitness because they differ in their degree of internal coordination. If we think of organisms as machines, then fitter genotypes in this context correspond to machines whose parts integrate better with each other. An example at the morphological level would be a variant with improved skeletal joint functioning; while at the cellular level an example would be a variant with improved intercellular communication.

2. Internal selection can relate to the degree of developmental coordination, the degree of functional coordination, or both (see Figure 9–2).

3. Fitness differences associated with internal selection are likely to remain approximately constant across a wide range of environments; unlike those caused by an external selective agent (e.g. a predator) which are likely to be highly environment specific. This is a key point which will be elaborated below.

4. In terms of biological causality, internal selection is quite distinct from external selection. There is not a continuum from one to the other – as we saw earlier that there was from 'micro' to 'macro' mutation. A selective agent is either part of the organism or it is not. (An endoparasite is an external selective agent, notwithstanding its physical location.) However, in terms of the degree of environment specificity (point 3 above), there *is* a continuum between the two: see below.

5. In elementary population genetic models of single-locus polymorphism using selective coefficients to measure the strength of selection, internal and external selective agents are formally indistinguishable. (The trajectories of population-level spread are determined by the numerical value of $s$, not by the nature of the agent that causes it.)

6. While *some* external selective agents may *co*evolve, others – the abiotic ones such as climate – may not. In internal selection, *all* selective agents may coevolve, as they are all genes or their direct or indirect products. Coevolution of developmentally interacting genes thus becomes a major focus of attention.

7. Because of this, epistatic effects on fitness should be expected to be the rule in cases of internal selection (though they are probably also very common in cases of external selection). From a population genetics perspective, coevolution of developmentally interacting genes can be considered a subcategory within epistatic interactions affecting fitness more generally. So some of the formal machinery for modelling it already exists, albeit not in an explicitly developmental context.

8. Because of points 5 and 7 above, it might be asserted, by those whose interest is focused at the population level, that internal selection is already implicitly taken into account by existing genetic models of the spread of variants within populations, and that little can be gained from a more explicit consideration. However, this

view is too limited. Explicit consideration of internal selection and developmental coevolution (a) is essential for understanding the origin of body plans; and (b) will help to unify developmental and population approaches to evolution, and will thus lead ultimately to a more comprehensive body of evolutionary theory.

These eight assertions represent a deliberate high-risk strategy. They are intended to be maximally visible, so that the reader cannot either ignore them or 'entangle' them, but rather is forced to confront each individually and to decide whether each is acceptable. I suspect that in the more developmentally enlightened climate of the 1990s (in contrast to the 1960s, when Whyte's book appeared), proposals 1–7 may be acceptable to a fairly wide range of evolutionary biologists. But proposal 8 is likely to remain controversial, especially given the lack of acceptance of the body-plan concept by some neo-Darwinians, for example Williams (1992), who describes the concept (or, to be more accurate, one particular version of it) as "misguided and dispensable".

Let us now return to the issue of the degree of environment specificity of the selective advantage of a new variant. Figure 9–3 shows three 'environmental profiles' representing three possible ways in which a selective advantage may vary across a range of environments. We will consider each of these in turn.

In (A), the fitness increment of a new variant is restricted to one ($f$) of a series of different environments. An example of this would be a variant with improved crypsis with regard to detection by a predator only found in environment $f$. In (B), the fitness increment varies, but in a more gradual way. An example would be a variant with improved temperature stability of a crucial enzyme (with $a$–$j$ now representing a range of environments with increasing mean temperatures). In (C), the fitness increase is general, and applies equally to all environments. This is characteristic of internal selection. As noted above, an example would be a variant that has improved intercellular communication, as in the case of a new allele that makes a new version of a ligand that interacts more efficiently with a receptor.

Of course, all three transenvironmental fitness profiles are very idealized. Much more complex patterns will occur in reality, and that inevitably means that the very 'peaky' patterns of pure external selection will intergrade with the 'flat' pattern of pure internal selection. But such intergradation, and the fact that no profiles will be completely flat, should not cause us to write off the idea of internal

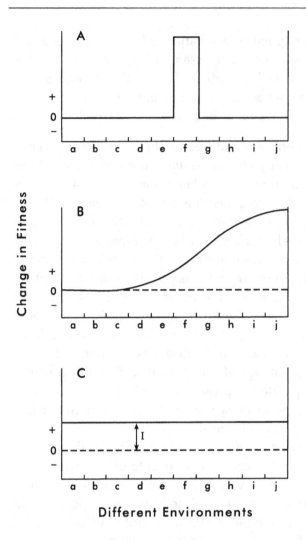

**Figure 9–3 Three possible environmental profiles of the selective advantage of a new variant**

a–j: ten environments that differ in some way. (A) new variant is advantageous in environment f only; (B) new variant is at increasing advantage through environments d to j; (C) new variant has a similar advantage in all environments. I, fitness increment of new variant.

selection and revert to thinking of all evolutionary changes as adaptations to particular environments, which is too limited a view. Perhaps a useful criterion for measuring the relative importance of internal and external factors in any particular case would be the ratio of the peak height to the general fitness increment (I in Figure 9–3). Cases close to the perfectly flat profile of C but with some slight undulation are then seen predominantly as internal developmental/functional improvements that happen to vary a little across different environments. This seems a less forced interpretation than to consider such profiles as external adaptations.

In the end, the relative fitnesses of individuals possessing alternative genotypes at any particular locus are determined by a range of external, ecological factors and also by a range of internal, genetic–developmental factors. If one of the former stands out as having a major effect, while none of the latter do, then the situation may be thought of as 'external selective agent with genetic background effects'. On the other hand, if one internal factor stands out as having a major effect on fitness, while none of the environmental ones do, then the situation may be envisaged as 'internal selective agent with environmental background effects'.

Fitness differences of alternative pigmentation morphs in a prey species broadly fit the former conceptual mould. In contrast, fitness differences associated with possession of different versions of a ligand, one of which interacts better with its receptor, fit better with the latter. Current evolutionary theory has been adequately influenced by the first of these ways of thinking. It has been inadequately influenced by the second, and by the recognition of a continuum of possibilities extending from one to the other. However, it seems likely that as data on comparative developmental genetics (Chapter 7) accumulate, the requisite shift in mind-set on the part of evolutionary biologists will become irresistible.

Throughout the above discussion, I have assumed that the new genetic variant causes a cellular/morphological alteration that is not influenced directly by the environment. However, phenotypic plasticity is commonplace, and adds a further complication to the overall picture. One possible evolutionary response to such plasticity is selection favouring increased autonomy of developmental systems from the environment (Schmalhausen 1949: see Chapter 4). However, another possibility is selection for patterns of plasticity that are themselves adaptive. This is essentially selection for particular 'reaction norms' (see Lewontin 1974b and review by West-Eberhard 1989). Such selection is clearly external rather than internal according to the above distinction, even though it is internal characteristics that are being modified.

Finally, I should stress that although I believe, like Whyte (1965), that positive internal selection has an important role to play in the origin of body plans, and negative internal selection an important role in their maintenance, I am *not* proposing that internal selection is all that is involved: far from it. An interplay between several distinct kinds of selection almost certainly 'drives' the origin and subsequent stabilization of any body plan. This interplay is described in the following section.

## 9.4  The Origin of Body Plans: a Population Perspective

### The Phylogenetic, Ecological and Ontogenetic Context

The following points are made very briefly, as they have all been made at greater length in earlier chapters. The animal kingdom is monophyletic (Slack *et al.* 1993). The first animals perhaps appeared in the Vendian, around 600 *my* ago (Valentine 1994), though a much earlier origin is also possible (Wray *et al.* 1996). There was an Ediacaran 'explosion' of body plans, followed by a second explosion around the base of the Cambrian (now revised to 545 *my*: Bowring *et al.* 1993) which produced the body plans on which all of today's phyla are based. Thus there appears to have been an early period of around 50 *my* during which nearly all major morphological novelties arose, even if it turns out that the main lineage divergences occurred much earlier.

The animal populations involved in these explosions were all marine, and it is probable that most inhabited the shallow waters of continental shelves. Their unicellular and algal food supplies were already well diversified (Lipps 1993), while established multicellular competitors, predators and parasites were absent. There was thus a situation of considerable ecological opportunity – or empty niche space – that has never been repeated in biospheric history. (The closest situation is that prevailing after a mass extinction, but even that is quite different, as most body plans continue to be represented by a few surviving species, from which rapid diversifications and recolonizations can occur.)

These early animals were all (by definition) multicellular, but consisted of relatively few cells compared with most present-day animals. Their ontogenetic trajectories were therefore much simpler than those to which most of our detailed embryological information relates. Also, since they had not been subject to a comparable history of selection for integration and canalization, their ontogenies were almost certainly more evolutionarily flexible than those of their later counterparts. Just as Weismannian separation of the germ plasm is an evolutionarily derived state (Buss 1987), so too is a highly coordinated development that is resistant to change. Mutations of genes controlling early development thus occurred, at the beginning of animal evolution, in a very different ontogenetic context to those occurring in *Caenorhabditis, Drosophila, Mus*, or any other present-day animal.

## Escape from Competition

The concept of fitness used both by Darwin (1859) and by present-day population geneticists is intrinsically comparative. A variant's probabilities of survival and reproduction are considered not in isolation but in terms of how they compare with the equivalent probabilities for other members of the same population. The comparison, therefore, is with organisms that are colocalized in space and time (sympatric and contemporary) and are *conspecific*. At first sight, this seems perfectly reasonable, as individual organisms do not exist in isolation from the rest of their species. However, it must be stressed that the biological criterion delimiting the group of organisms against which 'comparative fitness' should be measured is not those that interbreed with the variant in focus but rather those that compete with it (in the broad sense of competition, as defined below).

Two examples of selective scenarios serve to illustrate this point. First, in a population of an organism that reproduces asexually, a new variant that is fitter may spread – in the form of a clone – through the overall population, eventually displacing the other clone(s). Directional selection leading to fixation has occurred despite the lack of interbreeding between the types that are being 'compared'. Second, the ecological process of competitive exclusion (Gause 1934) wherein one species spreads to 'fixation' in a locality at the expense of another, is formally identical to this asexual selective scenario (Arthur 1987a). Yet again, no interbreeding occurs if the populations concerned represent 'good' biological species.

What all three scenarios – sexual populations, asexual populations and some ecological guilds – have in common is competition between the organisms concerned; though we need to be careful here, as 'competition' has proved remarkably difficult to define (Birch 1957;

Milne 1961) and the word is used in many different senses. In this case, the appropriate usage is *sharing a limiting factor*. Where a population is resource limited, the competition is exploitative (though interference-type mechanisms may subsequently evolve). But this need not be the case. Where a population is regulated around a lower equilibrium than resources would permit, due to the density-dependent action of a parasite, for example, then there is still 'competition' in a broader sense (Williamson 1957).

Given a population of a single type of organism – be it a clone of asexual individuals or an isogenic line of sexual ones – there is inevitably *intra*type competition, in the above sense, among the organisms concerned. Such competition may be temporarily suspended where the population is 'effectively unlimited' (e.g. after the colonization of a resource-rich environment by a small propagule of immigrating individuals). But such transitory 'unlimited' situations will rapidly give way to limited ones.

Given a mixture of types of organisms – conspecific genetic variants or (closely-related) species – that share a limiting factor, *inter*type competition is equally inevitable; and the disappearance, fixation or intermediate equilibrium of any one of the types depends on its fitness relative to the other(s). Also, complex scenarios involving a mixture of intergenotype and interspecies competition are possible, and indeed are likely to occur often in nature. In a closed community of species A and B, where the inferior species (B) is polymorphic, a gene conferring a selective advantage may spread through a population of B at the same time as that overall population declines to zero because of its competitive inferiority to A. If the fitter genotype provides a sufficient benefit, its spread through B might of course reverse the interspecific competitive relationship with A, and put the mixed-species population dynamics into reverse. Many theoretical models of these complex situations have been formulated; for examples see Lawlor and Maynard Smith (1976) and Slatkin (1980). These show that under certain conditions non-trivial equilibria can occur in both the genetic and ecological dynamics.

Just as competition causes the fitnesses of two types to be 'assessed' (by nature) in a comparative way, so does the absence of competition render such comparative assessment inoperable. In the absence of intertype competition, a variant that would otherwise be eliminated may be able to persist. The pioneering genetic model of this situation of 'multiple niche polymorphism' was formulated by Levene (1953);

many related models followed (e.g. Maynard Smith 1970; Strobeck 1974; Yokoyama and Schaal 1985; Reeves 1992). The ecological equivalent is 'resource partitioning' – a misleading phrase meaning incompletely overlapping niches of *species* leading to stable coexistence (see Schoener 1974).

Levene's (1953) model was a classic piece of population genetics theory: application was intended to be quite general, so the nature of the gene was unspecified, and the question of whether/how it altered development went unasked and unanswered. Effectively, such questions were irrelevant to the issue in focus at the time. However, as a result the model tends to be most readily interpreted as dealing with 'routine' intraspecific polymorphism. For example, one variant may possess an allozyme that enables it to utilize one resource more efficiently, while the alternative allozyme assists better in the utilization of another resource. In such situations, within-niche competition causes the overall system to be frequency dependent, and so the variation is maintained within the population. There is no explicit link with speciation, much less with the origin of a body plan. It may be, however, that a rather similar dynamics operating on radical developmental variants in primitive small-cell-number animals occupying the relatively 'uncluttered' niche space of the Vendian (or earlier) could produce much more dramatic results; and it is worthwhile to think our way through such a scenario, in conjunction with Figures 9–4 and 9–5.

The genes of Vendian animals were no doubt subject to a great variety of mutations, just like the genes of their descendants today. However, I would like to focus on a particular kind of mutation, which may be especially relevant to the origin of body plans. This involves a gene that has a controlling role in early development. The mutation causes an altered ontogenetic trajectory which leads to an adult morphology that is well outside the normal range of continuous variation. We will assume that such a mutation occurs in an early germline cell in a single individual, and that it is recessive. This situation can give rise to a cluster of progeny carrying the mutant allele, as suggested by Sinnott, Dunn and Dobzhansky (1958); and matings between these can give rise to a cluster of phenotypically mutant progeny: see Figure 9–4, top two panels.

Such mutations, when they occur in present-day organisms such as *Drosophila*, have two very predictable characteristics. First, by 'scrambling' normal development, they lower the degree of coadaptation of the mutant organism, sometimes to such an extent that the

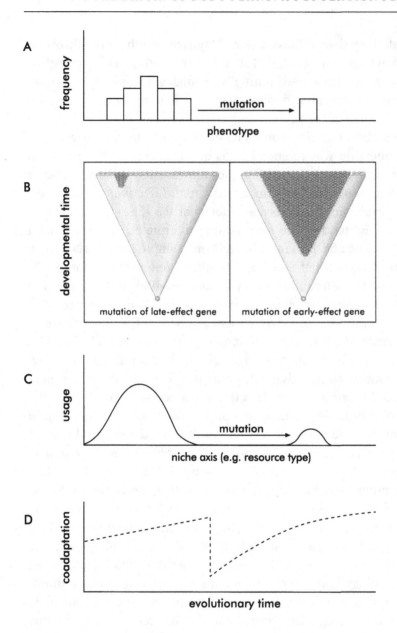

**Figure 9–4    A model of the invasion of a new niche in early animal evolution**

(A) Production of a cluster of mutant progeny by mutation; (B) effect of this mutation on development (right) contrasted with the effect of a less major mutation (inverted cone model, section, affected cells shaded); (C) niche shift that is brought about fortuitously by the mutation; (D) pattern of change in internal coadaptation over time (see also Figure 9–6).

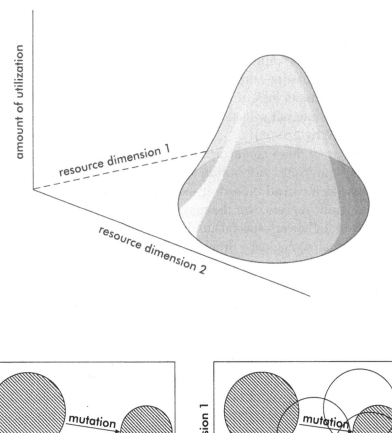

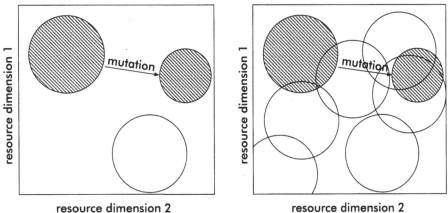

**Figure 9–5   The effect of empty niche space on the chance of success of an ill-coadapted mutant animal**

Top: Standard representation of a population's niche as a resource utilization 'dome'. (In reality many more resource dimensions are involved, see Pianka 1994.) Bottom: Radical mutants may have faced a low probability of coming into competition with heterospecific wild types in early animal evolution (left), while the corresponding probability now is much higher (right). Circles represent the bases of domes like that shown above.

animal dies before reaching adulthood. Second, if the animal survives to adulthood, then it is very unfit when in direct competition with the 'wild type', and rapidly disappears from any experimental polymorphic population that is established. Inasmuch as the fitness decrease is due to lack of internal integration, what is being observed is 'negative internal selection' (in the sense of Lewontin and Caspari 1960 and Whyte 1965).

Because of these characteristics, it is unlikely that such mutations make a significant contribution to present-day evolution, either intra- or transspecific. And indeed, it seems equally unlikely that they have had a significant role over the majority of animal evolutionary history. However, in the very special conditions – both internal/developmental and external/ecological – prevailing in early animal evolution, the fate of some such mutations may have been very different, and this may have had important consequences.

Consider first the issue of internal integration. Because of the shorter history over which prior 'integrative' selection was able to act (about 600 *my* shorter), the starting point for our mutation – some unspecified 'Vendian wild-type' – is likely to have had a lower level of coadaptation than a typical present-day animal. The degree of decrease in the level of coadaptation caused by a mutation of a gene controlling early development is thus likely to be smaller. So the chances of lethality are correspondingly lower.

This in itself will not achieve much. A group of mutant organisms may survive and reproduce (both with each other and with their progenitor stock), but if they are subject to competition with the wild type they will still disappear. However, in some cases they may *not* be subject to such competition. This is where we need to consider the role of Vendian 'empty niche space'.

Any major change in the ontogenetic trajectory and adult morphology of an animal is likely to alter its pattern of resource use and indeed its ecological characteristics generally. In a few cases, such changes will by chance result in utilization of a novel resource, and in Vendian times there was a reasonable probability that no other multicellular consumer was already using that resource (Figure 9–5). In such a situation, a badly coadapted but viable mutant animal may be *more* fit than its well-coadapted progenitor in the sense that on average it leaves more surviving progeny, simply because of the lack of competition for its new-found resource. (An alternative approach to visualizing this situation is simply to consider the fitnesses of the two types independently, which was behind my earlier label of $n$ selection – the $n$ deriving from the *net* reproductive rate of the new mutant: see Arthur 1984; also Jongeling 1996.)

In this situation, the new type is (a) ecologically and perhaps microspatially isolated, but (b) geographically *not* isolated and (c) reproductively *not* isolated from the wild type. What happens next? One possibility is that there will be strong selection, in the mutant subpopulation, for variation that decreases reproductive compatibility with the progenitor stock, as own-type matings will on average leave more offspring, because of the comparative lack of competition for the new-found resource. Such a situation can lead to a variety of modes of sympatric speciation (Maynard Smith 1966), including 'micro-allopatry', which of course is not true sympatry at all.

The scenario pictured is not a highly probably one, but that is hardly the point. I have already argued that probability/frequency on the one hand and evolutionary importance on the other cannot be readily equated. This kind of major developmental divergence only needs to happen in a tiny fraction of Vendian animal populations to provide a mechanism for the rapid invasion of novel morphospace. And in its favour it requires no unknown mechanisms – the process works entirely at the level of mutation and selection occurring within local populations.

Like most other evolutionary ideas that might at first sight seem new, the above ideas are really not. Three main themes are intertwined in the above proposal: less entrenched ontogenies in early evolution; the possible detachment, in some circumstances, of (co)adaptation from reproductive fitness; and the evolutionary 'permissiveness' (and hence 'creativity') associated with a temporary reduction or suspension of competition. All of these have been suggested before: respectively by Erwin and Valentine (1984), Burian (1983) and Frazzetta (1975), among many others. There are also echoes of Wright's (1932) adaptive landscapes and Simpson's (1944) quantum evolution.

The sudden appearance of a morphological novelty in the manner described above does not, in itself, constitute the origin of a new body plan. Body plans incorporate many deep-rooted and interlocked features, and the origin of any body plan is inevitably a multistage process. In particular, internal selection (first positive, then negative) has a role to play, and it is to that that we should now turn.

## Internal Selection and Developmental Coevolution

Let us consider, as a starting point, an early (e.g. Vendian) animal that has arisen through radical developmental change, invasion of a new niche, and sympatric or micro-allopatric speciation, as outlined above. Any such animal will be characterized by a low level of internal co-adaptation (see Figure 9–4D). This may manifest itself in various ways, but one obvious possibility is that the ontogenetic trajectory is rather variable – that is, there is a high variance in its outcome between different individuals – and this is likely to result in a high pre-adult mortality (see Gould, Young and Kasson 1985 for a relevant study of ontogeny in the mollusc *Cerion*).

In a population of such animals, there will inevitably be allelic variation in the genes downstream of the early acting one whose muta-tion led to the origin of this lineage in the first place. The pre-adult mortality referred to above is likely to be strongly selective on this variation: those alleles that interact more effectively with the new mutant upstream gene will be favoured. As suggested in Chapter 8, it may be that there is a net downstream flow in this process of 're-coadaptation' (Figure 9–6). However, any such trend is likely to be statistical rather than absolute, and will be complicated by the reuse of developmental genes at two or more stages in ontogeny, which is now known to be a common occurrence (see for example Salser and Kenyon 1996).

The very simplified arrangement shown in Figure 9–6 suggests that although the alleles change, and hence the nature of each upstream/ downstream link changes, the overall pattern of such links is (a) linear and (b) unchanging. Neither of these is likely to be the case. Indeed, selection may well favour an increasing density of linkages (especially cross-linkages), thus simultaneously increasing redundancy of control and repeatability of ontogenetic trajectory between individuals, and hence cause decreased evolutionary flexibility for the future.

The picture that begins to emerge, then, of the developmental evo-lution of early animals, is a *race between two processes, both selec-tively driven*. On the one hand, selection may favour occasional radical alterations to ontogeny, but, on the other, it favours increased canal-ization, cross-linking, generative entrenchment and so on, all of which act to prevent subsequent major changes from being successful.

I envisage the process of 're-coadaptation' being driven by selection that is (a) directional, (b) positive (in the sense of spreading a new

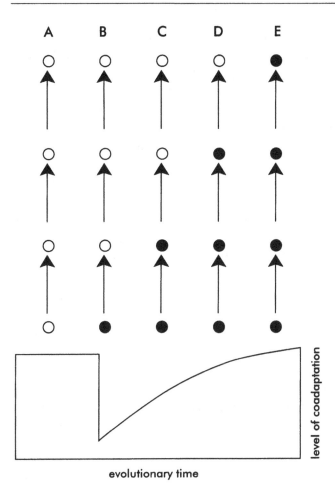

**Figure 9–6  A simplified picture of the process of 're-coadaptation'**
A, Original developmental pathway of interacting genes; B, mutant early gene, reducing the level of coadaptation; C, selection on next-level-down modifier increases coadaptation; D/E, the process continues further downstream. (Open circles, original alleles; shaded circles, new alleles.)

variant from a negligible frequency to fixation) and (c) *internal* in the sense of Whyte (1965), as discussed in the previous section. Critics might respond – as some did to Whyte – that all selection is really external, because all organisms inhabit particular environments and it is ultimately their relative probability of survival (and reproduction) *in that environment* that matters. The counter-view – that we really are dealing with internal selection – does not rely on proposing that developmental stages occur *in utero* or are otherwise protected from the rigours of the external environment. After all, no Vendian animals had anything even remotely similar to the mammalian level of embryo protection. Rather, this view is based on the idea that *the selective changes involved were not environment specific*, as was illustrated in Figure 9–3C. They would have occurred – albeit with slight variations – in virtually any viable environment for the organism concerned, as they are driven by variation in internal integration, not by variation in ability to deal with any external threat.

Once a new body plan has become established through a mixture of (and interaction between) positive external and positive internal selection, it then becomes stabilized and largely maintained through *negative* internal selection. That is, once the new suite of embryological characters becomes successfully 'interlocked', further mutations will tend to damage this new-found developmental integration. In some respects, this *maintenance* of the body plan is of more interest than its origin: as noted in Chapter 1, all phenotypic characters have an evolutionary origin, but only a very few of them – body-plan characters – are maintained for hundreds of millions of years.

None of the above should be taken as playing down the importance of external selection. Although I have had little to say throughout this chapter about external selection on quantitative phenotypic variation, about adaptation to environments, or about the divergence of sister species through allopatric speciation, I believe that these processes are ubiquitous, and were occurring throughout the phase of body-plan proliferation on which I am focusing, as well as before it and ever since. I do not subscribe to Willis's (1940) view that evolution made body plans first, and only later began to tinker with the details. But because so much has been written about adaptation, and because we have so many good examples of it, there is a danger that we will see adaptation as constituting the whole of evolution. I have attempted to show why I think such a view is incorrect.

Finally, a comment is in order on the extent to which the overall view proposed herein is 'uniformitarian'. This term was introduced by Whewell (1832) to refer to (geological) theories that did not invoke mechanisms other than those that can be observed at work today to explain events in the distant past.

Darwin's views on evolutionary mechanism were clearly uniformitarian, in that he envisaged (external) selection acting on (continuous) variation as driving all stages of evolution. A theory invoking a Lamarckian mechanism for the origin of body plans but admitting panselective modification thereafter would clearly be non-uniformitarian. What of a theory (as herein) that is based on mutation and selection throughout, but envisages statistical trends in the importance of different *kinds* of mutation and selection as evolutionary time proceeds, with progressively fewer successful 'radical' mutations and a diminishing importance of positive internal selection? Clearly, it is somewhere in between the two extremes, but it seems rather closer to the uniformitarian end of the spectrum. I will consider this point further in Chapter 10.

## 9.5   Types of Genetic Change

In Chapter 1, I drew attention to a point made by Coyne (1983) regarding the question of whether the kinds of genetic change underlying 'ordinary' speciation events (for example those underlying interspecific differences in genital arch morphologies in *Drosophila*) were the same as those underlying major evolutionary divergences such as those leading to different body plans. Coyne states (p.1115): "there is currently no evidence that major and minor taxonomic bifurcations involve different mechanisms of genetic change."

A good approach is to use this proposed lack of difference as a null hypothesis. In doing so, I will consider "mechanisms of genetic change" in a broad sense, to include both the issue of the *number* of genes involved (with which Coyne was primarily concerned) and other aspects, such as the ontogenetic period of gene expression and the mechanism of the spread of a new gene at the population level (e.g. by internal or external selection).

It now looks likely, from the extensive body of data at our disposal on comparative developmental genetics (Chapter 7), that there is at least one reason for rejecting our null hypothesis. Present-day representatives of different body plans (e.g. insect versus crustacean; chordate versus echinoderm) diverge from a very early ontogenetic stage in the expression patterns of key developmental genes and, in consequence, in their developmental trajectories. Although we cannot be certain, it seems almost inevitable that Vendian or Cambrian representatives of different body plans diverged equally early in their ontogenies. That is, early ontogenetic separation was a feature of the original body-plan divergence, even if it has been supplemented by further changes occurring in one or both lineages since that divergence. Comparing this state of affairs with the typically much later stage in ontogeny at which the developmental trajectories of sister species diverge, it seems likely that the "mechanisms of genetic change" are different – albeit we have yet to establish the relative contribution from different *alleles* versus different *loci*. However, it is important to note that this is a statistical difference rather than an absolute one. (Its likely form will be considered in Chapter 11.)

At the population level, the views advanced in Sections 9.3 and 9.4 above would also argue for the rejection of our null hypothesis, on at least two grounds. The origins of body plans almost certainly involved a greater role for: (a) emptier niche space and less competition than generally prevails in the present-day biosphere; and (b) positive internal selection associated with the coevolution of developmentally interacting genes. However, these ideas are themselves still only hypotheses, so we should tread carefully.

The overall view that emerges, then, is two-fold. First, major and minor taxonomic bifurcations probably do involve different sorts of change at both individual and population levels. But second, this difference takes the form of a number of related statistical patterns; it is neither 'singular' nor 'absolute'. The further elucidation of these patterns will be considered in Chapter 11. As will become apparent therein, this issue is central to the incorporation of a developmental component into the 'modern synthesis'.

## 9.6 Drift, Drive and Directed Mutation

While the foregoing account has been unconventional in some respects, it has been conventional in emphasizing the importance of selection in spreading developmental changes at the population level. But three other agents can potentially bring about such a spread, and these require brief mention.

The role of genetic drift in evolution was seriously underestimated – especially in the context of large populations – prior to the 'neutral theory' of Kimura (1968, 1983). However, mutations causing developmental change are very unlikely to be neutral, as Kimura himself stated (but see Wray and Bely 1994). The fate of transient polymorphism in a developmental gene is thus likely to be determined primarily by selection – but with the general proviso made earlier that drift may be influential, despite selection, when either allele is at a very low frequency.

Twenty years ago, discussions of 'meiotic drive' were restricted to a few examples in mice and *Drosophila* in which progeny ratios suggested that some unknown mechanism was distorting the normal Mendelian segregation of alleles in favour of one of them. Subsequently, the more mechanistically explicit idea of 'molecular drive' arose (Dover 1982, 1986), where processes of genomic flux such as biased gene conversion might consistently drive the frequency of one variant up and another down, both within gene families and within populations. Perhaps the most compelling molecular mechanism with inherent directionality is the duplicative transposition of mobile elements; yet this may often be opposed by selection (see reviews by Charlesworth and Langley 1989, and Charlesworth, Sniegowski and Stephan 1994). At present, there is considerable evidence of a role for transposable elements in generating mutation; but very little evidence of a role for them in causing the population-level spread of developmental novelties.

The claimed detection of 'directed' or selection-induced mutations in unicellular eukaryotes (Hall 1992) raises the question of whether it may occur in multicells and contribute to the evolution of development. Currently there is no evidence for this; and in general the idea of directed mutation remains controversial. All of the putative examples so far involve the apparent induction of specifically adaptive mutations by external agents such as nutrients. The equivalent, in a scenario of internal selection, would be the induction by one gene of mutational changes in another, interacting one, that resulted in improved compatibility between the two. If this sort of 'directed coevolution' were possible, it would be an evolutionary mechanism with astounding potential. However, there is currently no evidence for this, and it seems prudent to assume that no such mechanism operates.

# CHAPTER TEN

# Creation Versus Destruction

## 10.1 A Fourth 'Eternal Metaphor'?

We have now considered both mutation (Chapter 8) and selection (Chapter 9) from an evolution-of-development perspective. Clearly, both are important. Equally clearly, they are complementary: mutation gives rise to an altered developmental pathway and hence to a morphologically different individual, while selection governs the fate of such individuals in populations. Yet despite their obvious complementarity, the relative importance of the two processes – particularly in relation to their influences on the *direction* of evolution – has given rise to considerable controversy, which has dominated discussion of evolutionary mechanisms more than once over the past century. Let us now examine why.

If we contrast the present-day biosphere with its equivalent two or three billion years ago, what is the most striking difference? Undoubtedly, it is the existence now, but not then, of a diverse array of complex multicellular organisms, each internally coadapted as well as being adapted to its external environment. It is of course true that many lineages have *not* undergone a significant increase in visible complexity; both prokaryotic and eukaryotic unicells are still abundant and successful. It is also true that neither coadaptation nor adaptation are ever perfect for a variety of reasons, including 'constraints' and environmental variation. But such points to not detract from the fact that the explanation of "adaptive complexity" represents a central task for evolutionary theory (Maynard Smith 1972).

How strange it might seem, then, that mainstream neo-Darwinian theory has come to regard natural selection as the primary mechanism causing evolutionary change. Selection is a destructive force, which acts only to eliminate. Of course, positive directional selection differs from its negative (or 'purifying') counterpart in that it eliminates the old type, not the new. But even then, it does not create the new type in the first place. We have come to accept a theory of evolution that explains the origin and diversification of exquisitely-engineered organisms on the basis of the selective destruction of genetic/developmental variants whose initial production has been treated, for the most part, as a 'black box'.

Even the groups of evolutionary biologists who have given credence to agencies other than selection have fallen into the same trap of over-emphasizing destructive forces, as we saw in Chapter 1. In the early days of population genetics, Wright (1931) paid particular attention to genetic drift, and the power of drift to explain many aspects of molecular evolution has been elucidated by Kimura (1983). But, like selection, drift acts only to destroy. Many palaeontologists have drawn attention to the huge impact, in evolution, of both 'background' extinction rates and, superimposed upon these, occasional mass extinctions in each of which a substantial part of the world's biota was eliminated (see Stanley 1987). Nothing could be a more clearly destructive agency than extinction.

In some situations, however, selection acting in conjunction with other processes can act in a creative way. A good example is provided by Waddington's (1953) concept of genetic assimilation, where a phenotypic character initially produced only as a response to some environmental 'cue' becomes constitutive. Also, paradoxically, selection on those lineages that survive a mass extinction event (i.e. a combination of two destructive agencies) can lead to explosive radiations of novel types. But even in these cases, the mechanism that *produces* both the new genotypes and new phenotypes lies elsewhere – again, selection is merely modifying frequencies and eliminating 'old' and transitional variants.

As will by now be apparent, the theme that I am returning to here is the lopsided nature of mainstream evolutionary theory (Section 1.4), wherein selection is very thoroughly dealt with, while the developmental effects of mutations that contribute to evolution have been comparatively neglected. For the most part, it is simply assumed that unspecified mutations occur at random, and that these will provide whatever variation is necessary for selection to act upon.

How do mutations modify developmental pathways? There are many fundamental questions here, to which we need answers. For any particular mutation, we can ask: at what ontogenetic stage does it begin to modify the pattern of cell proliferation/differentiation?; how extensive are any such effects in developmental time and space?; how are they achieved, in terms of both gene-switching patterns and the types of gene product involved (adhesion molecules, morphogens, long-distance hormones, and so on)? In Chapter 8, I attempted a sort of 'developmental classification' of mutations, but this was very elementary and much remains to be done over the next couple of decades.

Why has this pronounced lopsidedness of evolutionary theory, with its emphasis on destructive forces, been allowed to develop? Or, to put it another way, why has the role of mutation in creating novel developmental trajectories contributed so little to the prevailing neo-Darwinian world view? There would appear to be at least two reasons for this unfortunate state of affairs.

First, as I have emphasized on several occasions, there has been a tendency to divide mutations into 'micro' and 'macro' on the basis of whether or not their developmental effects are individually recognizable, and to consider only the former to be involved in evolutionary change. For example, Dover *et al.* (1982) state: "Natural selection has no difficulty in increasing the frequencies of micromutations, leading to a gradual, small-step adaptation of a population. Mutations with larger phenotypic effects (macromutations) cannot enjoy the services of [positive] natural selection." This stance has its origins in the world view of the selectionists or biometricians as opposed to that of the mutationists (see Section 10.2). Clearly, if the 'continuum approach' adopted herein is accepted, then the need to give detailed consideration to the developmental effects of mutations becomes greater; and indeed it also becomes more possible, because in the case of some genes that contribute to evolution these effects can be directly studied.

Second, Darwinian theory triumphed, in the second part of the last century, over two alternative world views: that of 'special creation' (with 'special' meaning each species individually) and that of Lamarck, wherein organisms' own efforts (e.g. to acquire food) had a creative effect on the developmental pathways of their offspring. Perhaps we have all been unnecessarily resistant to explicit consideration of creative forces because of a residual concern that a creative approach (or 'generative paradigm': Goodwin 1984) may lead to a resurgence of such views.

Evolutionary theory should now be mature enough to disregard these old adversaries. Also, the recent advances in comparative developmental genetics (Chapter 7) provide just the right context for renewed interest in the evolutionary origins and genetic/mutational causality of new and 'successful' ontogenetic pathways. The time is ripe for a major drive to increase our understanding of the creative side of evolution, and to correct the above-mentioned lopsidedness in existing theory. The whole of this book is, of course, intended to be a contribution towards this drive.

In a much-referred-to work, Gould (1977b) drew attention to three "eternal metaphors" of palaeontology – essentially axes along which different theories occupied different positions. These were: internal/ environmental; gradual/punctuational and progressional/steady state. I have already drawn attention to the difference between Evolutionary Developmental Biology and conventional neo-Darwinism along the internal/external axis. However, the creation/destruction axis seems just as important and deep-rooted philosophically as the other three – perhaps even more so. Certainly, the degree to which mutation is able to contribute to the direction of evolutionary change has been a deeply contentious issue, as we will shortly see.

## 10.2 Mutationists versus Selectionists: a Protracted Debate

Starting around the beginning of this century, and continuing for several years, there was intense disagreement on the relative importance of mutation and selection in driving evolutionary change. This was linked to the question of whether selection usually acts upon the continuous phenotypic variation observed in all natural populations or on occasional bigger-effect mutations whose phenotypic consequences are such as to be individually recognizable and to put the organisms carrying them well outside the normal distribution of phenotypes characterizing the rest of the population.

The 'selectionists' such as Weldon (1901) and Poulton (1908) took the former view, and allied themselves with Darwin and Wallace (especially the latter, some of whose evolutionary essays (1870) are very strongly in favour of the ubiquity of imperceptible changes). The 'mutationists', such as Bateson (1894), de Vries (1910) and Gates

(1915) took the alternative view, and allied themselves with Mendel, since they saw evolution as being based on the same sort of discrete mutants that Mendel used to observe his famous ratios.

It is conventional to view the outcome of this debate in rather simplistic terms – namely, that the selectionists 'won'. After all, the current mainstream evolutionary theory is neo-Darwinism, whose roots lie firmly in the selectionist camp; and the views of latter-day mutationists, such as Goldschmidt (1940), were rejected. And it is true that some of the more extreme views of the mutationists were simply wrong. For example, we now know that continuous phenotypic variation is *not* wholly caused by direct environmental effects but rather by a combination of these and many small-effect gene differences among the individuals concerned (see Fisher 1918). Also, Bateson's (1894) assertion that "*the Discontinuity of Species results from the Discontinuity of Variation*" (italicized in the original, p.568) is incorrect, and the similar assertion made by de Vries that "The new species arises all at once" (1910, p.3) was based on an over-generalization from some special features of the genetic system of his study organism – the evening primrose *Oenothera lamarckiana*.

Another perspective on the debate is that *both* camps were wrong, and for two reasons. First, there is no clear distinction between mutants of the sort Mendel studied and the continuous variation emphasized by Darwin and Wallace. Every gradation exists between extreme 'macromutants' (such as homeotically transformed insects) and imperceptible differences in body length or other 'quantitative' characters. To put it another (and somewhat ironic) way, there is a continuum from 'continuous' to 'discontinuous'. Second, *neither* school of thought incorporated a 'directional' element into the nature of evolutionary change. Selectionists thought that virtually all speciation events at *every stage in evolution* involved only selection on continuous variation; while mutationists thought that virtually all such events, again *throughout evolutionary time*, were based on the occurrence of discrete mutant phenotypes. Thus while the view advanced herein (especially in Chapters 8, 9 and 11) is in one way intermediate between the selectionist and mutationist views, in another way it is fundamentally different from both (see Figure 10–1).

The widespread perception that the selectionists 'won' has had a very unfortunate side effect: the neglect of the creative role of mutation. Few biologists would deny that all morphological novelties arise ultimately from mutational changes in developmental programmes.

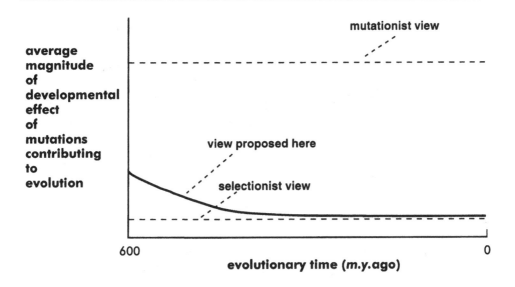

**Figure 10–1  Presence or absence of a directional trend in the kind of mutations contributing to the evolution of development under different theories**

The measurement of 'magnitude of effect' (vertical axis) is problematic, as noted at various stages in the book. The curvilinear pattern shown assumes that this magnitude is measured in a way that takes into account the size and complexity of the organism and the length of its developmental period. For example, we could use '*proportion* of cells whose developmental fate is altered'. If we used '*number* of cells' instead, then the curve might well be reversed, as cell number has on average increased over evolutionary time.

But if all such mutations whose effects are pronounced enough to study produce evolutionarily irrelevant 'monsters', and if all evolutionary change is driven by mutations which cause imperceptible deviations from the previously prevailing course of ontogeny, then it is easy to be pessimistic about both the importance of the creative side of evolution and our ability to elucidate it.

If the view of the evolution of development taken herein is correct, then this creativity becomes more important, but its elucidation gets only a little easier. If the main phase of evolutionary incorporation of mutations with non-negligible phenotypic effects was in the Vendian and Cambrian, we are hardly in a position to study it directly.

However, several less direct approaches are possible; these will be discussed in Chapter 12.

There is little doubt that the 'mutationists' were driven largely by a desire to elucidate the creative side of the evolutionary process. Commenting on Darwin's proposed evolutionary mechanism, natural selection, Bateson (1894, p.5) states that "this solution does not aim at being a complete solution like Lamarck's, for as to the *causes* of Variation it makes no suggestion." De Vries (1910, p.609) makes a similar distinction: "How the struggle for existence sifts is one question; how that which is sifted arose is another." And Morgan (1932, p.131) notes that "even without natural selection evolution might have taken place. What the theory does account for is the absence of many kinds of living things". Unfortunately, their interest in evolutionary creativity led the mutationists to over-emphasize the importance of big-effect mutations. We are now in a position to take a more enlightened view.

One major aspect of this overall debate concerns the question of what determines the *direction* of evolutionary change. Wallace (1870, p.290) represents one extreme view here: "Universal variability – small in amount but in every direction, ever fluctuating about a mean condition until made to advance in a given direction by 'selection', natural or artificial – is the simple basis for the indefinite modification of the forms of life." This idea of universally available variation in all directions, with selection being in sole charge the choice of one of these, can be traced through the work of the early population geneticists (e.g. Fisher 1930, chapter 1) and the ecological genetics school (e.g. Ford 1971) to present-day neo-Darwinians such as Dawkins (1986). (Ford concludes his text by stating that "if ever it could have been thought that mutation is important in the control of evolution, it is impossible to think so now".)

The opposite extreme is represented by Bateson (1894) and Goldschmidt (1940). Clearly, if transspecific evolution was based exclusively on the very rare occurrence of advantageous macromutations, then the direction of change would be primarily determined by the direction in which such a mutation led off from the previous ontogenetic programme.

Looking back at the debate from this perspective of the determinants of evolutionary direction, my conclusions are the same as before. The selectionists were much nearer to the truth in relation to the typical speciation event. But again, the whole debate is misleading, as there is no clear distinction between micro and macromutation; and both camps failed to recognize that the relative contributions of mutation and selection to the morphological direction taken by a lineage may have changed systematically over the course of evolutionary time.

Recent approaches to this issue have centred around the idea of quantifying 'morphospace'. I will examine such approaches in the following section.

## 10.3   The Structure of Morphospace

The use of $n$-dimensional hyperspace is familiar to students of both morphology ($n$ characters – morphospace) and ecology ($n$ environmental variables – Hutchinsonian niche space). In the case of morphospace, the characters can be very varied, and might include, for example: number of cells, number of cell types, form of symmetry, number and types of segments and appendages, and so on.

While any adult form is a single point under this conceptualization, a developing organism passes through a series of forms and so represents a trajectory of a particular length and shape. Evolutionary changes in morphology can be represented as departures from an initial trajectory (see Figure 10–2).

An important distinction is that between 'occupied' and 'occupiable' morphospace. The former – much the easier of the two to characterize – can be empirically defined by the cloud of points representing either all extant morphologies at a particular moment in phylogenetic history (e.g. the present) or all morphologies that have existed up to now (i.e. effectively an accumulation of all the 'instantaneous' hypervolumes from a starting point which for our present purposes should be the immediate unicellular ancestor of all animals (see Gould 1991)).

What constitutes occupiable morphospace is a much more difficult issue, because it depends on the frame of reference, that is, the assumptions we make about what is possible. Different sets of assumptions inevitably lead to different occupiable morphospaces. For example, we might specify that occupiable morphospace should encompass all morphologies that are possible given that they must be (a) composed of cells, (b) produced in the short term by a single-cell-start ontogenetic process and (c) produced in the long term through an evolutionary process driven by mutation and selection. With these specifications,

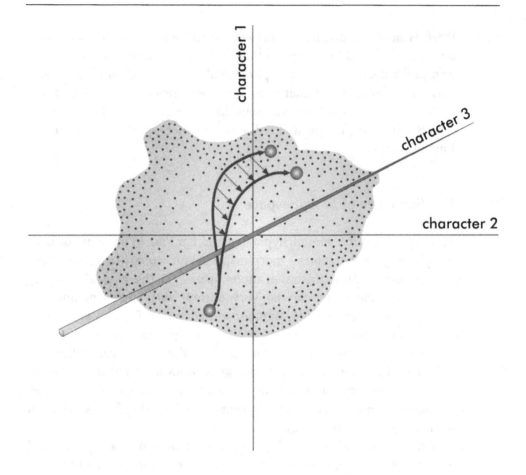

**Figure 10–2  Trajectories through animal morphospace**
LH curved line: ontogenetic trajectory of a typical individual of a given species (ignoring sexual dimorphism and other developmental/morphological polymorphisms). Dotted arrows: evolutionary shift in ontogenetic trajectory (resulting in RH ontogenetic trajectory); typically, these shifts take the form of divergences, but trajectories that are already divergent may evolve convergence. Number of characters reduced to three for ease of presentation.

occupiable morphospace will be much larger than that currently or previously occupied. Its additional 'volume' will include yet-to-be-produced morphologies of the evolutionary future and morphologies which were never produced because of the historical route that phylogeny took; itself a partly 'accidental' thing.

If we drop any of the three above specifications, then occupiable morphospace becomes larger still. For example, if the selection criterion is removed (or even suspended) there may be whole ranges of morphologies that could be reached – i.e. forms that are integrated and (co)adapted but to which there is no route except through morphological intermediates which are very unfit. If the cellular and developmental criteria are removed, then almost anything becomes possible, and there is no reason why there should not be 'kingdoms with wheels' (see Gould 1983).

Thus while for any point in evolutionary time there is a single hypervolume representing occupied morphospace (under either the 'instantaneous' or 'cumulative' approaches), there is a whole *range* of different hypervolumes representing occupiable morphospace under different assumptions.

An obvious question to ask at this stage is why occupied morphospace takes the form that it does. However, we need to be careful here, because this is really a way of asking how occupied and occupiable differ, and the question is clearly complicated by the range of different occupiable morphospaces available. We should avoid futile arguments about whether the occupied/occupiable difference is predominantly due to history or constraint, when in fact either may be true given an appropriate way of defining 'occupiable'.

I would like to make one specific suggestion about the structure of occupied morphospace. As the number of cells and cell types increases, the range of possible constructions becomes enormous, but the proportion of these that is 'prohibited' because of being either completely non-functional or partially so (and hence selected against) also becomes enormous. This is in contrast to the situation that applies when there is (say) just a single cell type and a small number (say 5–50) of cells. Here, the number of possible arrangements is relatively small – linear, branching, sheet, sphere, irregular mass, etc. – and most of these possible arrangements are actually found in one or more species. (Not all, though: for example, there are no precisely cubic animals.)

If we consider the vertical axis through morphospace to be 'cell number', then the above view leads to a picture of occupied morphospace as a series of upwardly pointing divergent 'fingers', each of which, for example, might represent a phylum (see Figure 10–3). It seems likely that part of each 'gap' is caused by history – that is, routes that were not taken but perhaps could be if the 'tape were replayed', to use Gould's (1989) analogy. But also, it seems likely that a part of each

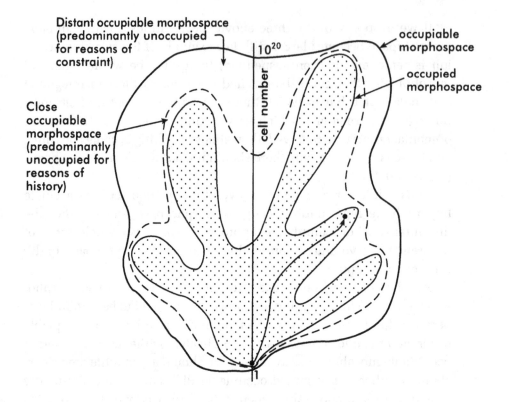

**Figure 10–3   A possible structure for animal morphospace**
Each 'finger' represents a different body plan. In the case of 'higher' animals, each individual traverses a considerable stretch of morphospace in the course of its development (arrow). Note that the vertical dimension is cell number. The horizontal 'character' is unidentified but is best thought of as a compound variable. The diagram is of course highly simplified: in reality we might envisage thirty-five 'fingers', each representing a different phylum.

gap represents morphologies that either cannot be made ontogenetically or are subject to negative internal selection regardless of the environment. Statistically, the regions of morphospace not invaded for historical regions might be expected to be closer to occupied morphospace than those which are unoccupied as a result of various kinds of constraint.

Since potentially occupiable morphospace is so hard to define, an alternative approach is to contrast some areas of occupied morphospace with others. In particular, we can ask why some areas are very densely populated, with many species representing variants on an oft-repeated theme, while others are much more sparsely populated, with few species and bigger gaps between them.

Goodwin (1994a) has taken this approach in relation to the phyllotactic (i.e. leaf-arrangement) patterns of higher plants. He notes that out of a quarter of a million species more than 80 percent have a spiral pattern; and he goes on to postulate (p.119) that "the frequency of the different phyllotactic patterns in nature may simply reflect the relative probabilities of the morphogenetic trajectories of the various forms and have little to do with natural selection."

This suggestion is important because it directs our attention to the 'missing link' in evolutionary theory noted earlier, namely the question of how mutations modify developmental programmes. In this case, we focus on the genes that set in place the morphogenetic 'self-organizing process' of phyllotaxis (Douady and Couder 1996). If a large proportion of all the possible mutations of these genes give rise to variant ontogenies *which still have a basic spiral arrangement*, then the material upon which selection acts has an intrinsic bias, and this may be reflected in the array of species found.

In one respect, I believe that Goodwin (1994a) is right: variation is not phenotypically random and perpetually available in all directions, as suggested by Wallace (1870). In another respect, however, I think Goodwin is wrong: natural selection does *not* have a negligible role. If 99 percent of available variations were spiral, but all were very unfit, and 1 percent were non-spiral but fit, then selection would override the mutational/morphogenetic bias.

In the end, we return to the conclusion we reached in the previous section: that overemphasizing the role of mutation *or* selection as a determinant of evolutionary direction is unhelpful. The probability of a lineage evolving in any particular morphological direction is a compound of two probabilities: the probability of mutations modifying developmental trajectories in that direction and the probability of those variants being selectively advantageous. Also, if density of occupation of morphospace is measured in terms of species numbers, then the probability of populations of variant ontogenies speciating needs to enter the picture too. We are still a long way from being able to put the three of these together in a form with sufficient explanatory power to solve the puzzle of why occupied morphospace takes the particular form that it does.

It might be thought (contrary to Figure 10–3) that animal morphospace consists of several disjunct hypervolumes, and that some sort of 'jump' is occasionally required to get from one to another. This view seems especially relevant in the case of Bateson's (1894) 'meristic' characters, such as the number of body segments. However, it must be recalled that morphospace is a whole-lifecycle developmental space, *not* an 'adults-only' finalized morphology space. So, no invisible jumps can occur; rather, altered ontogenetic trajectories are the only possible means of getting from one 'area' to another. And this fact serves to re-emphasize the lack of distinction between 'micro' and 'macro' changes. There is a continuum of possible ontogenetic divergence times and magnitude and direction of divergences. The continuity of morphospace even extends beyond animals to unicells and the other two great experiments in multicellularity: plants and fungi (see Figure 10–4). Generally speaking, if a taxon is monophyletic, and its constituent species have separated through ontogenetic divergence, then the morphospace for that taxon, whatever shape it takes, must be continuous. This is probably true for Animalia, for Eukaryota, and for terrestrial life in general.

Kauffman (1993) has postulated that occupied morphospace is located "at the edge of chaos". Inasmuch as evolutionary inflexibility of ontogeny (too ordered) and ontogenetic fragility (too chaotic) are likely to be disadvantageous, the location of occupied morphospace in between these extremes seems inevitable. However, if the views formulated herein are correct, ontogenies become more entrenched over evolutionary time, and so there is a gradual long-term shift away from the chaotic domain.

Finally, we should ask whether it is possible to progress beyond the rather general ideas about morphospace presented above to something more akin to a *quantification*. I suspect that Gould (1991) is correct in suggesting that this may be "logically intractable, not merely difficult" for Animalia, but that quantification may be possible for "creatures with comparable body plans and joint possession of sufficient homologies". And attempts have indeed been made to quantify morphospace for particular taxa: e.g. gastropods (Raup and Michelson 1965; see also Stone 1996). However, while such endeavours may help in elucidating patterns of morphological disparity within such taxa, they are of little help in furthering our understanding of the origins of the different body plans themselves.

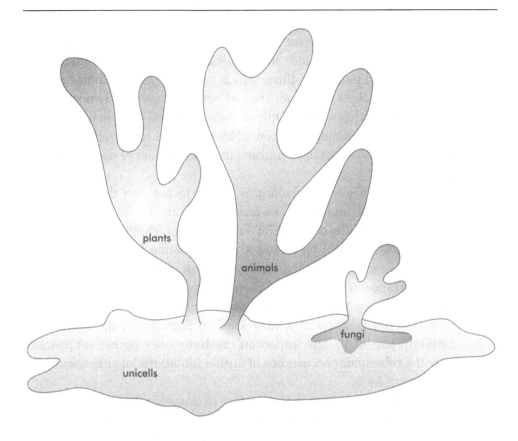

**Figure 10–4  Continuity of morphospace within a monophyletic taxon**
Note that animal, plant and fungal 'experiments' with multicellularity show no regions of overlap in all dimensions of morphospace. Although some animals have very plant-like growth forms, there are always some developmental/ morphological axes on which they separate.

## 10.4  Creation *and* Destruction

The contrasting of opposites is often useful in sharpening a debate; however, it can also be misleading. Although I feel that the creative role of mutation in shaping new ontogenetic trajectories has been comparatively neglected by evolutionary theory, and needs intensive study, the simple contrasting of mutation as a creative process and selection as a destructive one is an oversimplification.

Although mutations are generally creative, in that they produce novel DNA sequences and, in many cases, novel gene products and developmental pathways, those which cause the production of a truncated gene product and/or an aborted ontogeny are clearly destructive from a different viewpoint. Only mutations which are creative in both respects (i.e. producing both new DNA sequences and new/viable phenotypes) contribute to the evolutionary modification of developmental pathways.

The creative side of selection is harder to see than the destructive side of mutation, and requires a longer timescale. In the short term, the only possible results of selection – regardless of which stage of the life-cycle it acts at – are (a) changes in the relative frequencies of pre-existing variants and (b) the local extinction of one of them.

In the longer term, however, there is a creative aspect to this process of sweeping new mutations from low to high frequency; and it is more than just the creation of a new kind of population, though that is certainly part of it. The important creativity only becomes apparent after the subsequent occurrence of further mutations (at either the same or different loci).

Effectively, selection acts as a sort of mutation–accumulation system. If an initial mutation occurs and is selectively neutral, then the most probable outcome is that it will drift erratically through very low frequencies for several generations and then disappear again. If a second mutation occurs some generations after one which behaves in this way, then there are three possibilities: (a) the earlier mutation has already been lost; (b) the earlier mutation is still present, but the pattern of interbreeding is such that the two do not come together within an individual before one of them is lost; (c) they do by chance come together. The combination formed under (c) might conceivably be advantageous despite the fact that neither mutation is individually so, but the probability of (c) is rather low, and is a declining function of the length of period separating the occurrence of the two mutations.

If, on the other hand, the first mutation is advantageous and is spread through the population by positive directional selection, the subsequent mutation is much more likely to occur in an individual already carrying the first. Indeed, the probability of the two coming together within an individual is now an *increasing* function of their time of separation, rising to a maximum after fixation of the earlier mutation.

This 'accumulation system' will continue indefinitely, building up combinations of many mutations at many loci over long periods of evolutionary time. Thus Morgan's (1932) view that selection merely accounts for the *absence* of certain types of organism is too pessimistic. In the long term, selection results in combinations of mutations that might not otherwise occur. Selection is thus truly both creative and destructive.

There is, however, a twist in the tail. All of the mutations that selection accumulates are ones which were advantageous at the time of their occurrence and for some generations thereafter. So inasmuch as well-adapted and coadapted 'compound phenotypes' can be produced by accumulating individually fit mutations, selection provides an effective way of building these. But equally, it *cannot* build any beneficial compound phenotypes that might result from other kinds of mutational combinations. This connects with the point made in the previous section that there are likely to be areas of potentially occupiable morphospace that are never 'invaded' because of a lack of viable routes to them. In this context, the order of occurrence of different mutations may be very important (see Gillespie 1984; Barton 1989; Clarke, Shelton and Mani 1988; Mani and Clarke 1990).

As we have seen, when new areas of morphospace *are* invaded, the only way in which this can happen is through a new ontogenetic trajectory, with both mutational and selective processes underlying it. Statistical patterns in the production of such trajectories on both developmental and evolutionary timescales are thus of considerable interest. These can be examined through comparisons of developmental pathways between taxa of various ranks. Such comparisons provide the focus of the following chapter.

# CHAPTER ELEVEN

# Ontogeny and Phylogeny Revisited

## 11.1 Mapping the Two Hierarchies

Both ontogeny and phylogeny are, in certain respects, hierarchical processes. A theme running through the work of most contributors to what can now be described as evolutionary developmental biology is the relationship between these two hierarchies. This is true throughout the history of the subject (see Chapter 4) from the pioneering studies of von Baer (1828) and Haeckel (1866, 1896) to recent work, notably that of Thomson (1988) and Hall (1992). Goodwin (1994a, p.234) states: "Developmental processes are hierarchical. So are biological classification schemes. I shall follow a well-trodden route in attempting to relate them." And a possible *way* in which they may be related is given by Thomson (1988, p.92): "In principle we should be able to reconstruct for any species or any higher group a sequence of levels of morphological characteristics that define all the higher groups to which the taxon belongs, and to match these up with particular points in the hierarchy of morphogenesis." I would now like to extend Thomson's proposed relationship by superimposing on it, in the following paragraph, a series of further proposals. As we will shortly see, some of these are oversimplified or even incorrect. However, the aim for the moment is to build an extreme picture that will serve as a means to an end.

My 'extreme picture', then, is as follows. Within any overall ontogenetic trajectory, the developmental fates of different groups of proliferating cells diverge in a hierarchical manner. Early divergent fates lead to major differences such as contributing to anterior or posterior ends of the animal, while later divergent fates result in more subtle differences – such as contributing to one particular patch of wing tissue rather than another adjacent patch (see Wieschaus and Gehring 1976 and Section 1.7). The genes controlling these different developmental divergences are different; and so there is a hierarchy of 'early', 'middle' and 'late' genes operating over different periods of ontogenetic time. *Evolutionary* divergence in earlier acting genes occurred earlier in phylogenetic history and produced the major body plans which we recognize as characterizing different phyla; while evolutionary changes in later genes occurred subsequently, and gave rise to less major developmental changes which characterize taxa of lower rank nested hierarchically within phyla (see also Willis 1940). The picture of a neat inter-hierarchy correspondence that emerges has a certain beauty, but it is almost certainly wrong – or at least an over-simplification – for reasons that I discuss below.

I have deliberately chosen to paint a rather extreme picture in order to define one end of a spectrum of possible views. The other end of the spectrum is represented by the neo-Darwinian view, though there is a danger in painting this as homogeneous. Some neo-Darwinians explicitly deny inter-hierarchy correspondence. Recall, for example, the null hypothesis of 'no difference between the genetic changes underlying major and minor taxonomic bifurcations' that was derived from Coyne (1983) and discussed in Chapter 9. Others appear to be 'agnostic' in this respect, being simply more interested in short-term patterns of genetic change in populations than in long-term patterns of relationship between ontogeny and phylogeny.

My own view, which is influenced by both my neo-Darwinian training and my current advocacy of the nascent discipline of evolutionary developmental biology, is that there *is* a relationship between the developmental and evolutionary hierarchies, and that it *is* informative about the nature of evolutionary mechanisms. However, as will become apparent below, I think that the relationship is both *complex* in that it involves six hierarchies, not just two (see Table 11–1), and *messy* in that it takes the form of a statistical pattern (or several interconnected patterns) rather than a neat, clean correspondence.

In the next section, I will explain why we need to distinguish six different hierarchies. Before doing this, I should state what I mean by 'hierarchy', as there are many different versions of this concept (see, for

**Table 11-1. The six hierarchies**

| Realm | Type | Description | Kind of hierarchy |
|-------|------|-------------|-------------------|
| Ontogeny | Morphological | Sequence of forms, e.g. blastula → gastrula → neurula → pharyngula → adult | Linear |
| | Lineage | Branching pattern of cell lineage from zygote to adult | Divergent, exclusive |
| | Genetic | Cascade/network of interactions between developmental genes | Partial (i.e. partly hierarchical but with other kinds of organization such as feedback loops superimposed) |
| Phylogeny | Morphological | Groups-within-groups pattern based on phenetic analysis of morphological disparity | Divergent, inclusive |
| | Lineage | Branching pattern of phyletic lineages, as pictured in a cladogram | Divergent, exclusive |
| | Genetic | Genetic changes underlying successive lineage branching events (speciations) | Partial (and currently unclear: see text) |

example, Eldredge 1985; Salthe 1985; Panchen 1992). *Always* implicit in the term hierarchy is the idea of several levels – in some sense from 'high' to 'low' – in a definite ordered sequence. *Sometimes* superimposed upon this is the idea that any single higher category includes several lower ones. This is true, for example, of the taxonomic hierarchy of 'groups within groups', which can thus be described as a divergent, inclusive hierarchy (see Panchen 1992). A variant on this is represented by regimental structure within an army, where several privates report to one NCO and so on up to General, but several privates do not collectively 'form' an NCO, in the way that several species collectively form a genus. Such a hierarchy can be described as divergent but exclusive. In the case of the genes controlling development

within a particular organism: if there is indeed a series of 'levels' –
however messy – of early, mid and late acting genes (see below), then
the system is hierarchical in at least one sense. Whether it is also
divergent, in that one earlier gene typically controls several later
ones remains to be seen. (I have argued previously that this is usually
the case (Arthur 1984, 1988) but I am now less convinced of the generality of such a pattern.)

We should not expect this endeavour to be easy. It has taxed some
of the best minds in the biological sciences for at least two centuries –
starting with the German *Naturphilosophen* (see Chapter 4). However,
recent advances in systematics and developmental biology – respectively the advent of cladistics and the upsurge of data on comparative
developmental genetics – shed new light, and enable the problem to be
approached from a somewhat different perspective.

## 11.2    From Two Hierarchies to Six

An excellent account of the rationale underlying the cladistic approach
to evolution and taxonomy is given in the first chapter of Hennig
(1981). In this account, Hennig brings into sharp focus the potential,
and in some cases actual, non-correspondence between genealogical
and morphological change in evolution. Lineages that separated earlier
in phylogeny might be expected, typically, to be represented today by
species whose morphological disparity is greater than that between
species whose lineages separated more recently. But as Hennig pointed
out, this is not *necessarily* the case; and there are many examples of a
lack of morphological/genealogical correspondence. Perhaps the most
famous of these is the pronounced morphological restructuring associated with the origin of the birds, despite Aves being a relatively
recently diverged clade within the tetrapods (see Chiappe 1995).

At the level of major animal body plans there *is* a good correspondence between genealogy and morphology. As we saw in earlier
chapters, the body plans underlying the thirty-five or so animal
phyla all originated deep within the animal clade, at least 500 *my*
ago, and perhaps much earlier. In plants, the correspondence is less
good; I drew attention in Chapter 1 to the rather late appearance in
phylogeny (~130 *my* ago) of the angiosperm body plan. But regardless
of whether in any particular case there is or is not a clear correspondence between genealogy and morphology, the fact that there *need not*

*be one* forces us to distinguish them. Such a distinction caused Hennig to opt for a clear focus on one side of the divide, and the rest, as they say, is history; notwithstanding its 'transformation', cladistics is firmly based on the rigorous analysis of genealogy. Such a purist stance was understandable, given the realization that conventional taxa were based on an uncomfortable and ill-thought-out mixture of genealogical and morphological criteria. However, from the perspective of evolutionary developmental biology, *both* hierarchies are of interest; as are (a) cases of good correspondence between them (e.g. the origin of animal body plans) and (b) cases of marked mismatch (which suggest recent evolutionary change involving genes with an important role in controlling development). Also, in relating evolution to development we need to make clear whether on the evolutionary side we are referring to genealogy, morphology or both.

Although it has not featured prominently in the literature (but see Wray and McClay 1989), a broadly parallel distinction – morphological versus genealogical – can be made within the realm of the ontogeny of an individual. Any multicellular organism is produced through a branching pattern of cell lineages. (I will assume simple life histories for now; discussion of larvae will be deferred to Section 11.4.) Such a branching pattern is, of course, a divergent genealogical hierarchy of cells. Like its phylogenetic equivalent, it may or may not have a simple relationship with its corresponding hierarchy of morphological change.

Strangely, we find a broadly parallel pattern of correspondence and non-correspondence within the ontogenetic realm and within its phylogenetic equivalent. There is a period of intense morphogenetic creativity associated with early branching of the cell lineage hierarchy extending from cleavage and gastrulation up to early organogenesis. This has a broad parallel with early body-plan creativity associated with early genealogical branching of phyletic lineages. But equally, there are exceptions: pronounced morphological change can sometimes occur relatively late in ontogeny, as in the case of metamorphosis in holometabolous insects; this constitutes the embryological equivalent of the evolutionary non-correspondence that we saw exemplified by birds.

So far, by distinguishing between morphology and genealogy within both ontogenetic and phylogenetic realms, we have gone from two hierarchies to four (see Table 11–1). The remaining two emerge when we focus our attention on genes, and in particular on (a) the hierarchy of genes controlling the development of an individual organism and (b) the hierarchy of genetic changes underlying successive phylogenetic divergences from the Neoproterozoic to the present. It is at this point that we need to question whether a hierarchy (of any kind) is still an appropriate model. *This question lies at the heart of differences of opinion about the overall relationship between development and evolution.* I hope that the remainder of this chapter will explain why I believe that this question is so crucial.

Let us consider the genetic control of individual ontogeny first; and again it is useful to proceed by considering alternative 'extreme' views to identify the spectrum of possibilities. My own deliberately extreme picture, which was based on a statement by Thomson 1988 (see Section 11.1), incorporated the (incorrect) idea that each temporal phase in development involves its own unique set of genes controlling the pattern of cell proliferation (and differentiation). This is pictured, in simplified form, in Figure 11–1a. The alternative extreme view is illustrated by Figure 11–1b. Here, there is no overall correspondence between genes and developmental stages. Rather, each gene that is involved in the control of ontogeny has its own characteristic period of expression, but in many cases, these periods are either long or intermittent, thus occurring through a range of developmental stages. Under this view, there is *no* 'genetic hierarchy' underlying the cellular hierarchy of ontogeny.

I will now propose – and it is merely a hypothesis at this stage – that the true pattern lies somewhere in between these two extremes, and that it can be characterized, albeit in simplified form, as shown in Figure 11–1c. According to this view, there *is* a genetic hierarchy underlying development, but it is a messy one with a good deal of 'noise'. Some genes are exclusively 'early', others 'late', but many others have intermittent and/or prolonged expression.

The evidence, so far, eliminates one extreme view (Figure 11–1a) but not the other. We saw in Chapters 5–7 that some developmental genes are re-used at various stages in ontogeny. Increasingly, this looks like a commonplace occurrence (see Salser and Kenyon 1996). It is now clear that some of the variation in quantitative traits that appear late in development is due to segregation of alleles at loci that are already known to have a role in early developmental processes. For example, the *achaete-scute* complex, which is involved in the development of the embryonic central nervous system in *Drosophila,* is also involved in minor continuous variation in adult bristle number (Mackay and Langley 1990; Mackay 1996).

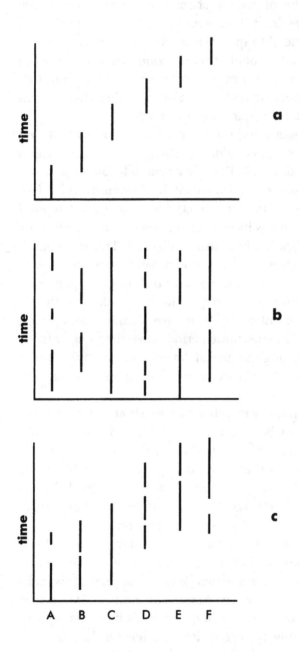

**Figure 11–1 Three views of the genetic control of ontogeny:
(a) strict hierarchy; (b) no hierarchy; (c) intermediate view**
A–F represent six separate genetic loci; bar indicates period of
expression.

The other extreme view (Figure 11–1b) cannot be eliminated with certainty at this stage. If we think of development in terms of localized cellular interactions, it seems entirely possible that at every stage in development where a particular phenomenon recurs, the *same* genes are involved. For example, we saw in Chapter 5 that cadherin genes stop being expressed prior to cell movements; and since cadherins form a sort of 'intercellular velcro', this makes sense. The switching off of these genes (and the switching on of others) may well occur prior to cell movement regardless of whether it occurs at gastrulation or during a very late stage of ontogeny. The same may apply to other genes controlling other cellular processes.

The position, then, is clear. Inasmuch as the 'messy hierarchy' proposed here (Figure 11–1c) is a compound of a time trend (Figure 11–1a) and 'noise' (Figure 11–1b), there is much evidence for the noise, but so far rather less evidence for the time trend. Indeed, the *nature* of the evidence that is needed to confirm the existence of a time trend is problematic in that it is essentially negative. We need to be certain that *some* early genes have *no* late expression and vice versa. I will, however, proceed on the assumption that such evidence will eventually be forthcoming. This is simply an act of faith: I cannot believe that such a directional process as ontogeny is underlain by a genetic architecture that has *no* hierarchical component.

The question of whether the genetic changes underlying the *evolution* of development also have a hierarchical component is even more difficult to answer. For example, we might ask whether genes with an early role in ontogeny were more involved in early (e.g. Vendian/Cambrian) evolutionary divergences than in recent ones; but the question is complicated by the fact that what constitutes an early acting gene may itself be subject to evolutionary change. Such changes can be brought about either by evolutionary changes in a gene's period of expression within a fixed ontogenetic period, or by changes in the length of the ontogenetic period itself. The former category of changes includes those accompanying a gene's cooption for a new developmental role. Roth (1988) has given the name 'genetic piracy' to this phenomenon; it is possible that the *bicoid* gene in *Drosophila* (Chapter 6) represents an example.

Despite such complicating factors, some aspects of the genetic basis of phylogenetic divergence do appear to differ between early and recent evolution. For example, we saw in Chapter 7 that the earliest stages in the proliferation of animal body plans must have involved duplication and divergence of the *Hox* genes, which have a controlling role in early development. Such processes must have characterized some Neoproterozoic speciation events which, following further divergence, led to today's higher taxa. In contrast, present-day speciation events typically do not include genetic changes of this kind. However, genetic divergence of a more routine allelic sort (and duplications at other kinds of loci) must have characterized speciation events at *all* stages of animal evolution. So again, there is a time trend, but it is only part of the overall picture.

As discussed above, and illustrated in Table 11–1, there are *six* hierarchies in a (3 × 2) pattern – (morphological, genealogical, genetic) × (ontogeny, phylogeny) – even if the genetic ones are partially obscured by other, non-hierarchical patterns and/or 'noise'. It thus becomes clear that the question 'what is the relationship between ontogeny and phylogeny?' cannot be expected to have a simple answer. There is a parallel here with the often-asked ecological question 'what is the relationship between complexity and stability?' Pimm (1984) showed that both 'complexity' and 'stability' are compound variables with several components, and that it is thus necessary to ask more specific questions. I hope I have succeeded in making the same point about the relationship between ontogeny and phylogeny.

## 11.3   An Important General Pattern

Of the many more specific questions that could be asked, I will now focus on one: did the origin of animal body plans in the Neoproterozoic and/or Cambrian (and the subsequent proliferation of *Unterbaupläne*) involve developmental changes that were distributed differently through ontogeny than those characterizing present-day speciation events? This deceptively simple question has a rather complex answer. I will approach it from a von Baerian starting point, but the sequence of reasoning that we will follow will ultimately lead (in Section 11.4) to a rejection of von Baer's laws – or perhaps more accurately to a pronounced taxonomic restriction in their applicability.

Let us start, then, by considering the classic example of the initially similar but progressively diverging, embryonic trajectories of different vertebrates, as were shown in Figure 2–7. We will ignore the early diversity made explicit in the 'egg-timer' concept (see Duboule

1994), and concentrate on the pharyngula-to-adult phase of the life cycle, where clear von Baerian divergence can be seen. (The explanation of this may lie in the simple asymmetry of causality in development, whereby some early changes alter later stages too, while no late changes – except maternal effects – alter early stages; but greater constraint of earlier-acting genes is probably also responsible: see Morgan 1932; Arthur 1988.)

This classic example of embryonic divergence is based on a particular overall taxonomic group (subphylum Vertebrata), and within that involves comparisons most of which are between different classes: usually Pisces, Aves and Mammalia. (Note (a) the use of 'traditional' non-cladistic classes here; also (b) the omission of Anura, where the picture is complicated by metamorphosis.)

We will now investigate what happens when we make taxonomic rank a variable and compare (1) more distantly-related embryos (from different phyla) and (2) much more closely-related embryos (for example from congeneric species or confamilial genera).

1. It is clear that phyla typically diverge from a very early developmental stage. Depending on which phyla are compared, divergence in the pattern of cell proliferation either begins immediately with different patterns of cleavage, or, where cleavage patterns are shared (e.g. between chordates and echinoderms) then immediately afterwards with different patterns of gastrulation, neurulation and so on. There is no between-phylum equivalent of a shared within-phylum pharyngula (vertebrates), germband (insects) or rudiment (echinoderms) stage, although among phyla with complex life histories there are a few cases of shared larvae (e.g. the trochophore that is found in many marine molluscs and annelids).

2. In contrast, the developmental trajectories of congeneric and even confamilial vertebrates are typically very similar until quite late in ontogeny. These thus exhibit a von Baerian pattern, but with divergence usually occurring long after the pharyngula stage. This would be the pattern found, for example, in comparisons of congeneric chimpanzees (*Pan troglodytes* and *P. paniscus*) or titmice (*Parus caeruleus* and *P. ater*). This pattern is also likely to hold in congeneric comparisons within invertebrate taxa with simple life histories.

I now want to perform a *gedanken* experiment. Suppose we were able to compare the overt morphology of all the developmental stages in many cross-congener comparisons, and then in many cross-class and cross-phylum comparisons, in each case noting the point in ontogenetic time at which overt divergence begins to be readily apparent. We then plot the distributions of these divergence times and examine their locations and shapes. (Convergences will be considered later.)

While such an exercise has never been undertaken on a sufficiently large scale, particularly in relation to acquiring an adequate sample size and broad taxonomic distribution of congener comparisons, there is sufficient accumulated information on 'comparative descriptive embryology' to allow us to have a rough idea of the shapes of the distributions (see Figure 11–2). The clear contrast in location (and shape) between the 'congeneric' and 'higher taxon' curves has important implications for evolutionary mechanisms, and these will be explored below. However, it is necessary first to discuss some complications and exceptions to the pattern shown in Figure 11–2.

Even in cases of direct development, examples are known where reasonably closely related species of quite recent evolutionary lineage divergence diverge very early in their ontogenies. For example, the onychophorans *Opisthopatus* and *Peripatopsis*, whose lineages appear (on the basis of allozyme data) to have separated only some 30 *my* ago, diverge developmentally from the gastrula stage onwards (with the latter genus having the more typical onychophoran pattern: Walker 1992, 1995). Such cases are unusual, but occur sporadically within many higher taxa. They give rise to the long tail of the skewed distribution of 'congeners' (although in the example given the taxonomic distance is a little greater).

Although it is not shown in Figure 11–2, there is probably a 'blip' in the congener curve at the beginning of developmental time, reflecting the known variability of the very earliest stages. Examples include different egg morphologies (e.g. in *Coelopa* seaweed flies: Phillips *et al.* 1995) and the modification of cleavage patterns by different amounts of yolk. This variability is incorporated into the egg-timer concept, as noted above.

In some cases, there is pronounced variability in one or more 'intermediate' developmental stages (anywhere between gastrula and subadult) which is *not* reflected in a diversity of adults (i.e. in some sense developmental trajectories diverge and then converge again). This is most pronounced in the case of complex life-histories. These deserve full and separate consideration, not least because they represent the *majority* of the animal kingdom, notwithstanding our inevit-

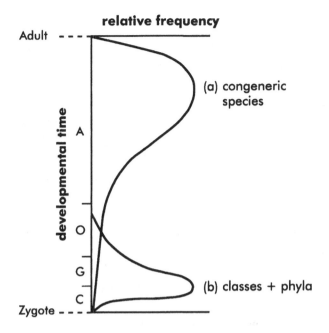

Figure 11–2  **Hypothetical distribution curves for developmental divergence times in comparisons of (a) congeners and (b) higher taxa**

C, cleavage; G, gastrulation; O, organogenesis; A, allometric growth.

able familiarity with direct development, given its occurrence in many of our 'model systems' (chick, mouse, *Caenorhabditis*) and indeed in humans. I will consider complex life histories in the following section.

One problem in producing the 'higher taxon curve' in Figure 11–2 is that *within* each phylum or class there is considerable variability in developmental trajectory. Thus whether we are comparing, for example, annelid with mollusc or bird with mammal, the point in ontogenetic time at which these higher taxa appear to diverge is affected by our choice of a particular 'representative' species of each. However, the curve illustrated is probably quite robust. For example, *no* choice of mollusc or annelid will result in a late ontogenetic separation. (Congener comparisons are less robust in the sense that some choices of species to be compared will give rise to an early ontogenetic separation – hence the pronounced skew of the congener curve, as noted above.)

Another problem that affects the production of *both* curves is that of alterations in ontogenetic period. There can be substantial changes in the zygote-to-adult timespan across both large and small taxonomic distances; and indeed there can be substantial within-species variation in many cases as a result of exposure to different environmental conditions. For this reason, no meaningful pattern would have emerged in Figure 11–2 had the time axis been calibrated in absolute units such as days or years. Division into common developmental stages allows what might be called a proportionate view of developmental time, which seems intuitively a more reasonable approach.

Finally, it is important to recognize that the overt morphology of various developmental stages, which is what we are dealing with here, does not necessarily connect in a simple way with underlying cell lineages or with the spatiotemporal expression patterns of developmental genes. Macroscopic morphological similarity can be underlain by microscopic cellular and genetic differences, as noted in relation to homology by Roth (1988) and Panchen (1994). This brings us back to the distinction between the different hierarchies: morphological, genealogical and genetic hierarchies *may* correspond in ontogeny (or in evolution), but they often do not, and so we should be reluctant to assume such correspondence.

Notwithstanding these various complications and exceptions, or indeed the *gedanken* nature of this whole exercise, the contrasting curves of Figure 11–2, or at least patterns not too dissimilar from them, apply widely through most direct-developing taxa and probably to selected life stages within most complex life histories (see next section). It is therefore important to consider possible explanations, and the implications these may have for evolutionary mechanisms. There are three main possibilities, as follows:

**1.** The pattern is retrospective; it does not reflect the nature of speciation events in early evolution.

This possibility is best considered by reference to Figure 11–3. In a comparison between present-day representatives (A, C) of two higher taxa (X, Y), any observed differences between A and C may have arisen during the early speciation event that produced species E and F; or may be due to subsequent changes in one or both lineages. Sometimes, the pattern of taxonomic distribution among many present-day species is clear enough to favour one or other of these possibilities, but often the pattern is not readily diagnostic.

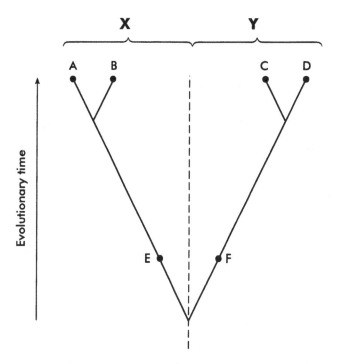

Figure 11–3  Evolutionary tree with four present-day species
(A–D) belonging to two higher taxa (X, Y) whose initial divergence
had its origin in an ancient speciation event (separating species E
and F)

A pattern of early ontogenetic divergences between higher taxa but
late ontogenetic divergences between congeners could arise
'retrospectively' by (a) increasing complexity occurring in the diverging clades through a lengthening of the ontogenetic sequence by
'terminal additions' and/or by (b) heterochronic processes (see
McKinney and McNamara 1991) pushing certain developmental
events back into earlier ontogenetic stages.

2. The pattern is caused by the nature of early speciation events:
   higher taxa were originated by a highly non-random sample of the
   kinds of speciation events that can still be found occurring today – a
   sample in which early ontogenetic divergences prevail (i.e. those in
   the tail of the congener curve in Figure 11–2).

In this case (unlike **1** above) 'deep' ontogenetic divergence *does* characterize species E and F. It is not brought about subsequently by term-

inal addition and/or heterochrony, though of course it may be enhanced by those processes. However, under this view, early speciations through which body plans arose are different to recent and current speciations in only a statistical, not an absolute, sense.

**3.** The pattern is caused by the nature of early speciation events: those early speciations through which body plans arose were fundamentally different from any speciations occurring in recent evolutionary history.

If body plans arose through experimental probing of morphospace by 'radical' mutations, and were enhanced and stabilized by internal selection (as discussed in Chapter 9), then some key early speciation events may have been fundamentally different from any that we can observe today.

Although **1–3** above have been presented as sharply-contrasting possibilities, there may be an element of each contributing to the overall pattern shown in Figure 11–2. Undoubtedly, the suites of embryological/morphological homologies that we recognize as body plans did not arise in single speciation events. Phylum-level body plans probably arose through many such events, perhaps tightly packed together in relatively short periods of geological time in one or more early creative phases of evolution in the Neoproterozoic and/or Cambrian characterized by uncluttered niche space and relatively unconstrained ontogenies. *Unterbaupläne*, such as those underlying the traditional vertebrate classes, arose later, and again involved a succession of changes, not just one. In most lineages, additions (both terminal and intercalated) and heterochronies have been occurring intermittently ever since.

It seems likely that the embryological differences between vertebrate classes with which we started, and their counterparts within other phyla, are to be explained by a combination of **1** and **2** above: atypical but not unique speciation events and subsequent additions and heterochronies. I would hypothesize (as in Chapter 9) that the origins of the phylum-level body plans also involved an input from **3** above (i.e. a distinct kind of speciation event). But whether these suggested combinations are correct, and if so then what the magnitudes of the different contributions are, remains to be resolved. This is a difficult if not intractable problem; but its fundamental scientific importance is hard to overemphasize.

## 11.4    Larval Forms and Complex Life Histories

As noted earlier, complex life histories are the rule rather than the exception within the animal kingdom (see Conn 1991). Any proposed generalization on the evolution of development that does not take them into account will ultimately fail. That is essentially the case with von Baer's laws, as will become clear below. Arguably, no single pattern in comparative embryology is sufficiently general to be elevated to the status of a law.

At the outset, we should try to be clear about what constitutes both a 'complex life cycle' and a 'larva'. There is no universally accepted definition of either. It is interesting to compare the suggestions made by Istock (1967) and Wilbur (1980). The former defines complex life cycles in *ecological* terms: two or more distinct niches or habitats over the life cycle, especially in relation to factors limiting population abundance. In contrast, Wilbur (1980) uses a *morphological* criterion, and states that a complex life cycle exists when there is "an abrupt onto-genetic change in an individual's morphology". McEdward and Janies (1993) take a comparable view to Wilbur (1980), and consider a larva to be a morphological stage that "is eliminated by the metamorphic transition to the juvenile".

Although the two approaches – morphological and ecological – will often go together, they need not necessarily do so. I will adopt Wilbur's (1980) view here, partly because it connects better with the idea of a larva (as opposed to a juvenile) being a morphologically distinct entity, quite unlike the adult. Of course, there are degrees of 'unlikeness' and thus we cannot in the end cleanly separate larvae from juveniles. Lepidopteran caterpillars, polychaete trochophores and anuran tadpoles are all clearly larvae, while young birds or mammals are not. An intermediate situation prevails in hemimetabolous insects. The so-called larvae of some teleost fish (see Young 1962) are probably better regarded as juveniles.

Both taxonomic and ecological patterns are found in (a) the occurrence of larvae/ complex life histories (as opposed to direct development); and (b) the occurrence of different *types* of larvae. Typically, marine animals with sedentary or limited-range adults have a dispersive planktonic larval stage (see Figure 11–4), some of which (planktotrophic) feed, while others (lecithotrophic) are sustained by a supply of yolk. Highly mobile marine animals typically do not have a planktonic larva – for example cephalopods and fish. The small,

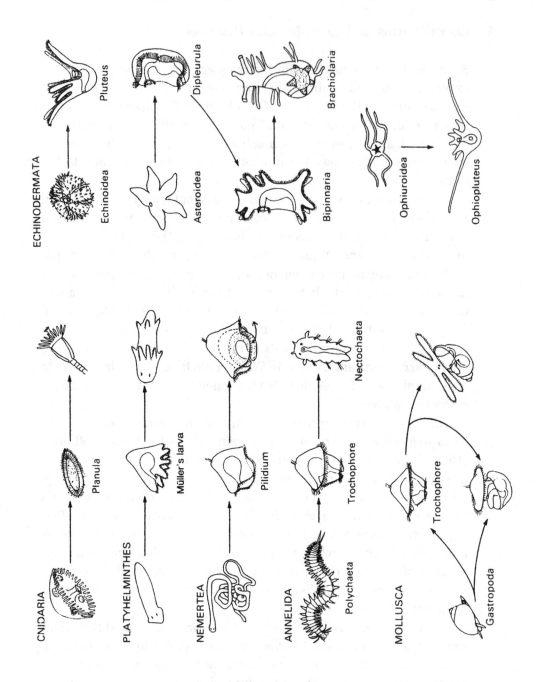

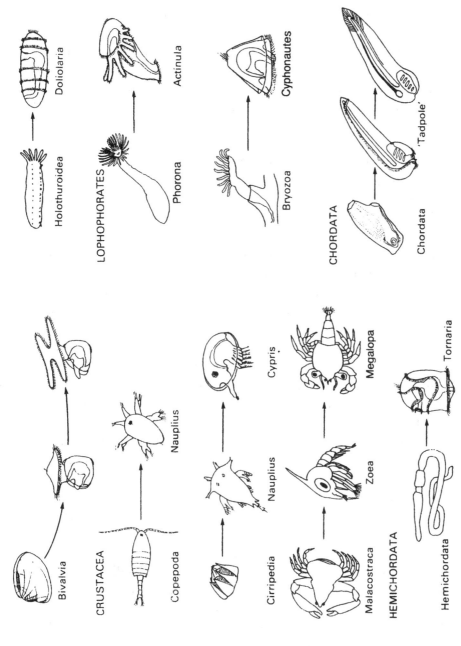

**Figure 11–4  Examples of the characteristic larvae of the major groups of marine invertebrates**
The sketches show the adult forms and the corresponding larvae. Note that the adults are drawn to a much smaller scale than the larvae. From Barnes *et al.* 1988. *The Invertebrates: A New Synthesis.* Blackwell, Oxford, with permission.

fragile, early-stage larvae found in originally marine groups such as molluscs and annelids tend to be dispensed with in favour of direct development in those clades stemming from invasion of freshwater and/or land, and there are obvious selective reasons for this.

Some marine larvae occur widely (but not universally) across a phylum, and indeed some are shared between phyla. The use of these larvae to infer a genealogical relationship between the phyla concerned is well known, particularly in the case of annelids and molluscs (trocophore), and chordates and echinoderms (dipleurula). (A recent study by Young *et al.* (1996) uses the same rationale to investigate the phylogenetic position of the enigmatic taxon Vestimentifera.) While the possibility of convergent 'invention' of these relatively simple forms means that this kind of homology inference on its own cannot be conclusive (see Chapter 7; also Willmer 1990), there seems little doubt that many of these marine larval morphologies have very ancient origins and represent a primitive life cycle that has subsequently been modified in some descendant groups (Strathmann 1985).

In groups that have left the marine environment, larval stages are again generally associated with the exploitation of quite distinct habitats/niches. Obvious examples include freshwater/land (frogs), freshwater/air (mayflies), land/air (butterflies) and primary/secondary hosts (parasitic Platyhelminthes). In these cases it is usually the adult rather than the larva that constitutes the dispersive phase of the life cycle. This has implications for the relative lengths of life-history phases: on average the relative length of the larval phase is greater in terrestrial than in marine animals.

Despite the existence of these various broad-scale taxonomic and ecological patterns, there are many cases of particular species or genera which show a marked variation from their close relatives. A good example is the anuran genus *Cornufer*, which exhibits direct development (Dent 1968). Such cases of unusual ontogenies tend to be related to the ecology of the species/genus concerned.

It is sometimes the case that a von Baerian pattern of early similarity giving way to later divergence applies *within* a particular lifestage of a complex life cycle. This is generally true, for example, in the embryo-to-larva development of Diptera. However, in other cases, a pattern of convergence rather than divergence can be found – rather like the 'base of the egg timer' that applies to the egg-to-pharyngula phase of vertebrate development. A good example is provided by com-

*Heliocidaris tuberculata*

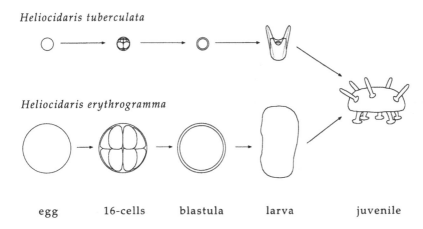

*Heliocidaris erythrogramma*

| egg | 16-cells | blastula | larva | juvenile |

**Figure 11–5   Developmental convergence in echinoid species**
The larva of *H. tuberculata* is planktotrophic while that of *H. erythrogramma* is lecithotrophic. The scale is the same for upper and lower embryos: that is, early stages of *H. tuberculata* are much smaller than those of *H. erythrogramma*. From Wray and Raff 1991. *TREE*, **6**, 45–50, with permission from Elsevier Science Ltd.

parison of the early ontogeny of two congeneric echinoids (Wray and Raff 1991); this is illustrated in Figure 11–5.

This is an appropriate point at which to recall the fact that both ontogeny and phylogeny consist of three hierarchies, not just one: morphological, genealogical and genetic. Within ontogeny, morphological similarity between two early-stage embryos does not necessarily imply similarity of cell lineages or the spatiotemporal expression patterns of developmental genes (see Wray and McClay 1989; Wray 1994). Since the different hierarchies may behave differently, this may affect the generality of any particular pattern, such as von Baerian divergence.

Let us now focus again on that particular pattern and attempt to delineate its 'realm' – that is, the set of conditions under which it is found. The primary example (Figure 2–7) applies to (a) vertebrates, (b) explicit morphology and (c) post-pharyngula development. No comparative developmental studies on other groups have provided equally clear and impressive examples; but in *all* other instances where a broadly similar pattern is found, there must again be restrictions of these three kinds: in terms of taxonomy, hierarchy and ontogenetic period. Thus von Baerian divergence is a taxonomic statement (in the sense of Panchen 1992), *not* a generally-applicable 'law'.

Inasmuch as there is a genetic hierarchy (recall Figure 11–1), it may well be that this is where a generally von Baerian pattern has the broadest scope. Comparative studies of *Hox* genes (Chapter 7) suggest that there may be a difference in the degree of evolutionary conservation of development-controlling genes which are primarily expressed in different ontogenetic stages: the earlier the initial expression, the higher the degree of conservation. It seems ironic that von Baer may have been in a sense more broadly correct about genes (of whose existence he was unaware) than about morphology (of which he was one of the leading students of his time).

Considering overall life cycles, why is the von Baerian pattern so much more pronounced in vertebrates than elsewhere? It seems likely that the high degree of embryo protection typical of vertebrates (and especially amniotes) considerably reduces the strength of external selection pressures compared with, for example, the case of a free-living juvenile stage of a marine invertebrate. Although positive internal selection may be involved in the origin of phylum-level body plans (Chapter 9), subsequent internal selection is likely to be negative (i.e. to act against departures from the status quo); in other words, it takes the form of a 'constraint'. If there is strong positive external selection for a new variant of an early developmental stage, the constraining influence of negative internal selection may be overcome; but if there is not, then there is no reason why it should be. Thus the greater the level of protection, the greater the conformity to a von Baerian pattern. This is clearly a testable prediction for the many groups whose comparative embryology is not yet well described. (This overall view is similar to that of Wray and Raff (1991) who suggest that "selective rather than developmental constraints govern changes in developmental strategy"; but recall Section 2.5 where the difficulty in distinguishing these two kinds of constraint was discussed.)

Where external selection 'wins' and elaborate larval forms evolve, these may well provide a poor morphological basis from which to develop an adult; but selection can weaken the explicit morphological link between larva and adult. One way to achieve this is by extensive loss of larval structures through cell death, and construction of the adult from relatively few small pieces of larval tissue, such as insect imaginal discs or the echinoderm rudiment. There appears, however, to be no direct equivalent of this 'separation' in the genetic domain: larvae and adults share a common genome, and, as we saw in Chapter 7, there is much reuse of developmental genes between lifestages. The wider applicability of a von Baerian pattern to genes than to morphology noted above is clearly compatible with this observation.

What we need, to make further progress in this area, is a comprehensive body of life history theory applied to complex life cycles (with direct development as just a special case). As Stearns (1992) points out, however, most existing quantitative life history theory is restricted to simple life cycles; though the studies by Istock (1967), Werner (1988) and Ebenman (1992) represent exceptions to this generalization (see also review by Wilbur 1980). In general, the interface between life history theory and evolutionary developmental biology is likely to be fertile ground for future advances in our overall understanding of the evolution of developmental systems.

## 11.5  Phenotypic Complexity and Evolutionary 'Explosions'

So far, we have seen examples of both von Baerian divergence, and its opposite, developmental convergence. However, both these patterns emerged from comparison of organisms whose level of phenotypic complexity was broadly similar. A third comparative embryological pattern emerges when we bring evolutionary increase in complexity explicitly into the picture. This pattern, as we will see, is generally 'Haeckelian', though it falls some way short of strict recapitulation.

We briefly considered possible ways of measuring phenotypic complexity in Chapter 7. Here, I will consider phenotypic complexity to be measureable by the number of different cell types an organism is made up of. In cases of complex life cycles, it seems appropriate to apply this criterion to whichever life stage has the most cell types. This is usually, but not always, the adult.

Throughout animal evolutionary history, maximum and median complexities have increased, while the minimum has remained at the unicellular level. Thus the *range* of complexities has massively expanded. However, this expansion over time has been far from linear. Marked elevations of maximum complexity occurred (a) with the appearance of the first eukaryotic cells, (b) with the advent of multi-cellularity, and (c) accompanying the Vendian/Cambrian 'explosion' of body plans in organisms of macroscopic size (see Figure 11–6).

In between these three 'steps', and following the last of them (i.e. since about 500 *my* ago), Figure 11–6 depicts a much slower but continuous increase in maximum complexity over time. This must, of

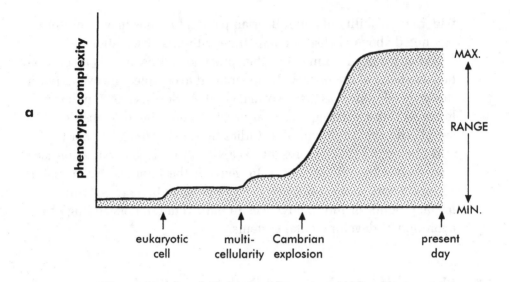

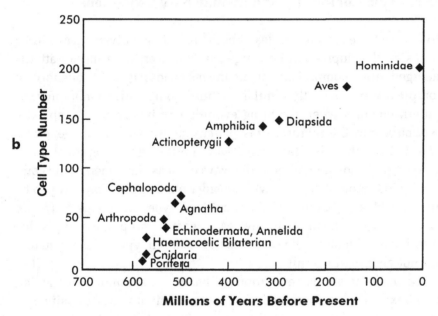

**Figure 11–6  Increasing complexity over evolutionary time**
(a) Approximate pattern of increase in phenotypic complexity,
showing range expansion as main phenomenon. (Time scale non-
linear and not correspondent with that in (b).) (b) The number of
cell types in various animal taxa plotted against their estimated
times of origin. From Valentine *et al.* 1994. *Paleobiology*, **20**,
131–142, with permission.

course, be a simplification of the true situation. Given that the overall range is only affected by changes in complexity occurring in the most phenotypically complex higher taxon existing at any one time (currently vertebrates), there must have been long periods when the range remained constant. Also, decreases are possible, particularly during mass extinctions. However, a decrease will only occur when the most complex higher taxon is completely eliminated. It is more usual, in mass extinctions, for higher taxa to suffer drastic reductions in diversity but for sporadic representative species to persist. So perhaps reductions in maximum complexity are indeed rare, as suggested by Valentine, Collins and Meyer (1994).

The pattern of increasing maximum and constant minimum – and hence increasing range – applies to virtually any starting point throughout evolutionary time that is used as the basis for a study of subsequent changes in complexity. The reason for this is that in most (possibly all) divergences from which one lineage leads to more complex forms, a lineage representing forms of approximately constant complexity also persists (see Figure 11–7). This is the case whether the divergence is simple (i.e. a dichotomy) or multiple (i.e. with one ancestral lineage simultaneously giving rise to many descendant lineages).

The Haeckelian pattern emerges when we consider the ontogenies of the sequence of forms represented in Figure 11–7: from eukaryotic unicell to teleost fish. Since the ontogenies of ancestral forms are usually very imperfectly known, it makes sense to consider present-day representatives of each 'level' of organism: for example *Paramecium*, *Volvox*, *Polycelis*, *Branchiostoma* and *Danio*.

The beginning of *Volvox* ontogeny is, of course, a single eukaryotic cell. Cleavage in flatworms gives rise initially to a tiny cluster of such cells. In early embryogenesis the amphioxus is a small unsegmented bilaterian. Early fish embryos have only rudimentary head and appendage features. In other words, in each comparison we can see a resemblance between the embryo of the more complex form and the *adult* of the less complex one. Of course, the resemblance is far from perfect, which is why 'recapitulation' is not a helpful term. Even 'Haeckelian pattern' is an imperfect label, given some of Haeckel's own remarks (see Chapter 4), but it does serve to emphasize the distinction between this pattern and the patterns of divergence and convergence that we saw above. Its existence reinforces the point made earlier: different

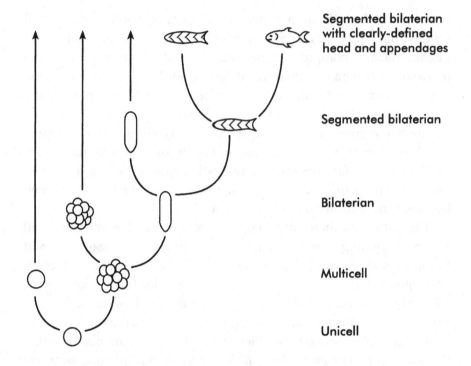

**Figure 11–7  Persistence of simpler forms alongside their more complex derivatives**

comparisons give rise to different embryological patterns. No one pattern rules supreme – each has its own 'domain'.

Two unrealistic aspects of Figure 11–7 are (a) the equal spacing of divergences and (b) their exclusively dichotomous nature. In reality, morphological evolution is characterized by an intermittent series of 'explosions'. Although the Vendian/Cambrian explosion of body plans is the best known, many subsequent explosions have occurred in various taxa, including such distinct groups as crinoids (Foote 1994) and mammals (see Benton 1990: Chapter 9). This pattern of explosive or bush-like proliferation of lineages is not, of course, an alternative to the much-discussed patterns of punctuated equilibrium and gradualism. Rather, it applies to a higher level (see Figure 11–8).

Why does morphological evolution take the form of an intermittent series of explosions? We cannot be certain at this stage, but the following 'three-plank hypothesis' readily suggests itself. The sudden diversification of an ancestral lineage into many descendant lineages of

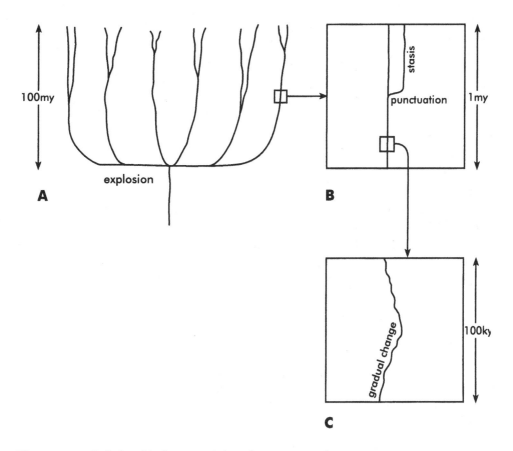

**Figure 11–8** Relationship between (A) evolutionary 'explosions', (B) species-level punctuations and (C) gradually shifting population mean character values under directional selection

morphologically and ecologically distinct organisms requires three things: (i) ecological opportunities, (ii) an easily-modifiable (= 'generalized'?) base form and (iii) a supply of genetic and developmental variants at least some of which are useful in invading novel ecospace. Consequently, explosions tend to occur only in unspecialized lineages following the occupation of new habitats/niche space (e.g. invasion of the land), or following the freeing-up of niche space by a mass extinction or geographical fragmentation (see Hedges *et al.* 1996). Such combinations of circumstances, when combined with availability of appropriate variants, lead to explosions, while other combinations tend to result in less dramatic change or even stasis.

Let us now consider the relationship between different evolutionary explosions over the phylogenetic history of the Animalia. In particular,

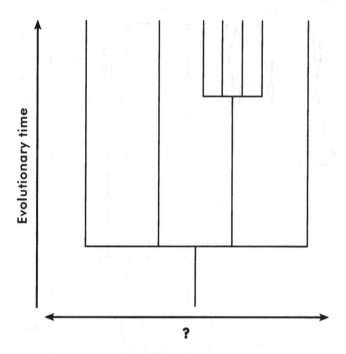

**Figure 11–9   Diminishing explosions over evolutionary time**
The interpretation of such patterns (and indeed their validity)
depends on what is being measured along the horizontal axis: see
text for details.

successive explosions are often pictured as diminishing in size (see
Figure 11–9). We need to enquire whether this apparent diminution
is real, and if so then exactly what is diminishing. That is, we need to
be clear about what the horizontal axis in Figure 11–9 represents.

This issue brings us back to Hennig's (1981) distinction between
genealogy and morphology. Clearly, from the former perspective evolutionary explosions 'diminish' in the sense that later ones are nested
within deeper lineage-splitting events. But the less obvious, and more
interesting, question relates to the morphological perspective: are the
developmental changes occurring in later explosions (e.g. mammals) in
some sense less dramatic than those that occurred earlier (e.g. in the
Vendian/Cambrian)?

The material covered in the last two sections suggests that the
answer is 'yes', providing that the horizontal axis of Figure 11–9 represents 'proportion of the overall developmental programme that is

altered'. As we have seen, it is not necessarily the earliest ontogenetic stages that are the most resistant to change – the exact pattern varies between taxa. However, the tendency towards increased 'proportional inertia' in development appears to be a general one, even if the precise way in which it is manifested varies.

The reason for this increased inertia is not clear. While it must be harder for mutation/selection to alter the whole developmental programme in a large, complex metazoan (e.g. a mammal) than in a small, simple one (e.g. a flatworm), we should recall that the evolutionary increase in complexity has been in the maximum, not the minimum. There have been simple flatworms available to form the basis of evolutionary radiations for *at least* 520 *my*. Yet since the Cambrian explosion this stock has not given rise to radically different (i.e. non-Platyhelminth) designs. This may be due to internal and/or external causes. Perhaps prolonged selection for increased cross-links or redundancy of control mechanisms has made the development of recent flatworms (and other animals) less readily modifiable than their distant ancestors – although clearly some kinds of modification are 'permitted', since some groups of present-day flatworms have rather unusual ontogenetic features. Perhaps also the ecological opportunities have been fewer and shorter-lived. It seems likely that a mixture of internal and external causes is responsible; determining their relative importance may well be an intractable problem.

We should be careful not to take this perceived directionality in the evolution of developmental systems too far. In particular, the views of Willis (1940) and, more recently, Kauffman (1993, 1995) are too extreme. Willis thought that evolution made phyla first, then classes and so on down to species. Kauffman (1995) expresses a similar view, in relation to Gould's (1989) metaphor of 're-playing the tape of life'. Kauffman states (p.14): "The particular branchings of life, were the tape played again, might differ, but the *patterns* of the branching, dramatic at first, then dwindling to twiddling with details later, are likely to be lawful."

The problem here is that evolution has *always* 'twiddled' (or 'tinkered', to use Jacob's (1977) famous metaphor). It has always produced close congeners. This is not a feature only of recent phylogenetic history. So the directional trend is not from major changes to minor ones, but rather takes the form of a progressive shift in the ratio of major-to-minor changes in favour of the latter, even though these already constituted the majority of changes at the outset. This is not such a neat, clean pattern, but that does not detract from its importance.

Finally, do all the various evolutionary explosions in morphology from the Neoproterozoic to the present – regardless of scale – constitute 'adaptive radiations'? At one level, the answer has to be 'yes' in that ultimately maladapted organisms perish and those whose radiation we observe are necessarily those whose level of adaptedness was sufficient to ensure their survival. But there has been a tendency for studies of adaptive radiation to overemphasize ecology at the expense of ontogeny. As Skelton (1991) has pointed out, adaptive radiations have both morphogenetic and environmental cues. Essentially, this book is one long argument that we should not neglect the former. Also, we should not neglect the interaction between the two. Rather than the environment passively selecting between genetically determined invariant ontogenies, selection may often favour organisms whose developmental pathways respond to environmental factors in particular ways (see Section 9.3). That is, there may be selection for a particular 'norm of reaction' (Lewontin 1974b), with genetic assimilation (Waddington 1953) being just one of many kinds of example of this process.

# CHAPTER TWELVE

# Prospect: Expanding the Synthesis

## 12.1 Neither Boredom nor Heresy

Scientific theories, or interconnected groups of them, are often likened to buildings – as in the phrase 'theoretical edifice'. There is a lot of sense in this analogy. General theories are often built up painstakingly from many different components. They are underpinned by various 'foundations'. Different parts of a theoretical edifice are often interdependent – if one part turns out to be wrong, the whole structure may eventually collapse.

We can picture neo-Darwinian evolutionary theory, or the 'modern synthesis', as one particular theoretical edifice. In this context, criticisms of the theory tend to be reacted to in a bimodal way. When a mild-mannered critic points out that the theory is fine as far as it goes, but that it lacks a developmental component (e.g. Horder 1994), a reaction verging on boredom is sometimes engendered: something along the lines of 'yes, a few bricks are missing, but they will eventually be discovered and inserted'. In contrast, more damning criticism, which implies that the whole edifice needs to be demolished and replaced with a different one with greater explanatory power (e.g. Rosen 1984), is rightly regarded as 'heresy'.

It should by now be apparent that the view of existing evolutionary theory taken herein constitutes a criticism of intermediate scale. There is not just a brick or two missing, but rather a whole section of the building. However, this can be – indeed is being – built and integrated into the existing edifice without causing it to collapse; though there may well be some stresslines that require attention.

## 12.2 Completing the Evolutionary Circle

At the risk of overdoing the analogies, I will now use another one. I will picture the various disciplinary contributions towards an overall understanding of evolution as forming a circle. This has the advantage of allowing us to visualize the size of the 'missing slice'. Figure 12–1 is an illustration of this evolutionary circle; and it indicates that the slice that is yet to be fully incorporated represents about one-third of the whole. While this estimate is clearly a very subjective one, it is certainly not compatible with the notion of a single 'missing brick' in a large edifice.

Let us now meander around this conceptual circle – clockwise from the starting point marked X – to get a more detailed feel for what it represents. At the end of this tour, I will indicate some of the circle's deficiencies.

We start, then, with classical studies of mutation. Mutations are, as is often said, the 'raw material' of evolution. They provide the genetic and developmental variation without which selection could achieve nothing. Some of their properties – notably dominance/recessiveness – can be elucidated through the standard laboratory crosses of transmission genetics. These can be performed either with mutations that are not generally found segregating in natural polymorphisms (e.g. *vestigial* in *Drosophila*) because of their deleterious effects, or on those that do contribute to such polymorphisms (e.g. *Adh* in *Drosophila*, pigmentation in *Cepaea* or blood groups in mammals).

Moving 'up', then, from the behaviour of variant genes in families to their more complex dynamics in whole populations, we enter the territory of classical population genetics. Here we observe the ability of selection and drift to move allelic frequencies away from Hardy–Weinberg equilibrium, and to cause evolutionary change by driving some alleles to fixation.

Most adaptively important phenotypic characters (e.g. body size) do not, of course, have a genetic basis consisting of just a single locus. Rather, each has a complex underlying genetic causality in which the varied effects of many QTLs (quantitative trait loci) combine, often in a non-additive, epistatic manner. The mapping and general elucidation

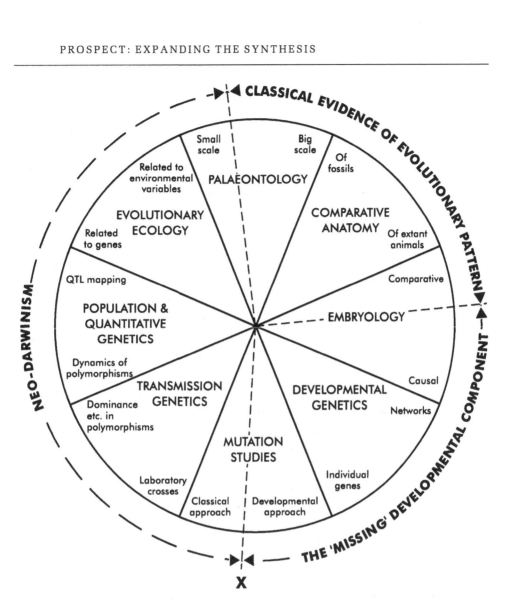

**Figure 12–1   The 'circle' of evolutionary theory**
Each contribution, represented by a segment, has a 'main heading'
(upper case). The lower-case descriptors refer to variation within
each segment, and are arranged so as to bring conceptually-related
topics in adjacent segments nearest to each other. QTL, quantitative
trait loci; X, departure point for 'tour': see text.

of these – a highly complicated task – is pursued by quantitative
geneticists.

Patterns of variation of quantitative traits within and between natural populations provide a focus of attention for evolutionary ecology.
Some of the traits that have been frequently studied are morphological

(e.g. beak size, head width, overall body size), some behavioural (e.g. habitat selection, foraging behaviour, territoriality), while others (e.g. niche breadth and location) have elements of both. A difficulty pervading this whole area of endeavour is that environmental factors sometimes alter phenotypes directly (i.e. not via the gene pool). Such ecophenotypic variation is often difficult to separate from its genetically-based counterpart (but see Arthur 1982c for an example; van Noordwijk (1984) for a general exhortation towards a more genetically-explicit approach to evolutionary ecology; and West-Eberhard (1989) for the paradoxical evolutionary importance of ecophenotypic variation (plasticity)).

Patterns of variation in body size, shape and associated characters often map to patterns of variation in environmental factors, either in space or time. A well-known example is 'Cope's law': the generalization that, within species, body size tends to increase with latitude (see James, Azevedo and Partridge (1995) for an interesting and developmentally explicit example). Whether certain long-term patterns – for example punctuated equilibrium within a lineage – are also associated with environmental change remains to be seen.

Detailed 'small-scale' palaeontological studies of individual lineages (e.g. Williamson 1981) are a relatively recent pursuit. Classical palaeontology more typically involved bigger-scale studies of broader taxa (e.g. the origin and diversification of mammals). A comparative approach to the morphology of both extant and fossil animals produced the crucial concept of homology (Owen 1848), and indeed this extends into the phenotype's fourth dimension in the form of comparative embryology. Von Baer's 'law' (not a law in fact: see Chapter 11) is a classic example of a comparative embryological pattern.

While such comparative patterns can be partly explained through population-level processes, such as different likelihoods of different kinds of selection, a complete explanation must also incorporate an adequate account of the 'target' upon which selection is acting. In other words, we need to understand development in terms of its underlying causality: both cellular and genetic. It is here – through part of the embryology slice, all of the developmental genetics slice, and back into mutation (where we started) that we lack an adequate overall picture of the processes at work. The 'genetic architecture of development' and its phylogenetic variation are the most important things that we now need to understand more thoroughly in order to achieve a more complete explanation of comparative embryological patterns (see Figure 12–2). In particular, the extent to which that architecture includes a hierarchical component has major implications for the ways in which developmental systems evolve – yet this issue remains elusive, as was discussed in Chapter 11.

Let us now briefly examine some of the evolutionary circle's deficiencies as a conceptual model. First, systematics is absent. That is actually not a problem, because in fact systematics pervades the whole circle. Whether we are looking at developmental genes, patterns of cell proliferation, morphology, body size, niches or polymorphisms, a systematic (i.e. phylogenetic) approach can be taken.

Second, in the form given in Figure 12–1, the circle has been influenced by my focus in this book on morphological characters (and their unfolding through ontogeny). Physiological and behavioural characters are not explicitly dealt with, except that we briefly encountered some niche-related behaviours in the above account. Again, though, this is not a problem. A particular physiological process or behaviour pattern has a genetic architecture, an ontogeny, a phylogenetic history (though without direct fossil evidence) and so on. There is no reason why a slightly modified version of the circle should not be capable of depicting our various approaches to such characters.

Finally, the 'circle model' runs the risk of down-playing important links between areas that do not occupy adjacent segments, of which there are many. A particularly interesting example of such a link is provided by gene loci that appear both as 'individual genes' in the developmental genetics segment and as quantitative trait loci in the population and quantitative genetics segment (see, for example, Mackay and Langley 1990). The way in which many genes have dual or multiple developmental roles is an important component of the overall genetic architecture of egg-to-adult ontogeny.

As the labels around the periphery of Figure 12–1 indicate, the various segments of the circle can be grouped under three broad headings: (a) classical evidence of large-scale evolutionary patterns; (b) neo-Darwinism (family- and population-level dynamic mechanisms); and (c) the 'missing' developmental component (of evolutionary theory, and indeed of biological theory more generally).

The evolutionary importance of the developmental component is, as I have stressed on many occasions in this book from the Preface

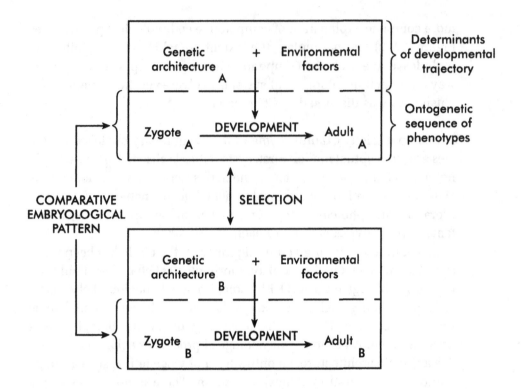

**Figure 12–2  Relationships between genetic architecture, development, selection and comparative embryological patterns**
Within any one organism (e.g. A, upper box) the genetic architecture of development and certain environmental factors combine to produce a predictable ontogenetic sequence of phenotypes. Selection (internal and/or external) alters the relative frequencies of different variants (A, B) within a species. As a clade diversifies, the *interspecific* distribution of variants (A, B: now interpreted differently) is what we observe as a comparative embryological pattern (e.g. von Baerian 'deviation').

onwards, that it has the potential to build a bridge between the other two. More specifically, the explanation of comparative embryological patterns must, as Figure 12–2 indicates, be based on the *interaction* between natural selection and its 'target' – the dynamic sequence of phenotypes whose basis is the genetic architecture of development. There are many fascinating comparative patterns: origins of body plans, divergence, convergence, increase in complexity, evolution of larval forms and so on. As noted in Chapter 11, none of these is universal; rather, each occurs within a particular domain. But attempts to explain *any* of them purely in selective terms, without reference to the underlying genetic architecture, will ultimately fail. Hence the need for the new discipline of evolutionary developmental biology; or, if you prefer to think of it this way, the need for a developmentally explicit, not a developmentally agnostic, neo-Darwinism.

Characterization of the developmental component as 'missing', is, of course, too extreme. If it were entirely absent, I could hardly have written a book about it. Rather, our knowledge of development is fragmentary, and, because of that, it has not yet made the contribution that it eventually will to evolutionary theory. Another way to put this point is that its current theoretical contribution underestimates the importance of ontogenetic processes in phylogeny – but this situation is rapidly changing.

## 12.3 The Main Themes of Evolutionary Developmental Biology

I was originally planning to make this section a summary of the whole book – that is, a summary of my attempt to sketch the form that the missing developmental component of evolutionary theory should take. However, a brief summary cannot do justice to this endeavour and is more likely to mislead than to inform. Instead, I decided merely to list (in Table 12–1) the main themes of Evolutionary Developmental Biology as I see them. Some of these already feature adequately in mainstream evolutionary theory, but most are under-emphasized, some are virtually absent and a few (e.g. internal selection) are treated with suspicion and indeed denied by 'pan-externalists'.

In the preceding chapters (especially Chapters 7–11), these themes have been woven together in a particular way. Whether that stands the test of time remains to be seen, but I am confident in predicting that the themes themselves – or at least the vast majority of them – will do so. That is not, of course, to say that they will replace the themes of current mainstream theory. Instead, there is likely to be a phase of 'tension', in which the degree to which the two sets of themes are compatible is gradually resolved, giving way eventually to some complementary combination of the two. (We have already entered the 'tension' phase: see, for example, Gilbert, Opitz and Raff 1996.)

The most crucial need at present is the identification of a research programme that will provide new data and test hypotheses. There has often been a tendency for study of the ontogeny/phylogeny interface to

**Table 12–1. The main themes of evolutionary developmental biology**

- Gene duplication and divergence in key 'developmental master genes' may have been particularly important in the earliest stages of body plan proliferation (Chapter 7).
- Changes in the interactional architecture and spatiotemporal expression patterns (STEPs) of *all* development-controlling genes have been important throughout evolution (Chapter 7).
- A more sophisticated classification of the developmental effects of mutation needs to be developed. The false distinction between 'micro' and 'macro' mutations should be dropped (Chapter 8).
- Evolutionary theory needs to acknowledge the role of internal selection, and the coevolution of developmentally-interacting genes. Associated with this, internal factors are not always negative 'constraints', but can also be positive evolutionary forces (Chapter 9).
- There should be a greater emphasis on creative processes: how variant ontogenies are mutationally produced is as important as how they are selectively sieved (Chapter 10).
- The nature of the evolutionary process has changed progressively over time: a few ontogenetic divergences in early evolution were more dramatic (in proportional terms) than any of their phylogenetically later counterparts – although even in early evolution, *most* ontogenetic divergences were probably subtle ones (Chapters 9 and 11).
- Associated with this, the genetic/developmental changes that were involved in the ancient origins of animal body plans were not identical to those characterizing typical present-day speciations (Chapters 9 and 11).
- Uncluttered niche space, relatively simple but multicellular phenotypes and as-yet poorly canalized ontogenies were all vital ingredients of body plan radiation (Chapters 9 and 11).
- The genetic architecture of development within any particular animal has a hierarchical component, though not a straightforward one. Its structure can change in many different ways during evolution (Chapter 11).
- No single comparative embryological pattern is universally found or can be described as a 'law'. Von Baerian divergence, its antithesis (convergence) and a broadly Haeckelian (quasi-recapitulatory) pattern can all be found, depending on the comparison made (Chapter 11).
- At the interface between evolutionary developmental biology and mainstream 'external adaptation theory': selection on the ways in which developmental systems react to environmental variation ('norms of reaction') is of considerable importance (Chapter 11).

border on the metaphysical (e.g. the *Naturphilosophen*); and recent work in evolutionary developmental biology has sometimes been criticized for being too philosophical and *not* producing ideas for specific hypotheses and experiments to test them (see, for example, Calow

1992). We cannot confidently predict the imminent successful incorporation of a whole series of new themes into the evolutionary synthesis if those themes are simply untestable conjectures. But we need not worry on this count. All sorts of new experimental and observational approaches have been adopted in recent years (notably the 'comparative developmental genetics' of Chapter 7) and others may be anticipated for the next decade. These 'paths into the future' (see Pollard 1984 for a prior use of this phrase to cover some very different paths) form the focus for my concluding section.

## 12.4   Paths into the Future

In general, what we need are observations and experiments which help to confirm, modify or reject the themes of Table 12–1, or to add new themes. The eight paragraphs that follow cover eight specific suggestions of possible approaches. Most of these are simply exhortations for the elaboration of approaches that are already under way, but some (e.g. use of the inverted cone model of Chapter 8) are relatively new.

1. Perhaps the most obvious need is for further studies of comparative developmental genetics, such as those of Garcia-Fernàndez and Holland (1994) and Averof and Akam (1995). We need to extend such work to more genes and to more classes and phyla. Also, we need to look not just at single genes or gene complexes (e.g. *Hox*) but rather at pairs and groups of genes that interact functionally to determine the patterns of cellular behaviour that constitute developmental pathways. Comparative study of interacting genes should help to elucidate the extent and nature of the 'internal coevolution' that we considered in Chapters 8 and 9.

2. Understanding the overall genetic architecture of development in any particular animal is another obvious goal. Included here is the issue of multiple effects of particular development-controlling genes. Further studies like that on the *achaete-scute* complex by Mackay and Langley (1990) are clearly desirable. Multiple effects indicate that the genetic architecture of development is not a simple hierarchical cascade. An 'upstream' gene may switch on a 'downstream' target which activates yet other genes, but further downstream in this process the 'upstream' gene may be switched on again (see Salser and Kenyon (1996) for an example). Further

studies on other loci and other organisms should help to elucidate any regularities in the overall interactional network; and I would hazard a guess that some kind of hierarchical structure is one of these regularities, despite the above rejection of a *simple* hierarchy.

3.  At the population level, we need more studies of 'developmental polymorphisms' – that is, polymorphisms involving development-controlling, not terminal-target or housekeeping, genes; and, at the phenotypic level, polymorphisms involving altered patterns of cell proliferation, for example polymorphisms of segment number or symmetry. The link between genetic and phenotypic levels is especially important. Sometimes polymorphism at a major development-controlling gene causes a major phenotypic polymorphism (e.g. in the direction of shell coiling in gastropods: Murray and Clarke 1966). But sometimes polymorphism at a 'major' locus has a relatively inconspicuous phenotypic effect (e.g. 'minor' polymorphism in *Hox* genes: Gibson and Hogness 1996). And in most cases, the genetic and developmental bases of major phenotypic polymorphisms (e.g. variation in myriapod segment number: Minelli and Bortoletto 1988) remain to be fully elucidated.

4.  My concern about the excessive emphasis on external factors in mainstream evolutionary theory should not be taken to imply that I believe such factors to be unimportant – far from it. Indeed, the interplay between internal and external factors is likely to be a productive field of study. In particular, we need to elucidate the way in which a developmental programme comes under *direct* control from the environment – Van Valen's (1974) famous 'control of development by ecology' – but now in an *ecophenotypic* sense. The pattern of ecophenotypic response of a developmental trajectory to environmental variation – Lewontin's (1974b) 'norm of reaction' – may itself be selected upon. There have been many previous studies of this phenomenon, most notably Waddington's (1953, 1956) work on genetic assimilation, and several relevant reviews – for example that by West-Eberhard (1989). However, further work is necessary to establish both the underlying mechanisms (how do environmental factors directly alter the spatiotemporal expression patterns of particular genes? – see Pigliucci 1996) and the general importance of this kind of evolutionary change, which may be considerable.

5.  With regard specifically to the origins of the major, phylum-level body plans, the discovery and interpretation of new pre-Cambrian fossils is to be eagerly anticipated. Although fossil material from

early stages of metazoan history is sparse, there has been a steady flow of new finds over the last few years, with undoubtedly more to follow. All such discoveries help to refine our picture of body plan origins, and so help to clarify the nature of the morphological phenomena for which we seek a joint developmental/genetic and ecological explanation. There is also a need for further comparative sequence analysis to resolve the currently difficult issue of the timing of major lineage divergences: see Wray *et al.* (1996).

6. As noted earlier, comparative embryology reveals a variety of patterns, not a single, all-pervasive one. This opens up the possibility of being able to predict which pattern should characterize any particular higher taxon that has so far been little-studied from a developmental perspective. For example, as I indicated in Chapter 11, the internalizing of embryos, or any other way of protecting them from the environment, should reduce or eliminate external selection for adaptation of early stages to particular niches. Consequently, if we think of a 'battle' between purifying internal selection to maintain developmental coordination and diversifying external selection for the adoption of particular methods of feeding etc., the former should 'win' in an embryo-protecting group, giving a von Baerian pattern, but conversely the latter may 'win' in an embryo-exposing group. Comparative embryological study of the lesser-known phyla should enable such predictions to be tested.

7. Because of the sheer complexity of developmental processes, it makes sense to probe issues in the evolution of development using simplified models, whether conceptual/pictorial or mathematical, alongside studies of real animals. Mathematical models of development have been explored by Meinhardt (1982) among others. The inverted cone model used herein (Chapters 1 and 8) is so far merely a 'conceptual' model. However, it provides a useful context for considering variation in the degree of evolutionary constraint of mutations that alter developmental processes. A combination of this *organismic* model with population-level models such as those related to the 'Fisher principle' (Chapter 1) could provide a new and comprehensive kind of modelling approach that transcends the individual/population divide. Models in themselves cannot, of course, be used to test hypotheses: essentially their predictions *are* hypotheses, and they need to be tested by experiment.

**8.** I will remain at the conceptual level for my final suggested approach to furthering an overall understanding of the evolution of development. There is now a substantial body of theory relating to the ecology and evolution of life histories (see text by Stearns (1992) for an introduction). Evolutionary developmental biology and life-history theory have grown up largely independently of each other to date, yet they deal with very closely related issues. There is a clear need for a theoretical 'knitting together' of these two approaches.

It seemed appropriate to let Darwin (1859) begin this book, and it seems equally appropriate to let Whyte (1965) end it – given his influence on the ideas developed herein. He states (p.32): "The statistical theory of populations must be complemented by a structural theory of ontogenesis and of its influence on phylogeny." Towards the end of the book (p.72), he continues: "The signs point towards a comprehensive biological synthesis... perhaps within this century." Well, he was somewhat optimistic on the time scale – we have not yet arrived at a truly comprehensive synthesis of development and evolution – but we are certainly beginning to glimpse the form that this overall synthesis will take. The 'soul' of natural history will not elude us for much longer.

# References

Abouheif, E., Akam, M., Dickinson, W.J., Holland, P.W.H., Meyer, A., Patel, N.H., Raff, R.A., Roth, V.L. & Wray, G.A. 1997. Homology and developmental genes. *TiG*, **13**, 432–433.

Aguinaldo, A.M.A., Turbeville, J.M., Linford, L.S., Rivera, M.C., Garey, J.R., Raff, R.A. & Lake, J.A. 1997. Evidence for a clade of nematodes, arthropods and other moulting animals. *Nature*, **387**, 489–493.

Ahlberg, P.E. 1995. *Elginerpeton pancheni* and the earliest tetrapod clade. *Nature*, **373**, 420–425.

Akam, M. 1998a. *Hox* genes, homeosis and the evolution of segment identity: no need for hopeless monsters. *Int. J. Dev. Biol.*, **42**, 445–451.

Akam, M. 1998b. *Hox* genes: from master genes to micromanagers. *Current Biology*, **8**, R676–R678.

Akam, M. 1995. *Hox* genes and the evolution of diverse body plans. *Phil. Trans. Roy. Soc. Lond. B.*, **349**, 313–319.

Akam, M., Averof, M., Castelli-Gair, J., Dawes, R., Falciani, F. & Ferrier, D. 1994a. The evolving role of *Hox* genes in arthropods. In: *The Evolution of Developmental Mechanisms*, eds M. Akam, P. Holland, P. Ingham & G. Wray. Development 1994 Supplement, Company of Biologists, Cambridge, UK.

Akam, M., Holland, P., Ingham, P. & Wray, G. (eds) 1994b. *The Evolution of Developmental Mechanisms*. Development 1994 Supplement, Company of Biologists, Cambridge, UK.

Aparicio, S., Hawker, K., Cottage, A., Mikawa, Y., Zuo, L., Venkatesh, B., Chen, E., Krumlauf, R. & Brenner, S. 1997. Organization of the *Fugu rubripes* *Hox* clusters: evidence for continuing evolution of vertebrate *Hox* complexes. *Nature Genetics*, **16**, 79-83.

Appel, B. & Sakonju, S. 1993. Cell-type-specific mechanisms of transcriptional repression by the homeotic gene products Ubx and abd-A in *Drosophila* embryos. *EMBO J.*, **12**, 1099–1109.

Aroian, R.V., Koga, M., Mendel, J.E., Oshima, Y. & Sternberg, P.W. 1990. The *let-23* gene necessary for *Caenorhabditis elegans* vulval induction encodes a tyrosine kinase of the EGF receptor subfamily. *Nature*, **348**, 693–699.

Artavanis-Tsakonas, S., Matsuno, K. & Fortini, M.E. 1995. Notch signaling. *Science*, **268**, 225–232.

Arthur, W. 1982a. A developmental approach to the problem of variation in evolutionary rates. *Biol. J. Linn. Soc.*, **18**, 243–261.

Arthur, W. 1982b. The evolutionary consequences of interspecific competition. *Adv. Ecol. Res.*, **12**, 127–187.

Arthur, W. 1982c. Control of shell shape in *Lymnaea stagnalis*. *Heredity*, **49**, 153–161.

Arthur, W. 1984. *Mechanisms of Morphological Evolution: A Combined Genetic, Developmental and Ecological Approach*. Wiley, Chichester, UK.

Arthur, W. 1987a. *The Niche in Competition and Evolution*. Wiley, Chichester, UK.

Arthur, W. 1987b. *Theories of Life*. Penguin, Harmondsworth, UK.

Arthur, W. 1988. *A Theory of the Evolution of Development*. Wiley, Chichester, UK.

Arthur, W., Jowett, T. & Panchen, A.L. 1999. Segments, limbs, homology and co-option. *Evolution and Development*, **1**, 74–76.

Ashburner, M. 1971. Induction of puffs in polytene chromosomes of *in vitro* cultured salivary glands in *Drosophila melanogaster* by ecdysone analogues. *Nature New Biol.*, **230**, 222–224.

Ashburner, M., Chihara, C., Meltzer, P. & Richards, G. 1974. Temporal control of puffing activity in polytene chromosomes. *Cold Spring Harb. Symp. Quant. Biol.*, **38**, 655–662.

Auerbach, C. 1976. *Mutation Research: Problems, Results and Perspectives*. Chapman & Hall, London.

Averof, M. & Akam, M. 1993. HOM/*Hox* genes of *Artemia*: implications for the origin of insect and crustacean body plans. *Curr. Biol.*, **3**, 73–78.

Averof, M. & Akam, M. 1995. *Hox* genes and the diversification of insect and crustacean body plans. *Nature*, **376**, 420–423.

Averof, M. and Cohen, S.M. 1997. Evolutionary origin of insect wings from ancestral gills. *Nature*, **385**, 627–630.

Ayala, F.J., Rzhetsky, A. & Ayala, F.J. 1998. Origin of the metazoan phyla: molecular clocks confirm paleontological estimates. *Proc. Natl. Acad. Sci. USA*, **95**, 606–611.

Azpiazu, N. & Frasch, M. 1993. *tinman* and *bagpipe*: two homeobox genes that determine cell fates in the dorsal mesoderm of *Drosophila*. *Genes and Devel.*, **7**, 1325–1340.

Backeljau, T., Winnepenninckx, B. & De Bruyn, L. 1993. Cladistic analysis of metazoan relationships: a reappraisal. *Cladistics*, **9**, 167–181.

Balavoine, G. & Telford, M.J. 1995. Identification of planarian homeobox sequences indicates the antiquity of most *Hox*/homeotic gene subclasses. *Proc. Nat. Acad. Sci. USA*, **92**, 7227–7231.

Balinsky, B.I. 1981. *An Introduction to Embryology*, 5th edition. Saunders College, Philadelphia.

Barnes, R.D. 1987. *Invertebrate Zoology*, 5th edition. Saunders College, Philadelphia, PA.

Barnes, R.S.K. (ed.) 1984. *A Synoptic Classification of Living Organisms*. Blackwell, Oxford, UK.

Barnes, R.S.K., Calow, P. & Olive, P.J.W. 1988. *The Invertebrates: A New Synthesis*. Blackwell, Oxford, UK.

Bartels, J.L., Murtha, M.T. & Ruddle, F.H. 1993. Multiple *Hox*/HOM-class homeoboxes in Platyhelminthes. *Molec. Phylogenet. Evol.*, **2**, 143–151.

Barton, N.H. 1989. Founder effect speciation. In: *Speciation and its Consequences*, eds D. Otte & J.A. Endler. Sinauer, Sunderland, MA.

Basler, K. & Struhl, G. 1994. Compartment boundaries and the control of *Drosophila* limb pattern by *hedgehog* protein. *Nature*, **368**, 208–214.

Bate, M. & Martinez-Arias, A. 1991. The embryonic origin of imaginal discs in *Drosophila*. *Development*, **112**, 755–761.

Bate, M. & Martinez-Arias, A. (eds) 1993. *The Development of Drosophila melanogaster* (vols 1 & 2). Cold Spring Harbor Laboratory Press, New York.

Bateson, W. 1894. *Materials for the Study of Variation, Treated with Especial Regard to Discontinuity in the Origin of Species*. Macmillan, London.

Baumgartner, S., Martin, D., Hagios, C. & Chiquet-Ehrismann, R. 1994. *Tenm*, a *Drosophila* gene related to tenascin, is a new pair-rule gene. *EMBO J.*, **13**, 3728–3740.

Becker, H.J. 1959. Die Puffs de Speichedrusenchromosomen von *Drosophila melanogaster*. 1. Beobachtung zum verhalten des Puffmusters un Normalstamm und bei zwei Mutanten, *giant* und *lethal-giant-larvae*. *Chromosoma*, **10**, 654–678.

Beeman, R.W., Stuart, J.J., Brown, S.J. & Denell, R.E. 1993. Structure and function of the homeotic gene complex (HOM-C) in the beetle *Tribolium castaneum*. *Bioessays*, **15**, 439–444.

Behrens, J. 1994. Cadherins as determinants of tissue morphology and suppressors of invasion. *Acta Anatomica*, **149**, 165–169.

Behrens, J., von Kries, J.P., Kühl, M., Bruhn, L., Wedlich, D., Grosschedl, R. & Birchmeier, W. 1996. Functional interaction of ß-catenin with the transcription factor LEF-1. *Nature*, **382**, 638–642.

Bender, W., Akam, M., Karch, F., Beachy, P.A., Peifer, M., Spierer, P., Lewis, E.B. and Hogness, D.S. 1983. Molecular genetics of the bithorax complex in *Drosophila melanogaster*. *Science*, **221**, 23–29.

Benton, M.J. 1990. *Vertebrate Palaeontology: Biology and Evolution*. Unwin Hyman, London.

Benton, M.J. (ed.) 1993. *The Fossil Record 2*. Chapman & Hall, London.

Bergström, J. 1991. Metazoan evolution around the Precambrian–Cambrian transition. In: *The Early Evolution of Metazoa and the Significance of Problematic Taxa*, eds A.M. Simonetta & S. Conway Morris. Cambridge University Press, Cambridge, UK.

Berner, R.A. 1991. A model for atmospheric $CO_2$ over phanerozoic time. *Am. J. Sci.*, **291**, 339–376.

Berner, R.A. & Canfield, D.E. 1989. A new model for atmospheric oxygen over phanerozoic time. *Am. J. Sci.*, **289**, 333–361.

Berrill, N.J. 1961. *Growth, Development and Pattern*. Freeman, San Francisco, CA.

Berry, R.J. 1982. *Neo-Darwinism*. Edward Arnold, London.

Bhanot, P., Brink, M., Samos, C.H., Hsieh, J.-C., Wang, Y., Macke, J.P., Andrew, D., Nathans, J. & Nusse, R. 1996. A new member of the *frizzled* family from *Drosophila* functions as a Wingless receptor. *Nature*, **382**, 225–230.

Bienz-Tadmor, B., Smith, H.S. & Gerbi, S.A. 1991. The promoter of DNA puff gene II/9-1 of *Sciara coprophila* is inducible by ecdysone in late prepupal salivary glands of *Drosophila melanogaster*. *Cell Regul.*, **2**, 875–888.

Birch, L.C. 1957. The meanings of competition. *Am. Nat.*, **91**, 5–18.

Bishop, J.A. & Cook, L.M. 1980. Industrial melanism and the urban environment. *Adv. Ecol. Res.*, **11**, 373–404.

Blair, S.S. 1995. Compartments and appendage development in *Drosophila*. *Bioessays*, **17**, 299–309.

Blake, C.C.F. 1978. Do genes-in-pieces imply proteins-in-pieces? *Nature*, **273**, 267.

Bominaar, A.A., Kesbeke, F., Snaar-Jagalska, B.E, Peters, D.J., Schapp, P. & van Haastert, P.J. 1991. Aberrant chemotaxis and differentiation in *Dictyostelium* mutant *fgdC* with a defective regulation of receptor-stimulated phosphoinositidase C. *J. Cell Sci.*, **100**, 825–831.

Bomze, H.M. & Lopez, A.J. 1994. Evolutionary conservation of the structure and expression of alternatively spliced Ultrabithorax isoforms from *Drosophila*. *Genetics*, **136**, 965–977.

Boncinelli, E., Simeone, A., La Volpe, A., Faiella, H., Acampora, D. & Scotto, L. 1985. Human cDNA clones containing homeobox sequences. *Cold Spring Harb. Symp. Quant. Biol.*, **50**, 301–306.

Boncinelli, E., Somma, R., Acampora, D., Pannese, M., D'Esposito, M., Faiella, A. & Simeone, A. 1988. Organization of human homeobox genes. *Hum. Reprod.*, **3**, 880–886.

Bond, G.A., Kominz, M.A. & Devlin, W.J. 1989. An overview of late Proterozoic and earliest Cambrian passive margins: Implications for the formation and breakup of a supercontinent. *Abstracts of the 28th International Geological Congress*, **1**, 1–171.

Bonner, J.T. 1974. *On Development*. Harvard University Press, Cambridge, MA.

Boore, J.L., Collins, T.M., Stanton, D., Daehler, L.L. & Brown, W.M. 1995. Deducing the pattern of arthropod phylogeny from mitochondrial DNA rearrangements. *Nature*, **376**, 163–165.

Bowring, S.A., Grotzinger, J.P., Isachsen, C.E., Knoll, A.H., Pelechaty, S.M. & Kolosov, P. 1993. Calibrating rates of Early Cambrian evolution. *Science*, **261**, 1293–1298.

Brady, R.H. 1985. On the independence of systematics. *Cladistics*, **1**, 113–126.

Brasier, M.D. 1992. Global ocean-atmosphere change across the Precambrian–Cambrian transition. *Geol. Mag.*, **129**, 161–168.

Breen, T.R. & Harte, P.J. 1991. Molecular characterization of the *trithorax* gene, a positive regulator of homeotic gene expression in *Drosophila*. *Mech. Dev.*, **35**, 113–127.

Brenner, S.E., Hubbard, T., Murzin, A. & Chothia, C. 1995. Gene duplications in *H. influenzae*. *Nature*, **378**, 140.

Brieher, W.M. & Gumbiner, B.M. 1994. Regulation of C-cadherin function during activin induced morphogenesis of *Xenopus* animal caps. *J. Cell Biol.*, **126**, 519–527.

Briggs, D.E.G. & Fortey, R.A. 1989. The early radiation and relationships of the major arthropod groups. *Science*, **246**, 241–243.

Briggs, D.E.G., Fortey, R.A. & Wills, M.A. 1992. Morphological disparity in the Cambrian. *Science*, **256**, 1670–1673.

Bromham, L., Rambaut, A., Fortey, R., Cooper, A. & Penny, D. 1998. Testing the Cambrian explosion hypothesis by using a molecular dating technique. *Proc. Natl. Acad. Sci. USA*, **95**, 12386–12389.

Buick, R., Thornett, J.R., McNaughton, N.J., Smith, J.B., Barley, M.E. & Savage, M. 1995. Record of emergent continental crust ~ 3.5 billion years ago in the Pilbara craton of Australia. *Nature*, **375**, 574–577.

Bürglin, T.R. 1994. A comprehensive classification of homeobox genes. In: *Guidebook to the Homeobox Genes*, ed. D. Duboule. Sambrook & Tooze, Oxford, UK.

Bürglin, T.R. & Ruvkun, G. 1993. The *Caenorhabditis elegans* homeobox gene cluster. *Curr. Op. Gen. Dev.*, **3**, 615–620.

Bürglin, T.R., Ruvkun, G., Coulson, A., Hawkins, N.C., McGhee, J.D., Schaller, D., Wittman, C., Müller, F. & Waterston, R.H. 1991. Nematode homeobox cluster. *Nature*, **351**, 703.

Burian, R.M. 1983. Adaptation. In: *Dimensions of Darwinism: Themes and Counterthemes in Twentieth-Century Evolutionary Theory*, ed. M. Grene. Cambridge University Press, Cambridge, UK.

Buss, L.W. 1987. *The Evolution of Individuality*. Princeton University Press, Princeton, NJ.

Buss, L.W. & Seilacher, A. 1994. The phylum Vendobionta: a sister group of the Eumetazoa? *Paleobiology*, **20**, 1–4.

Bussemakers, M.J., van Bokhoven, A., Voller, M., Smit, F.P. & Schalken, J.A. 1994. The genes for the calcium-dependent cell adhesion molecules P- and E-cadherin are tandemly arranged in the human genome. *Biochem. Biophys. Res. Comm.*, **203**, 1291–1294.

Butlin, R.K. & Day, T.H. 1984. The effect of larval competition on development time and adult size in the seaweed fly, *Coelopa frigida. Oecologia*, **63**, 122–127.

Butlin, R.K., Collins, P.M., Skevington, S.J. & Day, T.H. 1982. Genetic variation at the alcohol dehydrogenase locus in natural populations of the seaweed fly, *Coelopa frigida. Heredity*, **48**, 45–55.

Butterfield, N.J., Knoll, A.H. & Swett, K. 1990. A Brangiophyte red alga from the Proterozoic of arctic Canada. *Science*, **250**, 104–111.

Byers, S., Amaya, E., Munro, S. & Blaschuk, O. 1992. Fibroblast growth factor receptors contain a conserved HAV region common to cadherins and influenza strain A hemagglutinins: a role in protein–protein interactions? *Dev. Biol.*, **152**, 411–414.

Cairns, J., Overbaugh, J. & Miller, S. 1988. The origin of mutants. *Nature*, **335**, 142–145.

Calow, P. 1992. Constraint and malleability: Review of *Evolutionary Developmental Biology* by B.K. Hall. Times Higher Education Supplement, London, 12 June.

Campbell, I.D. & Baron, M. 1991. The structure and function of protein modules. *Phil. Trans. Roy. Soc. Lond. B*, **332**, 165–170.

Carroll, S.B. 1995. Homeotic genes and the evolution of arthropods and chordates. *Nature*, **376**, 479–485.

Carroll, S.B., Weatherbee, S.D. & Langeland, J.A. 1995. Homeotic genes and the regulation and evolution of insect wing number. *Nature*, **375**, 58–61.

Cartwright, P., Dick, M. & Buss, L.W. 1993. HOM/*Hox* type homeoboxes in the chelicerate *Limulus polyphemus*. *Molec. Phylogenet. Evol.*, **2**, 185–192.

Casares, F. & Sanchez-Herrero, E. 1995. Regulation of the infraabdominal regions of the bithorax complex of *Drosophila* by gap genes. *Development*, **121**, 1855–1866.

Chan, S.K., Jaffe, L., Capovilla, M., Botas, J. & Mann, R.S. 1994. The DNA binding specificity of Ultrabithorax is modulated by cooperative interactions with extradenticle, another homeoprotein. *Cell*, **78**, 603–615.

Charlesworth, B. 1994. *Evolution in Age-structured Populations,* 2nd edition. Cambridge University Press, Cambridge, UK.

Charlesworth, B. & Langley, C.H. 1989. The population genetics of *Drosophila* transposable elements. *Ann. Rev. Genet.*, **23**, 251–287.

Charlesworth, B., Sniegowski, P. & Stephan, W. 1994. The evolutionary dynamics of repetitive DNA in eukaryotes. *Nature*, **371**, 215–220.

Chen, J.-Y., Dzik, J., Edgecombe, G.D., Ramsköld, L. & Zhou, G.-Q. 1995. A possible Early Cambrian chordate. *Nature*, **377**, 720–722.

Chen, Y., Huang, L. & Solursh, M. 1994. A concentration gradient of retinoids in the early *Xenopus laevis* embryo. *Dev. Biol.*, **161**, 70–76.

Chiang, A., O'Connor, M.B., Paro, R., Simon, J. & Bender, W. 1995. Discrete Polycomb-binding sites in each parasegmental domain of the bithorax complex. *Development*, **121**, 1681–1689.

Chiappe, L.M. 1995. The first 85 million years of avian evolution. *Nature*, **378**, 349–355.

Child, C.M. 1941. *Patterns and Problems of Development.* University of Chicago Press, Chicago, IL.

Chitnis, A., Henrique, D., Lewis, J., Ish-Horowicz, D. & Kintner, C. 1995. Primary neurogenesis in *Xenopus* embryos regulated by a homologue of the *Drosophila* neurogenic gene *Delta*. *Nature*, **375**, 761–766.

Chothia, C. 1994. Protein families in the metazoan genome. In: *The Evolution of Developmental Mechanisms*, eds M. Akam, P. Holland, P. Ingham & G. Wray. Company of Biologists, Cambridge, UK.

Cigliano, M.M. 1989. A cladistic analysis of the family Tristiridae (Orthoptera, Acridoidea). *Cladistics*, **5**, 379–393.

Clarke, B. 1970. Darwinian evolution of proteins. *Science*, **168**, 1009–1011.

Clarke, B. 1975a. The contribution of ecological genetics to evolutionary theory: detecting the direct effects of natural selection on particular polymorphic loci. *Genetics*, **79** (Suppl.), 101–113.

Clarke, B. 1975b. The causes of biological diversity. *Sci. Am.*, **233**, 50–60.

Clarke, B. & Murray, J. 1969. Ecological genetics and speciation in land snails of the genus *Partula*. *Biol. J. Linn. Soc.*, **1**, 31–42.

Clarke, B., Murray, J. & Johnson, M.S. 1984. The extinction of endemic species by a program of biological control. *Pacific Science*, **38**, 97–104.

Clarke, B., Shelton, P.R. & Mani, G.S. 1988. Frequency-dependent selection, metrical characters and molecular evolution. *Phil. Trans. Roy. Soc. Lond. B.*, **319**, 631–640.

Clayton, G.A., Morris, J.A. & Robertson, A. 1957. An experimental check on quantitative genetical theory. I. Short-term responses to selection. *J. Genet.*, **55**, 131–151.

Clever, U. & Karlson, P. 1960. Induktion von Puff-Veränderungen in den Speicheldrüsen-chromosomen von *Chironomus tentans* durch Ecdyson. *Exp. Cell Res.*, **20**, 623–626.

Cloud, P. & Glaessner, M.F. 1982. The Ediacaran period and system: Metazoa inherit the earth. *Science*, **217**, 783–792.

Coates, M.I. & Clack, J.A. 1990. Polydactyly in the earliest known tetrapod limbs. *Nature*, **347**, 66–69.

Cohen, S.M. 1993. Imaginal disc development. In: *The Development of Drosophila melanogaster*, vol. II, eds M. Bate & A. Martinez-Arias. Cold Spring Harbor Laboratory Press, NY.

Collignon, J., Varlet, I. & Robertson, E.J. 1996. Relationship between asymmetric *nodal* expression and the direction of embryonic turning. *Nature*, **381**, 155–158.

Collins, D.H. 1996. The "evolution" of *Anomalocaris* and its classification in the Arthropod class Dinocarida (nov.) and order Radiodonta (nov.). *J. Paleontol.*, **70**, 280–293.

Concordet, J.-P. & Ingham, P. 1995. Patterning goes Sonic. *Nature*, **375**, 279–280.

Conn, D.B. 1991. *Atlas of Invertebrate Reproduction and Development*. Wiley-Liss, New York.

Conway Morris, S. 1985. The Middle Cambrian metazoan *Wiwaxia corrugata* (Matthew) from the Burgess Shale and *Ogygopsis* Shale, British Columbia, Canada. *Phil. Trans. Roy. Soc. Lond. B*, **307**, 507–582.

Conway Morris, S. 1993. The fossil record and the early evolution of the Metazoa. *Nature*, **361**, 219–225.

Conway Morris, S. 1995. Ecology in deep time. *TREE*, **10**, 290–294.

Conway Morris, S. & Peel, J.S. 1995. Articulated halkieriids from the Lower Cambrian of North Greenland and their role in early protostome evolution. *Phil. Trans. Roy. Soc. Lond. B*, **347**, 305–358.

Cook, L.M. 1971. *Coefficients of Natural Selection*. Hutchinson, London.

Cooper, D.N. & Krawczak, M. 1993. *Human Gene Mutation*. BIOS Scientific Publishers, Oxford, UK.

Counce, S.J. & Waddington, C.H. (eds) 1972. *Developmental Systems: Insects*, Vol. 1. Academic Press, London.

Coyne, J.A. 1983. Genetic basis of differences among three sibling species of *Drosophila*. *Evolution*, **37**, 1101–1118.

Crane, P.R., Friis, E.M. & Pedersen, K.R. 1995. The origin and early diversification of angiosperms. *Nature*, **374**, 27–33.

Crick, F.H.C. 1970. Diffusion in embryogenesis. *Nature*, **225**, 420–422.

Crimes, T.P., Insole, A. & Williams, B.P.J. 1995. A rigid-bodied Ediacaran biota from Upper Cambrian strata in Co. Wexford, Eire. *Geol. J.*, **30**, 89–109.

Crow, J.F. & Kimura, M. 1970. *An Introduction to Population Genetics Theory*. Harper & Row, New York.

Currie, P.D. & Ingham, P.W. 1996. Induction of a specific muscle cell type by a hedgehog-like protein in zebrafish. *Nature*, **382**, 452–455.

Dahl, U., Sjödin, A. & Semb, H. 1996. Cadherins regulate aggregation of pancreatic β-cells *in vivo*. *Development*, **122**, 2895–2902.

Daniel, C.W., Strickland, P. & Friedmann, Y. 1995. Expression and functional role of E- and P-cadherins in mouse mammary ductal morphogenesis and growth. *Dev. Biol.*, **169**, 511–519.

Darnell, J.E. 1978. Implications of RNA–RNA splicing in evolution of eukaryotic cells. *Science*, **202**, 1257–1260.

Darwin, C. 1859. *On the Origin of Species by Means of Natural Selection, or the Preservation of Favoured Races in the Struggle for Life*. John Murray, London.

Darwin, F. 1887. *The Life and Letters of Charles Darwin*. John Murray, London.

Davidson, E.H. 1986. *Gene Activity in Early Development,* 3rd edition. Academic Press, Orlando, FL.

Dawkins, R. 1986. *The Blind Watchmaker*. Longman, London.

Day, T.H., Hillier, P.C. & Clarke, B. 1974. Properties of genetically polymorphic isozymes of alcohol dehydrogenase in *Drosophila melanogaster*. *Biochem. Genet.*, **11**, 141–153.

De Beer, G.R. 1930. *Embryology and Evolution*. Oxford University Press, Oxford, UK.

De Beer, G.R. 1940. *Embryos and Ancestors*. Clarendon Press, Oxford, UK.

De Beer, G.R. 1971. *Homology: An Unsolved Problem*. Oxford University Press, London.

De Celis, J.F., Barrio, R. & Kafatos, F.C. 1996. A gene complex acting downstream of *dpp* in *Drosophila* wing morphogenesis. *Nature*, **381**, 421–424.

De Queiroz, K. 1985. The ontogenetic method for determining character polarity and its relevance to phylogenetic systematics. *Syst. Zool.,* **34**, 280–299.

De Robertis, E.M. 1994. The homeobox in cell differentiation and evolution. In: *Guidebook to the Homeobox Genes*, ed. D. Duboule. Sambrook & Tooze, Oxford, UK.

De Robertis, E.M. 1997. The ancestry of segmentation. *Nature,* **387**, 25-26.

De Robertis, E.M. & Sasai, Y. 1996. A common plan for dorsoventral patterning in Bilateria. *Nature*, **380**, 37–40.

De Rosa, R., Grenier, J.K., Andreevas, T., Cook, C.E., Adoutte, A., Akam, M., Carroll, S.B. & Balavoine, G. 1999. *Hox* genes in brachiopods and priapulids and protostome evolution. *Nature*, **399**, 772–776.

De Vries, H. 1910. *The Mutation Theory: Experiments and Observations on the Origin of Species in the Vegetable Kingdom*. (Translated by J.B. Farmer &

A.D. Darbyshire; two volumes – II published in 1911.) Kegan Paul, Trench, Trübner & Co., London.

Degnan, B.M., Degnan, S.M., Giusti, A. & Morse, D.E. 1995. A *Hox*/hom homeobox gene in sponges. *Gene*, **155**, 175–177

Degnan, B.M. & Morse, D.E. 1993. Identification of eight homeobox-containing transcripts expressed during larval development and at metamorphosis in the gastropod mollusc *Haliotis rufescens. Molec. Mar. Biol. Biotechnol.*, **2**, 1–9.

Dekker, E.J., Pannese, M., Houtzager, E., Boncinelli, E. & Durston, A. 1993. Colinearity in the *Xenopus laevis Hox*-2 complex. *Mech. Dev.*, **40**, 3–12.

Demerec, M. (ed.) 1950. *Biology of Drosophila*. Wiley, New York.

Denell, R. 1994. Discovery and genetic definition of the *Drosophila* Antennapedia complex. *Genetics*, **138**, 549–552.

Dent, J.N. 1968. Survey of amphibian metamorphosis. In: *Metamorphosis: A Problem in Developmental Biology*, eds W. Etkin & L.I. Gilbert. Appleton Century Crofts, New York.

Diaz-Benjumea, F.J., Cohen, B. & Cohen, S.M. 1994. Cell interaction between compartments establishes the proximal–distal axis of *Drosophila* legs. *Nature*, **372**, 175–179.

Dick, M.H. & Buss, L.W. 1994. A PCR-based survey of homeobox genes in *Ctenodrilus serratus* (Annelida: Polychaeta). *Molec. Phylogenet. Evol.*, **3**, 146–158.

Dickson, B. 1995. Nuclear factors in sevenless signalling. *Trends Genet.*, **11**, 106–111.

Dietrich, S., Schubert, F.R. & Gruss, P. 1993. Altered *Pax* gene expression in murine notochord mutants: the notochord is required to initiate and maintain ventral identity in the somite. *Mech. Dev.*, **44**, 189–207.

Di Gregorio, A., Spagnuolo, A., Ristoratore, F., Pischetola, M., Aniello, F., Branno, M., Cariello, L. & Di Lauro, R. 1995. Cloning of ascidian homeobox genes provides evidence for a primordial chordate cluster. *Gene*, **156**, 253–257.

Dobzhansky, T. 1961. On the dynamics of chromosomal polymorphism in *Drosophila. Symp. Roy. Ent. Soc. Lond.*, **1**, 30–42.

Doolittle, W.F. 1978. Genes in pieces: were they ever together? *Nature*, **272**, 581–582.

Dorit, R.L. & Gilbert, W. 1991. The limited universe of exons. *Genet. Devel.*, **1**, 464–469.

Douady, S. & Couder, Y. 1996. Phyllotaxis as a dynamical self organizing process, part III: the simulation of the transient regimes of ontogeny. *J. Theor. Biol.*, **178**, 295–312.

Dover, G.A. 1982. Molecular drive: a cohesive mode of species evolution. *Nature*, **299**, 111–117.

Dover, G.A. 1986. Molecular drive in multigene families: how biological novelties arise, spread and are assimilated. *Trends Genet.*, **2**, 159–165.

Dover, G.A. 1992. Observing development through evolutionary eyes: a practical approach to molecular coevolution. *Bioessays*, **14**, 281–287.

Dover, G.A. 1993. Evolution of genetic redundancy for advanced players. *Curr. Opin. Genet. Dev.*, **3**, 902–910.

Dover, G.A., Brown, S., Coen, E., Dallas, J., Strachan, T. & Trick, M. 1982. The dynamics of genome evolution and species differentiation. In: *Genome Evolution*, eds G.A. Dover & R.B. Flavell. Academic Press, London.

Duboule, D. 1992. The vertebrate limb: a model system to study the *Hox*/HOM gene network during development and evolution. *Bioessays*, **14**, 375–384.

Duboule, D. 1994. Temporal collinearity and the phylotypic progression: a basis for the stability of a vertebrate Bauplan and the evolution of morphologies through heterochrony. In: *The Evolution of Developmental Mechanisms* (*Development* 1994 Supplement). Company of Biologists, Cambridge,UK.

Duncan, I. 1987. The bithorax complex. *Ann. Rev. Genet.*, **21**, 285–319.

Durkin, M.E., Wewer, U.M. & Chung, A.E. 1995. Exon organization of the mouse entactin gene corresponds to the structural domains of the polypeptide and has regional homology to the low-density lipoprotein receptor gene. *Genomics*, **26**, 219–228.

Ebenman, B. 1992. Evolution in organisms that change their niches during the life cycle. *Am. Nat.*, **139**, 990–1021.

Echelard, Y., Epstein, D.J., St-Jaques, B., Shen, L., Mohler, J., McMahon, J.A. & McMahon, A.P. 1993. Sonic hedgehog, a member of a family of putative signaling molecules, is implicated in the regulation of CNS polarity. *Cell*, **75**, 1417–1430.

Ede, D.A. 1978. *An Introduction to Developmental Biology*. Blackie, Glasgow, UK.

Ekis, G. 1977. Classification, phylogeny and zoogeography of the genus *Perilypus* (Coleoptera: Cleridae). *Smithsonian Contrib. Zool.*, no. 227.

Ekker, S.C., von Kessler, D.P. & Beachy, P.A. 1992. Differential DNA sequence recognition is a determinant of specificity in homeotic gene action. *EMBO J.*, **11**, 4059–4072.

Ekker, S.C., Jackson, D.G., von Kessler, D.P., Sun, B.I., Young, K.E. & Beachy, P.A. 1994. The degree of variation in DNA sequence recognition among four *Drosophila* homeotic proteins. *EMBO J.*, **13**, 3551–3560.

Eldredge, N. 1971. The allopatric model and phylogeny in Paleozoic invertebrates. *Evolution*, **25**, 156–167.

Eldredge, N. 1979. Cladism and common sense. In: *Phylogenetic Analysis and Paleontology*, eds J. Cracraft & N. Eldredge. Columbia University Press, New York.

Eldredge, N. 1985. *Unfinished Synthesis: Biological Hierarchies and Modern Evolutionary Thought*. Oxford University Press, New York.

Eldredge, N. & Gould, S.J. 1972. Punctuated equilibria: an alternative to phyletic gradualism. In: *Models in Paleobiology*, ed. T.J.M. Schopf. Freeman, San Francisco.

Ericson, J., Muhr, J., Placzek, M., Lints, T., Jessell, T.M. & Edlund, T. 1995. *Sonic hedgehog* induces the differentiation of ventral forebrain neurons: a common signal for ventral patterning within the neural tube. *Cell*, **81**, 747–756.

Errington, J. 1991. Possible intermediate steps in the evolution of a prokaryotic developmental system. *Proc. Roy. Soc. Lond B.,* **244**, 117–121.

Erwin, D.H. & Valentine, J.W. 1984. "Hopeful monsters", transposons, and the metazoan radiation. *Proc. Nat. Acad. Sci. USA,* **81**, 5482–5483.

Estabrook, G.F. 1972. Cladistic methodology: a discussion of the theoretical basis for the induction of evolutionary history. *Ann. Rev. Ecol. Syst.,* **3**, 427–456.

Fain, A. 1994. Adaptation, specificity and host-parasite coevolution in mites (Acari). *Int. J. Parasitol.,* **24**, 1273–1283.

Fainsod, A., Steinbeisser, H. & De Robertis, E.M. 1994. On the function of *BMP-4* in patterning the marginal zone of the *Xenopus* embryo. *EMBO J.,* **13**, 5015–5025.

Falconer, D.S. 1989. *Introduction to Quantitative Genetics,* 3rd edition. Longman, London.

Fan, C.-M. & Tessier-Lavigne, M. 1994. Patterning of mammalian somites by surface ectoderm and notochord: evidence for sclerotome induction by a *hedgehog* homolog. *Cell,* **79**, 1175–1186.

Fan, C.-M., Porter, J.A., Chiang, C., Chang, D.T., Beachy, P.A. & Tessier-Lavigne, M. 1995. Long-range sclerotome induction by Sonic hedgehog: direct role of the amino-terminal cleavage product and modulation by the cyclic AMP signalling pathway. *Cell,* **81**, 457–465.

Farkas, G., Gausz, J., Galloni, M., Reuter, G., Gyurkovics, H. & Karch, F. 1994. The *Trithorax-like* gene encodes the *Drosophila* GAGA factor. *Nature,* **371**, 806–808.

Fedonkin, M.A. 1985. Precambrian metazoans: the problems of preservation, systematics and evolution. *Phil. Trans. Roy. Soc. Lond. B.,* **311**, 27–45.

Field, K.G., Olsen, G.J., Lane, D.J., Giovannoni, S.J., Ghiselin, M.T., Raff, E.C., Pace, N.R. & Raff, R.A. 1988. Molecular phylogeny of the animal kingdom. *Science,* **239**, 748–753.

Fields, C., Adams, M.D., White, O. & Venter, J.C. 1994. How many genes in the human genome? *Nature Genet.,* **7**, 345–346.

Fietz, M.J., Concordet, J-P., Barbosa, R., Johnson, R., Krauss, S., McMahon, A.P., Tabin, C. & Ingham, P. 1994. The *hedgehog* gene family in *Drosophila* and vertebrate development. In: *The Evolution of Developmental Mechanisms,* eds M. Akam, P. Holland, P. Ingham & G. Wray. Company of Biologists, Cambridge, UK.

Fisher, R.A. 1918. The correlations between relatives on the supposition of Mendelian inheritance. *Trans. R. Soc. Edinburgh,* **52**, 399–433.

Fisher, R.A. 1930. *The Genetical Theory of Natural Selection.* Clarendon Press, Oxford, UK.

Fitch, W.M. 1970. Distinguishing homologous from analogous proteins. *Syst. Zool.,* **19**, 99–113.

Fletcher, J.C., Burtis, K.C., Hogness, D.S. & Thummel, C.S. 1995. The *Drosophila* E74 gene is required for metamorphosis and plays a role in the polytene chromosome puffing response to ecdysone. *Development,* **121**, 1455–1465.

Fletcher, J.C. & Thummel, C.S. 1995. The *Drosophila* E74 gene is required for the proper stage- and tissue-specific transcription of ecdysone-regulated genes at the onset of metamorphosis. *Development*, **121**, 1411–1421.

Foote, M. 1994. Morphological disparity in Ordovician–Devonian crinoids and the early saturation of morphological space. *Paleobiology*, **20**, 320–344.

Ford, E.B. 1971. *Ecological Genetics*, 3rd edition. Chapman & Hall, London.

Fortey, R.A. & Chatterton, B.D.E. 1988. Classification of the trilobite suborder Asaphina. *Palaeontology*, **31**, 165–222.

Fortey, R.A. & Jefferies, R.P.S. 1982. Fossils and phylogeny – a compromise approach. In: *Problems of Phylogenetic Reconstruction*, eds K.A. Joysey & A.E. Friday. Academic Press, London.

Fortey, R.A., Briggs, D.E.G. & Wills, M.A. 1996. The Cambrian evolutionary 'explosion': decoupling cladogenesis from morphological disparity. *Biol. J. Linn. Soc.*, **57**, 13–33.

François, V. & Bier, E. 1995. *Xenopus chordin* and *Drosophila short gastrulation* genes encode homologous proteins functioning in dorsal–ventral axis formation. *Cell*, **80**, 19–20.

François, V., Solloway, M., O'Neill, J.W., Emery, J. & Bier, E. 1994. Dorsal–ventral patterning of the *Drosophila* embryo depends on a putative negative growth factor encoded by the *short gastrulation* gene. *Genes Devel.*, **8**, 2602–2616.

Frazzetta, T.H. 1975. *Complex Adaptations in Evolving Populations*. Sinauer, Sunderland, MA.

Freeman, G. & Lundelius, J.W. 1982. The developmental genetics of dextrality and sinistrality in the gastropod *Lymnaea peregra*. *Wilhelm Roux's Archives*, **191**, 69–83.

Friedrich, M. & Tautz, D. 1995. Ribosomal DNA phylogeny of the major extant arthropod classes and the evolution of myriapods. *Nature*, **376**, 165–167.

Fryxell, K.J. 1996. The coevolution of gene family trees. *Trends Genet.*, **12**, 364–369.

Funch, P. & Mobjerg Kristensen, R. 1995. Cycliophora is a new phylum with affinities to Entoprocta and Ectoprocta. *Nature*, **378**, 711–714.

Gabbott, S.E., Aldridge, R.J. & Theron, J.N. 1995. A giant conodont with preserved muscle tissue from the Upper Ordovician of South Africa. *Nature*, **374**, 800–803.

Garcia-Bellido, A. 1975. Genetic control of wing disc development in *Drosophila*. In: *Cell Patterning, CIBA Found. Symp.* No. 29 (new series). Associated Scientific Publishers, Amsterdam.

Garcia-Bellido, A., Ripoll, R. & Morata, G. 1973. Developmental compartmentalization of the wing disk of *Drosophila*. *Nature New Biol.*, **245**, 251–253.

Garcia-Fernàndez, J. & Holland, P.W.H. 1994. Archetypal organization of the amphioxus *Hox* gene cluster. *Nature*, **370**, 563–566.

Gardiner, D.M., Blumberg, B., Komine, Y. & Bryant, S.V. 1995. Regulation of *Hox*A expression in developing and regenerating axolotl limbs. *Development*, **121**, 1731–1741.

Garstang, W. 1929. The origin and evolution of larval forms. *Brit. Ass. Adv. Sci. Rep. 96th Mtg.* 77–98.

Gates, R.R. 1915. *The Mutation Factor in Evolution, With Particular Reference to Oenothera*. Macmillan, London.

Gaunt, S.J. 1991. Expression patterns of mouse *Hox* genes: clues to an understanding of developmental and evolutionary strategies. *Bioessays*, **13**, 234–242.

Gaunt, S.J., Sharpe, P.T. & Duboule, D. 1988. Spatially restricted domains of homeo-gene transcripts in mouse embryos: relation to a segmented body plan. *Development*, **104**, 169–179.

Gause, G.F. 1934. *The Struggle for Existence*. Williams & Wilkins, Baltimore, MD.

Gehring, W.J., Affolter, M. & Bürglin, T. 1994. Homeodomain proteins. *Ann. Rev. Biochem.*, **63**, 487–526.

Geoffroy Saint-Hilaire, E. 1822. Considérations générales sur la vertèbre. *Mem. Mus. Hist. Nat.*, **9**, 89–119.

Gibson, G. & Hogness, D.S. 1996. Effect of polymorphism in the *Drosophila* regulatory gene *Ultrabithorax* on homeotic stability. *Science*, **271**, 200–203.

Gilbert, S.F. 1994. *Developmental Biology*, 4th edition. Sinauer, Sunderland, MA.

Gilbert, S.F. & Raunio, A.M. (eds) 1997. *Embryology: Constructing the Organism*. Sinauer, Sunderland, MA.

Gilbert, S.F., Opitz, J.M. & Raff, R.A. 1996. Resynthesizing evolutionary and developmental biology. *Dev. Biol.*, **173**, 357–372.

Gilbert, W. 1978. Why genes in pieces? *Nature*, **271**, 501.

Gilbert, W. & Glynias, M. 1993. On the ancient nature of introns. *Gene*, **135**, 137–144.

Gillespie, J.H. 1984. Molecular evolution over the mutational landscape. *Evolution*, **38**, 1116–1129.

Glaessner, M.F. 1958. New fossils from the base of the Cambrian in South Australia. *Trans. Roy. Soc. S. Aust.*, **81**, 185–188.

Glaessner, M.F. 1959. Precambrian Coelenterata from Australia, Africa and England. *Nature*, **183**, 1472–1473.

Glaessner, M.F 1969. Trace fossils from the Precambrian and basal Cambrian. *Lethaia*, **2**, 369–393.

Glaessner, M.F. 1971. Geographic distribution and time range of the Ediacara Precambrian Fauna. *Geol. Soc. Am. Bull.*, **82**, 509–514.

Glaessner, M.F. 1976. A new genus of polychaete worms from the Late Precambrian of South Australia. *Trans. Roy. Soc. S. Aust.*, **100**, 169–170.

Glaessner, M.F. 1984. *The Dawn of Animal Life: A Biohistorical Study*. Cambridge University Press, Cambridge, UK.

Goldschmidt, R. 1940. *The Material Basis of Evolution*. Yale University Press, New Haven, CT.

Goldschmidt, R. 1952. Homeotic mutants and evolution. *Acta Biotheoretica*, **10**, 87–104.

González-Reyes, A., Elliott, H. & St. Johnston, D. 1995. Polarization of both major body axes in *Drosophila* by *gurken-torpedo* signalling. *Nature*, **375**, 654–658.

Goodwin, B. 1984. Changing from an evolutionary to a generative paradigm in biology. In: *Evolutionary Theory: Paths into the Future*, ed. J.W. Pollard. Wiley, Chichester, UK.

Goodwin, B. 1994a. *How the Leopard Changed its Spots: The Evolution of Complexity*. Weidenfeld & Nicolson, London.

Goodwin, B. 1994b. Homology, development and heredity. In: *Homology: The Hierarchical Basis of Comparative Biology*, ed. B.K. Hall. Academic Press, New York.

Gould, A.P. & White, R.A. 1992. Connectin, a target of homeotic gene control in *Drosophila. Development*, **116**, 1163–1174.

Gould, S.J. 1977a. *Ontogeny and Phylogeny*. Harvard University Press, Cambridge, MA.

Gould, S.J. 1977b. Eternal metaphors of palaeontology. In: *Patterns of Evolution*, ed. A. Hallam. Elsevier, Amsterdam.

Gould, S.J. 1983. *Hen's Teeth and Horse's Toes: Further Reflections in Natural History*. Norton, New York.

Gould, S.J. 1989. *Wonderful Life: The Burgess Shale and the Nature of History*. Hutchinson Radius, London.

Gould, S.J. 1991. The disparity of the Burgess Shale arthropod fauna and the limits to cladistic analysis: why we must strive to quantify morphospace. *Paleobiology*, **17**, 411–423.

Gould, S.J. & Vrba, E.S. 1982. Exaptation – a missing term in the science of form. *Paleobiology*, **8**, 4–15.

Gould, S.J., Young, N.D. & Kasson, B. 1985. The consequences of being different: sinistral coiling in *Cerion. Evolution*, **39**, 1364–1379.

Goustin, A.S., Betsholtz, C., Pfeifer-Ohlsson, S., Persson, H., Rydnert, J., Bywater, M., Holmgren, G., Heldin, C.-H., Westermark, B. & Ohlsson, R. 1985. Coexpression of the *sis* and *myc* proto-oncogenes in developing human placenta suggests autocrine control of trophoblast growth. *Cell*, **41**, 301–312.

Graba, Y., Aragnol, D., Laurenti, P., Garzino, V., Charmot, D., Berenger, H. & Pradel, J. 1992. Homeotic control in *Drosophila*; the *scabrous* gene is an *in vivo* target of Ultrabithorax proteins. *EMBO J.*, **11**, 3375–3384.

Grant, P.R. 1986. *Ecology and Evolution of Darwin's Finches*. Princeton University Press, Princeton, NJ.

Grant, V. 1981. *Plant Speciation*, 2nd edition. Columbia University Press, New York.

Gupta, R.S. 1995. Evolution of the chaperonin families (Hsp60, Hsp10 and Tcp-1) of proteins and the origin of eukaryotic cells. *Molec. Microbiol.*, **15**, 1–11.

Gurdon, J.B., Harger, P., Mitchell, A. & Lemaire, P. 1994. Activin signalling and response to a morphogen gradient. *Nature*, **371**, 487–492.

Gurdon, J.B., Mitchell, A. & Mahoney, D. 1995. Direct and continuous assessment by cells of their position in a morphogen gradient. *Nature*, **376**, 520–521.

Gutjahr, T., Frei, E. & Noll, M. 1993. Complex regulation of early *paired* expression: initial activation by gap genes and pattern modulation by pair-rule genes. *Development*, **117**, 609–623.

Haeckel, E. 1866. *Generelle Morphologie der Organismen.* Georg Reimer, Berlin.

Haeckel, E. 1896. *The Evolution of Man: A Popular Exposition of the Principal Points of Human Ontogeny and Phylogeny.* Appleton, New York.

Hafen, E., Kuroiwa, A. & Gehring, W.J. 1984. Spatial distribution of transcripts from the segmentation gene *fushi tarazu* during *Drosophila* embryonic development. *Cell*, **37**, 833–841.

Haldane, J.B.S. 1932. *The Causes of Evolution.* Longman, London.

Haldane, J.B.S. 1964. A defence of beanbag genetics. *Perspectives in Biology and Medicine*, **7**, 343–359.

Hall, B.G. 1992. Selection-induced mutations occur in yeast. *Proc. Nat. Acad. Sci. USA*, **89**, 4300–4303.

Hall, B.K. 1992. *Evolutionary Developmental Biology.* Chapman & Hall, London.

Hallam, A. 1992. *Phanerozoic Sea-Level Changes.* Columbia University Press, New York.

Hamilton, L. 1976. *From Egg to Adolescent: Xenopus – A Model for Development.* English Universities Press, London.

Harland, W.B., Armstrong, R.L., Cox, A.V., Craig, L.E., Smith, A.G. & Smith, D.G. 1990. *A Geological Time Scale 1989.* Cambridge University Press, Cambridge, UK.

Hartl, D.L. & Clark, A.G. 1989. *Principles of Population Genetics.* Sinauer, Sunderland, MA.

Hartley, B.S., Brown, J.R., Kauffman, D.L. & Smillie, L.B. 1965. Evolutionary similarities between pancreatic proteolytic enzymes. *Nature*, **207**, 1157–1159.

Harvey, P.H. & Pagel, M.D. 1991. *The Comparative Method in Evolutionary Biology.* Oxford University Press, Oxford, UK.

Heberlein, U., Singh, C.M., Luk, A.Y. & Donohoe, T.J. 1995. Growth and differentiation in the *Drosophila* eye coordinated by *hedgehog*. *Nature*, **373**, 709–711.

Hedges, S.B., Parker, P.H., Sibley, C.G. & Kumar, S. 1996. Continental breakup and the ordinal diversification of birds and mammals. *Nature*, **381**, 226–229.

Hendrickson, J.E. & Sakonju, S. 1995. Cis and trans interactions between the *iab* regulatory regions and *abdominal-A* and *Abdominal-B* in *Drosophila melanogaster*. *Genetics*, **139**, 835–848.

Henfrey, A. & Huxley, T.H. 1853. *Scientific Memoirs, Selected from the Transactions of Foreign Academies of Science, and from Foreign Journals: Natural History.* Taylor & Francis, London.

Hennig, W. 1966. *Phylogenetic Systematics.* University of Illinois Press, Urbana, IL.

Hennig, W. 1981. *Insect Phylogeny.* Wiley, Chichester and New York.

Hickey, D.A. & Benkel, B. 1986. Introns as relict retrotransposons : implications for the evolutionary origin of eukaryotic mRNA splicing mechanisms. *J. Theor. Biol.*, **121**, 283–291.

Hickey, L.J. & Doyle, J.A. 1977. Early Cretaceous fossil evidence for angiosperm evolution. *Bot. Rev.*, **43**, 3–104.

Hill, J., Clarke, J.D.W., Vargesson, N., Jowett, T. & Holder, N. 1995. Exogenous retinoic acid causes specific alterations in the development of the midbrain and hindbrain of the zebrafish embryo including positional respecification of the Mauthner neuron. *Mech. Dev.*, **50**, 3–16.

Hoch, J.A. & Silhavy, T.J. 1995. *Two-Component Signal Transduction*. ASM Press, Washington DC.

Holland, L.Z., Kene, M., Williams, N.A. & Holland, N.D. 1997. Sequence and embryonic expression of the amphioxus *engrailed* gene (*AmphiEn*): the metameric pattern of the transcription resembles that of its segment-polarity homolog in *Drosophila*. *Development*, **124**, 1723-1732.

Holland, P.W.H. & Garcia-Fernàndez, J. 1996. *Hox* genes and chordate evolution. *Dev. Biol.*, **173**, 382–395.

Holland, P.W.H., Garcia-Fernàndez, J., Holland, L.Z., Williams, N.A. & Holland, N.D. 1994a. The molecular control of spatial patterning in Amphioxus. *J. Mar. Biol. Ass. U.K.*, **74**, 49–60.

Holland, P.W.H., Garcia-Fernàndez, J., Williams, N.A. & Sidow, A. 1994b. Gene duplications and the origins of vertebrate development. In: *The Evolution of Developmental Mechanisms*, eds M. Akam, P. Holland, P. Ingham & G. Wray. Company of Biologists, Cambridge, UK.

Holley, S.A., Jackson, P.D., Sasai, Y., Lu, B., De Robertis, E.M., Hoffmann, F.M. & Ferguson, E.L. 1995. A conserved system for dorsal–ventral patterning in insects and vertebrates involving *sog* and *chordin*. *Nature*, **376**, 249–253.

Horder, T.J. 1994. Partial truths: a review of the use of concepts in the evolutionary sciences. In: *Models in Phylogeny Reconstruction*, eds R.W. Scotland, D.J. Siebert & D.M. Williams. Clarendon Press, Oxford, UK.

Hörstadius, S. 1973. *Experimental Embryology of Echinoderms*. Clarendon Press, Oxford, UK.

Hotary, K.B. & Robinson, K.R. 1992. Evidence of a role for endogenous electric fields in chick embryo development. *Development*, **114**, 985–996.

Huet, F., Ruiz, C. & Richards, G. 1995. Sequential gene activation by ecdysone in *Drosophila melanogaster*: the hierarchical equivalence of early and early-late genes. *Development*, **121**, 1195–1204.

Hughes, C.P. 1975. Redescription of *Burgessia bella* from the Middle Cambrian Burgess Shale, British Columbia. *Fossils and Strata* (Oslo), **4**, 415–435.

Hunter, C.P. & Kenyon, C. 1995. Specification of anteroposterior cell fates in *Caenorhabditis elegans* by *Drosophila Hox* proteins. *Nature*, **377**, 229–232.

Huxley, J.S. 1924. Constant differential growth-ratios and their significance. *Nature*, **114**, 895.

Huxley, J.S. 1932. *Problems of Relative Growth*. Methuen, London.

Huxley, J.S. 1942. *Evolution, The Modern Synthesis*. Allen and Unwin, London.

Ingham, P.W. 1988. The molecular genetics of embryonic pattern formation in *Drosophila. Nature*, **335**, 25–34.

Ingham, P.W. 1995. Signalling by hedgehog family proteins in *Drosophila* and vertebrate development. *Curr. Opin. Genet. Dev.*, **5**, 492–498.

Ingham, P.W. & Martinez-Arias, A. 1986. The correct activation of Antennapedia and Bithorax complex genes requires the *fushi tarazu* gene. *Nature*, **324**, 592–597.

Ingham, P.W., Baker, N.E. & Martinez-Arias, A. 1988. Regulation of segment polarity genes in the *Drosophila* blastoderm by *fushi tarazu* and *even skipped. Nature*, **331**, 73–75.

Ingham, P.W., Taylor, A.M. & Nakano, Y. 1991. Role of the *Drosophila patched* gene in positional signalling. *Nature*, **353**, 184–187.

Ingram, V.M. 1961. Gene evolution and the haemoglobins. *Nature*, **189**, 704–708.

Irish, V.F. & Gelbart, W.M. 1987. The *decapentaplegic* gene is required for dorsal-ventral patterning of the *Drosophila* embryo. *Genes Dev.*, **1**, 868–879.

Irvine, K.D., Botas, J., Jha, S., Mann, R.S. & Hogness, D.S. 1993. Negative autoregulation by *Ultrabithorax* controls the level and pattern of its expression. *Development*, **117**, 387–399.

Irvine, R.F., Michell, R.H. & Marshall, C.J. (eds) 1996. Current understanding of intracellular signalling pathways. *Phil. Trans. Roy. Soc. Lond. B.*, **351**, 123–241.

Istock, C.A. 1967. The evolution of complex life cycle phenomena: an ecological perspective. *Evolution*, **21**, 592–605.

Jacob, F. 1977. Evolution and tinkering. *Science*, **196**, 1161–1166.

James, A.C., Azevedo, R.B. & Partridge, L. 1995. Cellular basis and developmental timing in a size cline of *Drosophila melanogaster. Genetics*, **140**, 659–666.

Jefferies, R.P.S. 1986. *The Ancestry of the Vertebrates*. British Museum (Natural History), London.

John, A., Smith, S.T. & Jaynes, J.B. 1995. Inserting the ftz homeodomain into engrailed creates a dominant transcriptional repressor that specifically turns off ftz target genes *in vivo. Development*, **121**, 1801–1813.

Johnson, F.B., Parker, E. & Krasnow, M.A. 1995. Extradenticle protein is a selective cofactor for the *Drosophila* homeotics: role of the homeodomain and YPWM amino acid motif in the interaction. *Proc. Nat. Acad. Sci. USA*, **92**, 739–743.

Johnson, M.S., Murray, J. & Clarke, B. 1993. The ecological genetics and adaptive radiation of *Partula* on Moorea. *Oxford Surveys in Evolutionary Biology*, **9**, 167–238.

Johnson, R.L., Laufer, E., Riddle, R.D. & Tabin, C. 1994. Ectopic expression of *Sonic hedgehog* alters dorsal–ventral patterning of somites. *Cell*, **79**, 1165–1173.

Jones, C.M., Lyons, K.M., Lapan, P.M., Wright, C.V.E. & Hogan, B.L.M. 1992. DVR-4 (Bone Morphogenetic Protein-4) as a posterior-ventralizing factor in *Xenopus* mesoderm induction. *Development*, **115**, 639–647.

Jones, C.W., Jahraus, C.D. & Tran, P.V. 1994. Interspecific comparisons of the unusually long 5′ leader of the *Drosophila* ecdysone-inducible gene E74A. *Insect Biochem. Molec. Biol.*, **24**, 875–882.

Jones, J.S. 1984. Developing theories of evolution. *Nature*, **312**, 386.

Jones, J.S., Leith, B.H. & Rawlings, P. 1977. Polymorphism in *Cepaea*: A problem with too many solutions? *Ann. Rev. Ecol. Syst.*, **8**, 109–143.

Jongeling, T.B. 1996. Self-organization and competition in evolution: a conceptual problem in the use of fitness landscapes. *J. Theor. Biol.*, **178**, 369–373.

Jonk, L.J., de Jonge, M.E., Vervaart, J.M., Wissink, S. & Kruijer, W. 1994. Isolation and developmental expression of retinoic-acid-induced genes. *Dev. Biol.*, **161**, 604–614.

Judd, B.H., Shen, M.W. & Kaufman, T.C. 1972. The anatomy of a segment of the X chromosome of *Drosophila melanogaster*. *Genetics*, **71**, 139–156.

Karim, F.D., Guild, G.M. & Thummel, C.S. 1993. The *Drosophila* Broad-Complex plays a key role in controlling ecdysone-regulated gene expression at the onset of metamorphosis. *Development*, **118**, 977–988.

Kauffman, S.A. 1993. *The Origins of Order: Self-Organization and Selection in Evolution*. Oxford University Press, New York.

Kauffman, S.A. 1995. *At Home in the Universe: The Search for Laws of Self-Organization and Complexity*. Viking-Penguin, London.

Kaufman, T.C., Seeger, M.A. & Olsen, G. 1990. Molecular and genetic organization of the Antennapedia gene complex of *Drosophila melanogaster*. *Adv. Gen.*, **27**, 309–362.

Kenyon, C. & Wang, B. 1991. A cluster of *Antennapedia*-class homeobox genes in a nonsegmental animal. *Science*, **253**, 516–517.

Kerney, M.P. & Cameron, R.A.D. 1979. *A Field Guide to the Land Snails of Britain and North-West Europe*. Collins, London.

Kerr, J.G. 1919. *Text-Book of Embryology, vol. II, Vertebrata with the Exception of Mammalia*. Macmillan, London.

Kimmel, C.B., Ballard, W.W., Kimmel, S.R., Ullman, B. & Schilling, T.F. 1995. Stages of embryonic development of the zebrafish. *Dev. Dynamics*, **203**, 253–310.

Kimura, M. 1968. Genetic variability maintained in a finite population due to production of neutral and nearly neutral isoalleles. *Genet. Res.*, **11**, 247–269.

Kimura, M. 1983. *The Neutral Theory of Molecular Evolution*. Cambridge University Press, Cambridge, UK.

Kimura, Y., Matsunami, H., Inoue, T., Shimamura, K., Uchida, N., Ueno, T., Miyazaki, T. & Takeichi, M. 1995. Cadherin-11 expressed in association with mesenchymal morphogenesis in the head, somite, and limb bud of early mouse embryos. *Dev. Biol.*, **169**, 347–358.

King, J.A. & Millar, R.P. 1995. Evolutionary aspects of gonadotropin-releasing hormone and its receptor. *Cell. Molec. Neurobiol.*, **15**, 5–23.

King, M. 1993. *Species Evolution: The Role of Chromosomal Change*. Cambridge University Press, Cambridge, UK.

Kintner, C. 1992. Regulation of embryonic cell adhesion by the cadherin cyto-plasmic domain. *Cell*, **69**, 225–236.

Kluge, A.G. 1985. Ontogeny and phylogenetic systematics. *Cladistics*, **1**, 13–27.

Kluge, A.G. & Farris, J.S. 1969. Quantitative phyletics and the evolution of Anurans. *Syst. Zool.*, **18**, 1–32.

Knoll, A.H. 1991. End of the Proterozoic Eon. *Sci. Am.*, **265**(4), 64–73.

Knoll, A.H. 1992. The early evolution of eukaryotes: a geological perspective. *Science*, **256**, 622–627.

Knoll, A.H. 1996. Daughters of time. *Paleobiology*, **22**, 1–7.

Koch, E.A., Smith, P.A. & King, R.C. 1967. The division and differentiation of *Drosophila* oocytes. *J. Morphol.*, **121**, 55–70.

Krauss, S., Concordet, J-P. & Ingham, P.W. 1993. A functionally conserved homolog of the *Drosophila* segment polarity gene *hh* is expressed in tissue with polarizing activity in zebrafish embryos. *Cell*, **75**, 1431–1444.

Kraut, R. & Levine, M. 1991. Spatial regulation of the gap gene *giant* during *Drosophila* development. *Development*, **111**, 601–609.

Kreitman, M. 1996. The neutral theory is dead. Long live the neutral theory. *Bioessays*, **18**, 678–683.

Kretsinger, R.H. & Nakayama, S. 1993. Evolution of EF-hand calcium-modu-lated proteins. IV. Exon shuffling did not determine the domain composi-tions of EF-hand proteins. *Molec. Evol.*, **36**, 477–488.

Kristensen, N.P. 1981. Phylogeny of insect orders. *Ann. Rev. Entomol.*, **26**, 135–157.

Kröner, A. 1981. Precambrian plate tectonics. In: *Precambrian Plate Tectonics*, ed. A. Kröner, Elsevier, Amsterdam.

Krumlauf, R. 1992. Evolution of the vertebrate *Hox* homeobox genes. *Bioessays*, **14**, 245–252.

Krumlauf, R. 1993. *Hox* genes and pattern formation in the branchial region of the vertebrate head. *Trends Genet.*, **9**, 106–112.

Kuma, K., Iwabe, N. & Miyata, T. 1993. Motifs of cadherin- and fibronectin type III-related sequences and evolution of the receptor-type-protein tyro-sine kinases: sequence similarity between proto-oncogene *ret* and cadherin family. *Molec. Biol. Evol.*, **10**, 539–551.

Labedan, B. & Riley, M. 1995. Widespread protein sequence similarities: ori-gins of *Escherichia coli* genes. *J. Bacteriol.*, **177**, 1585–1588.

Lacalli, T.C. 1995. Dorsoventral axis inversion. *Nature*, **373**, 110–111.

Lack, D. 1947. *Darwin's Finches: An Essay on the General Biological Theory of Evolution*. Cambridge University Press, Cambridge, UK.

Lake, J.A. 1990. Origin of the Metazoa. *Proc. Nat. Acad. Sci. USA*, **87**, 763–766.

Lambert, M.E., McDonald, J.F. & Weinstein, I.B. (eds) 1988. *Eukaryotic Transposable Elements as Mutagenic Agents*. Cold Spring Harbor Laboratory, New York.

Lankester, E.R. 1870. On the use of the term homology in modern zoology, and the distinction between homogenetic and homoplastic agreements. *Ann. Mag. Nat. Hist. [4]*, **6**, 34–43.

Lauffenburger, D.A. & Linderman, J.J. 1993. *Receptors: Models for Binding, Trafficking and Signaling.* Oxford University Press, New York.

Lawlor, R. & Maynard Smith, J. 1976. The coevolution and stability of competing species. *Am. Nat.*, **110**, 79–99.

Lawrence, P.A. 1992. *The Making of a Fly: The Genetics of Animal Design.* Blackwell, Oxford, UK.

Lawrence, P.A. & Johnston, P. 1989. Pattern formation in the *Drosophila* embryo: allocation of cells to parasegments by *even-skipped* and *fushi tarazu. Development*, **105**, 761–768.

Lawson, M.A. & Maxfield, F.R. 1995. $Ca^{2+}$- and calcineurin-dependent recycling of an integrin to the front of migrating neutrophils. *Nature*, **377**, 75–79.

Leather, S.R. 1996. The case for the passive voice. *Nature*, **381**, 467.

Lecuit, T., Brook, W.J., Ng, M., Calleja, M., Sun, H. & Cohen, S.M. 1996. Two distinct mechanisms for long-range patterning by Decapentaplegic in the *Drosophila* wing. *Nature*, **381**, 387–393.

Lemaire, P. & Gurdon, J.B. 1994. Vertebrate embryonic inductions. *Bioessays*, **16**, 617–620.

Lenski, R.E. 1989. Are some mutations directed? *TREE*, **4**, 148–150.

Lepage, T., Cohen, S.M., Diaz-Benjumea, F.J. & Parkhurst, S.M. 1995. Signal transduction by cAMP-dependent protein kinase A in *Drosophila* limb patterning. *Nature*, **373**, 711–715.

Levene, H. 1953. Genetic equilibrium when more than one ecological niche is available. *Am. Nat.*, **87**, 331–333.

Levi, G., Ginsberg, D., Girault, J.M., Sabanay, I., Thiery, J.P. & Geiger, B. 1991. EP-cadherin in muscles and epithelia of *Xenopus laevis* embryos. *Development*, **113**, 1335–1344.

Levin, M., Johnson, R.L., Stern, C.D., Kuehn, M. & Tabin, C. 1995. A molecular pathway determining left–right asymmetry in chick embryogenesis. *Cell*, **82**, 803–814.

Levin, M. & Mercola, M. 1998. The compulsion of chirality: toward an understanding of left–right asymmetry. *Genes and Development*, **12**, 763–769.

Lewin, B. 1994. *Genes V.* Oxford University Press, Oxford, UK.

Lewis, E.B. 1963. Genes and developmental pathways. *Am. Zool.*, **3**, 33–56.

Lewis, E.B. 1978. A gene complex controlling segmentation in *Drosophila. Nature*, **276**, 565–570.

Lewontin, R.C. 1974a. *The Genetic Basis of Evolutionary Change.* Columbia University Press, New York.

Lewontin, R.C. 1974b. The analysis of variance and the analysis of causes. *Am. J. Hum. Gen.*, **26**, 400–411.

Lewontin, R.C. & Caspari, E.W. 1960. Developmental selection of mutations. *Science*, **132**, 1688–1692.

Lillie, F. 1895. The embryology of the Unionidae. *J. Morph.*, **10**, 1–100.

Lipps, J.H. (ed.) 1993. *Fossil Prokaryotes and Protists.* Blackwell, Cambridge, MA.

Lodish, H., Baltimore, D., Berk, A., Zipursky, S.L., Matsudaira, P. & Darnell, J. 1995. *Molecular Cell Biology*, 3rd edition. Freeman, New York.

Logsdon, J.M. Jr., Tyshenko, M.G., Dixon, C., D-Jafari, J., Walker, V.K. & Palmer, J.D. 1995. Seven newly discovered intron positions in the triose-phosphate isomerase gene: evidence for the introns-late theory. *Proc. Nat. Acad. Sci. USA*, **92**, 8507–8511.

Long, J.A., Moan, E.I., Medford, J.I. & Barton, M.K. 1996. A member of the KNOTTED class of homeodomain proteins encoded by the *STM* gene of *Arabidopsis*. *Nature*, **379**, 66–69.

Løvtrup, S. 1974. *Epigenetics: A Treatise on Theoretical Biology*. Wiley, London.

Løvtrup, S. 1978. On von Baerian and Haeckelian recapitulation. *Syst. Zool.*, **27**, 348–352.

Lowe, L.A., Supp, D.M., Sampath, K., Yokoyama, T., Wright, C.V.E., Potter, S.S., Overbeek, P. & Kuehn, M.R. 1996. Conserved left-right asymmetry of nodal expression and alterations in murine *situs inversus*. *Nature*, **381**, 158–161.

Mabee, P.M. 1989. An empirical rejection of the ontogenetic polarity criterion. *Cladistics*, **5**, 409–416.

MacArthur, R.H. 1972. *Geographical Ecology: Patterns in the Distribution of Species*. Harper and Row, New York.

MacDonald, P.M., Ingham, P. & Struhl, G. 1986. Isolation, structure and expression of *even-skipped*: a second pair-rule gene of *Drosophila* containing a homeobox. *Cell*, **47**, 721–734.

MacFadden, B.J. 1992. *Fossil Horses: Systematics, Paleobiology and Evolution of the Family Equidae*. Cambridge University Press, New York.

Mackay, T.F.C. 1996. The nature of quantitative genetic variation revisited: lessons from *Drosophila* bristles. *Bioessays*, **18**, 113–121.

Mackay, T.F.C. & Langley, C.H. 1990. Molecular and phenotypic variation in the *achaete-scute* region of *Drosophila melanogaster*. *Nature*, **348**, 64–66.

Maddison,W.P., Donoghue, M.J. & Maddison, D.R. 1984. Outgroup analysis and parsimony. *Syst. Zool.*, **33**, 83–103.

Mahoney, P.A., Weber, U., Onofrechuk, P., Biessmann, H., Bryant, P.J. & Goodman, C.S. 1991. The fat tumor supressor gene in *Drosophila* encodes a novel member of the cadherin gene superfamily. *Cell*, **67**, 853–868.

Mani, G.S. & Clarke, B.C. 1990. Mutational order: a major stochastic process in evolution. *Proc. Roy. Soc. Lond. B.*, **240**, 29–37.

Manoukian, A.S. & Krause, H.M. 1993. Control of segmental asymmetry in *Drosophila* embryos. *Development*, **118**, 785–796.

Manton, S.M. 1974. Arthropod phylogeny – a modern synthesis. *J. Zool. Lond.*, **171**, 111–130.

Manton, S.M. 1977. *The Arthropods: Habits, Functional Morphology and Evolution*. Oxford University Press, Oxford, UK.

Margalef, R. 1968. *Perspectives in Ecological Theory*. University of Chicago Press, Chicago, IL.

Margulis, L. & Schwartz, K.V. 1988. *Five Kingdoms: An Illustrated Guide to the Phyla of Life on Earth*, 2nd edition. Freeman, New York.

Marigo, V., Davey, R.A., Zuo, Y., Cunningham, J.M. & Tabin, C.J. 1996. Biochemical evidence that Patched is the Hedgehog receptor. *Nature*, **384**, 176–179.

Markow, T.A. 1995. Evolutionary ecology and developmental instability. *Ann. Rev. Entomol.*, **40**, 105–120.

Martinez-Arias, A. & Lawrence, P.A. 1985. Parasegments and compartments in the *Drosophila* embryo. *Nature*, **313**, 639–642.

Mastick, G.S., McKay, R., Oligino, T., Donovan, K. & López, A.J. 1995. Identification of target genes regulated by homeotic proteins in *Drosophila melanogaster* through genetic selection of *Ultrabithorax* protein-binding sites in yeast. *Genetics*, **139**, 349–363.

Matsunami, H., Miyatani, S., Inoue, T., Copeland, N.G., Gilbert, D.J., Jenkins, N.A. & Takeichi, M. 1993. Cell binding specificity of mouse R-cadherin and chromosomal mapping of the gene. *J. Cell Sci.*, **106**, 401–409.

Maynard Smith, J. 1966. Sympatric speciation. *Am. Nat.*, **100**, 637–650.

Maynard Smith, J. 1970. Genetic polymorphism in a varied environment. *Am. Nat.*, **104**, 487–490.

Maynard Smith, J. 1972. *On Evolution*. Edinburgh University Press, Edinburgh, UK.

Maynard Smith, J. 1989. *Evolutionary Genetics*. Oxford University Press, Oxford, UK.

Maynard Smith, J. & Szathmáry, E. 1995. *The Major Transitions in Evolution*. Freeman, Oxford, UK.

Maynard Smith, J., Burian, R., Kauffman, S., Alberch, P., Campbell, J., Goodwin, B., Lande, R., Raup, D. & Wolpert, L. 1985. Developmental constraints and evolution. *Q. Rev. Biol.*, **60**, 265–287.

Mayr, E. 1960. Where are we? *Cold Spring Harb. Symp. Quant. Biol.*, **24**, 1–14.

Mayr, E. 1963. *Animal Species and Evolution*. Harvard University Press, Cambridge, MA.

Mayr, E. 1969. *Principles of Systematic Zoology*. McGraw-Hill, New York.

Mayr, E. 1974. Cladistic analysis or cladistic classification? *Zeitschrift für Zoologische Systematik und Evolutionsforschung*, **12**, 94–128.

Mayr, E. 1994. Recapitulation reinterpreted: the somatic program. *Q. Rev. Biol.*, **69**, 223–232.

McDonald, J.H. & Kreitman, M. 1991. Adaptive protein evolution at the *Adh* locus in *Drosophila*. *Nature*, **351**, 652–654.

McEdward, L.R. & Janies, D.A. 1993. Life cycle evolution in Asteroids: what is a larva? *Biol. Bull.*, **184**, 255–268.

McGinnis, W. & Krumlauf, R. 1992. Homeobox genes and axial patterning. *Cell*, **68**, 283–302.

McKinney, M.L. & McNamara, K.J. 1991. *Heterochrony: The Evolution of Ontogeny*. Plenum Press, New York.

McMenamin, M.A.S. (1989). The origins and radiation of the early metazoa. In: *Evolution and the Fossil Record*, eds K. Allen & D.E.G. Briggs. Bellhaven Press, London.

Medawar, P.B. & Medawar, J.S. 1977. *The Life Science: Current Ideas of Biology*. Wildwood House, London.

Meinhardt, H. 1982. *Models of Biological Pattern Formation*. Academic Press, London.

Meno, C., Saijoh, Y., Fujii, H., Ikeda, M., Yokoyama, T., Yokoyama, M., Toyoda, Y. & Hamada, H. 1996. Left-right asymmetric expression of the TGFβ-family member *lefty* in mouse embryos. *Nature*, **381**, 151–155.

Meyerowitz, E.M. & Somerville, C.R. 1995. *Arabidopsis*. Cold Spring Harbor Laboratory Press, New York.

Miettinen, P.J., Ebner, R., Lopez, A.R. & Derynck, R. 1994. TGF-beta induced transdifferentiation of mammary epithelial cells to mesenchymal cells: involvement of type I receptors. *J. Cell Biol.*, **127**, 2021–2036.

Milne, A. 1961. Definition of competition among animals. *Symp. Soc. Exp. Biol.*, **15**, 40–61.

Minelli, A. & Bortoletto, S. 1988. Myriapod metamerism and arthropod segmentation. *Biol. J. Linn. Soc.*, **33**, 323–343.

Molven, A., Wright, C.V., Bremiller, R., De Robertis, E.M. & Kimmel, C.B. 1990. Expression of a homeobox gene product in normal and mutant zebrafish embryos: evolution of the tetrapod body plan. *Development*, **109**, 279–288.

Monk, M. (ed.) 1987. *Mammalian Development: A Practical Approach*. IRL Press, Oxford, UK.

Moon, R.T., DeMarais, A. & Olson, D.J. 1993. Responses to Wnt signals in vertebrate embryos may involve changes in cell adhesion and cell movement. *J. Cell Sci. Supplements*, **17**, 183–188.

Moreno, G. 1994. Genetic architecture, genetic behaviour, and character evolution. *Ann. Rev. Ecol. Syst.*, **25**, 31–44.

Morgan, T.H. 1932. *The Scientific Basis of Evolution*. Faber & Faber, London.

Mullis, K.B., Ferré, F. & Gibbs, R.A. (eds) 1994. *The Polymerase Chain Reaction (PCR)*. Birkhauser Verlag, Basel.

Murphy, J.B. & Nance, R.D. 1991. Supercontinent model for the contrasting character of Late Proterozoic orogenic belts. *Geology*, **19**, 469–472.

Murphy-Erdosh, C., Yoshida, C.K., Paradies, N. & Reichardt, L.F. 1995. The cadherin-binding specificities of B-cadherin and LCAM. *J. Cell Biol.*, **129**, 1379–1390.

Murray, J. & Clarke, B. 1966. The inheritance of polymorphic shell characters in *Partula* (Gastropoda). *Genetics*, **54**, 1261–1277.

Murray, J., Murray, E., Johnson, M.S. & Clarke, B. 1988. The extinction of *Partula* on Moorea. *Pacific Science*, **42**, 150–153.

Needham, J. 1934. *A History of Embryology*. Cambridge University Press, Cambridge, UK.

Nelson, G.J. 1978. Ontogeny, phylogeny, paleontology and the biogenetic law. *Syst. Zool.*, **27**, 324–345.

Nielsen, C. 1995. *Animal Evolution: Interrelationships of the Living Phyla*. Oxford University Press, Oxford, UK.

Nohno, T., Kawakami, Y., Ohuchi, H., Fujiwara, A., Yoshioka, H. & Noji, S. 1995. Involvement of the *Sonic hedgehog* gene in chick feather formation. *Biochem. Biophys. Res. Comm.*, **206**, 33–39.

Nursall, J.R. 1962. On the origin of the major groups of animals. *Evolution*, **16**, 118–123.

Nüsslein-Volhard, C. & Wieschaus, E. 1980. Mutations affecting segment number and polarity in *Drosophila*. *Nature*, **287**, 795–801.

O'Hagan, A. 1988. *Probability: Methods & Measurement*. Chapman & Hall, New York.

Oberlender, S.A. & Tuan, R.S. 1994. Expression and functional involvement of N-cadherin in embryonic limb chondrogenesis. *Development*, **120**, 177–187.

Ohno, S. 1970. *Evolution by Gene Duplication*. Springer-Verlag, New York.

Ohta, T. 1996. The current significance and standing of neutral and nearly neutral theories. *Bioessays*, **18**, 673–677.

Orr, H.A. 1998. The population genetics of adaptation: the distribution of factors fixed during adaptive evolution. *Evolution*, **52**, 935–949.

Orr, H.A. & Coyne, J.A. 1992. The genetics of adaptation: a reassessment. *Am. Nat.*, **140**, 725–742.

Owen, R. 1847. Report on the archetype and homologies of the vertebrate skeleton. *Meet. Br. Assoc. Adv. Sci. Rep.*, **16**, 169–340.

Owen, R. 1848. *On the Archetype and Homologies of the Vertebrate Skeleton*. John van Voorst, London.

Panchen, A.L. 1982. The use of parsimony in testing phylogenetic hypotheses. *Zool. J. Linn. Soc.*, **74**, 305–328.

Panchen, A.L. 1992. *Classification, Evolution and the Nature of Biology*. Cambridge University Press, Cambridge, UK.

Panchen, A.L. 1994. Richard Owen and the concept of homology. In: *Homology: The Hierarchical Basis of Comparative Biology*, ed. B.K. Hall. Academic Press, New York.

Panganiban, G., Irvine, S.M., Lowe, C., Roehl, H., Corley, L.S., Sherbon, B., Grenier, J.K., Fallon, J.F., Kimble, J., Walker, M., Wray, G.A., Swalla, B.J., Martindale, M.Q. & Carroll, S.B., 1997. The origin and evolution of animal appendages. *Proc. Natl. Acad. Sci.*, **94**, 5162–5166.

Park, T. 1948. Experimental studies of interspecies competition. I. Competition between populations of the flour beetles, *Tribolium confusum* and *Tribolium castaneum* Herbst. *Ecol. Monogr.*, **18**, 265–307.

Park, T. 1954. Experimental studies of interspecies competition. II. Temperature, humidity, and competition in two species of *Tribolium*. *Physiol. Zool.*, **27**, 177–238.

Park, T. 1962. Beetles, competition, and populations. *Science*, **138**, 1369–1375.

Patten, B.M. 1971. *Early Embryology of the Chick*, 5th edition. McGraw-Hill, New York.

Patterson, C. 1982. Morphological characters and homology. In: *Problems of Phylogenetic Reconstruction*, eds K.A. Joysey & A.E. Friday. Academic Press, London.

Patterson, C. 1994. Null or minimal models. In: *Models in Phylogeny Reconstruction*, eds R.W. Scotland, D.J. Siebert & D.M. Williams. Clarendon Press, Oxford, UK.

Patterson, C. & Smith, A.B. 1987. Is the periodicity of extinctions a taxonomic artefact? *Nature*, **330**, 248–251.

Patthy, L. 1991. Exons – original building blocks of proteins? *Bioessays*, **13**, 187–192.

Peifer, M., Karch, F. & Bender, W. 1987. The Bithorax complex: control of segmental identity. *Genes Dev.*, **1**, 891–898.

Pendleton, J.W., Nagai, B.K., Murtha, M.T. & Ruddle, F.H. 1993. Expansion of the *Hox* gene family and the evolution of chordates. *Proc. Nat. Acad. Sci. USA*, **90**, 6300–6304.

Penton, A. & Hoffmann, F.M. 1996. Decapentaplegic restricts the domain of *wingless* during *Drosophila* wing patterning. *Nature*, **382**, 162–165.

Perrimon, N. 1994. The genetic basis of patterned baldness in *Drosophila*. *Cell*, **76**, 781–784.

Peterson, K.J. 1995. Dorsoventral axis inversion. *Nature*, **373**, 111–112.

Pflug, H.D. 1972a. Zur Fauna der Nama-Schichten in Südwest-Afrika. III. Erniettomorpha, Bau und Systematik. *Palaeontographica*, *Abt. A.* **139**, 134–170.

Pflug, H.D. 1972b. Systematik der jung-präkambrischen Petalonamae. *Paläontologische Zeitschrift*, **46**, 56–67.

Phillips, D.S., Arthur, W., Leggett, M. & Day, T.H. 1995. Differential use of seaweed species by British seaweed flies, *Coelopa* spp. (Diptera: Coelopidae) with a description of the egg morphology of the two species. *Entomologist*, **114**, 158–165.

Pianka, E.R. 1994. *Evolutionary Ecology*, 5th edition. Harper Collins, New York.

Pigliucci, M. 1996. How organisms respond to environmental changes: from phenotypes to molecules (and vice versa). *TREE*, **11**, 168–173.

Pimm, S.L. 1984. The complexity and stability of ecosystems. *Nature*, **307**, 321–326.

Placzek, M., Jessell, T.M. & Dodd, J. 1993. Induction of floor plate differentiation by contact-dependent, homeogenetic signals. *Development*, **117**, 205–218.

Placzek, M., Tessier-Lavigne, M., Yamada, T., Jessell, T. & Dodd, J. 1990. Mesodermal control of neural cell identity: floor plate induction by the notochord. *Science*, **250**, 985–988.

Platnick, N.I. 1977. Review of *Evolution and the Diversity of Life* by Ernst Mayr. *Syst. Zool.*, **26**, 224–228.

Platnick, N.I. 1979. Philosophy and the transformation of cladistics. *Syst. Zool.*, **28**, 537–546.

Platnick, N.I. 1987. An empirical comparison of microcomputer parsimony programs. *Cladistics*, **3**, 121–144.

Platnick, N.I. 1989. An empirical comparison of microcomputer parsimony programs, II. *Cladistics*, **5**, 145–161.

Pollard, J.W. (ed.) 1984. *Evolutionary Theory: Paths into the Future*. Wiley, Chichester, UK.

Popadíc, A., Rusch, D., Peterson, M., Rogers, B.T. & Kaufman, T.C. 1996. Origin of the arthropod mandible. *Nature*, **380**, 395.

Porter, J.A., von Kessler, D.P., Ekker, S.C., Young, K.E., Lee, J.J., Moses, K. & Beachy, P.A. 1995. The product of *hedgehog* autoproteolytic cleavage active in local and long-range signalling. *Nature*, **374**, 363–366.

Pouliot, Y. 1992. Phylogenetic analysis of the cadherin superfamily. *Bioessays*, **14**, 743–748.

Poulton, E.B. 1908. *Essays on Evolution 1889–1907*. Clarendon Press, Oxford, UK.

Pourquié, O., Coltey, M., Teillet, M.-A., Ordahl, C. & Le Douarin, N.M. 1993. Control of dorsoventral patterning of somitic derivatives by notochord and floor plate. *Proc. Nat. Acad. Sci. USA*, **90**, 5242–5246.

Purnell, M.A. 1995. Microwear on conodont elements and macrophagy in the first vertebrates. *Nature*, **374**, 798–800.

Qian, S., Capovilla, M. & Pirrotta, V. 1991. The bx region enhancer, a distant cis-control element of the *Drosophila Ubx* gene and its regulation by *hunchback* and other segmentation genes. *EMBO J.*, **10**, 1415–1425.

Raff, R.A. 1996. *The Shape of Life: Genes, Development and the Evolution of Animal Form*. Chicago University Press, Chicago, IL.

Raff, R.A. & Kaufman, T.C. 1983. *Embryos, Genes and Evolution: The Developmental–Genetic Basis of Evolutionary Change*. Macmillan, New York.

Raff, R.A., Marshall, C.R. & Turbeville, J.M. 1994. Using DNA sequences to unravel the Cambrian radiation of the animal phyla. *Ann. Rev. Ecol. Syst.*, **25**, 351–375.

Ramsköld, L. & Xianguang, H. 1991. New early Cambrian animal and onychophoran affinities of enigmatic metazoans. *Nature*, **351**, 225–228.

Randazzo, F.M., Cribbs, D.L. & Kaufman, T.C. 1991. Rescue and regulation of *proboscipedia*: a homeotic gene of the Antennapedia complex. *Development*, **113**, 257–271.

Randazzo, F.M., Seeger, M.A., Huss, C.A., Sweeney, M.A., Cecil, J.K. & Kaufman, T.C. 1993. Structural changes in the Antennapedia complex of *Drosophila pseudoobscura*. *Genetics*, **134**, 319–330.

Raup, D.M. & Michelson, A. 1965. Theoretical morphology of the coiled shell. *Science*, **147**, 1294–1295.

Raup, D.M. & Sepkoski, J.J. 1984. Periodicity of extinctions in the geologic past. *Proc. Nat. Acad. Sci. USA*, **81**, 801–805.

Ray, R., Arora, K., Nüsslein-Volhard, C. & Gelbart, W.M. 1991. The control of cell fate along the dorsal–ventral axis of the *Drosophila* embryo. *Development*, **113**, 35–54.

Reeves, P.R. 1992. Variation in O-antigens, niche-specific selection and bacterial populations. *FEMS Microbiol. Lett.*, **79**, 509–516.

Resnik, D. 1995. Developmental constraints and patterns: some pertinent distinctions. *J. Theor. Biol.*, **173**, 231–240.

Retallack, G.J. 1994. Were the Ediacaran fossils lichens? *Paleobiology*, **20**, 523–544.

Richardson, M.K., Hanken, J., Gooneratne, M.L., Pieau, C., Raynaud, A., Selwood, L. Wright, G.M. 1997. There is no highly conserved embryonic stage in the vertebrates: implications for current theories of evolution and development. *Anat. Embryol.*, **196**, 91–106.

Riddihough, G. & Ish-Horowicz, D. 1991. Individual stripe regulatory elements in the *Drosophila hairy* promoter respond to maternal, gap, and pair-rule genes. *Genes Devel.*, **5**, 840–854.

Riddle, R.D., Johnson, R.L., Laufer, E. & Tabin, C. 1993. *Sonic hedgehog* mediates the polarizing activity of the ZPA. *Cell*, **75**, 1401–1416.

Ridley, M. 1993. *Evolution*. Blackwell, Oxford, UK.

Riedl, R. 1978. *Order in Living Organisms: A Systems Analysis of Evolution*. Wiley, Chichester.

Rieppel, O. 1993. The conceptual relationship of ontogeny, phylogeny, and classification: the taxic approach. In: *Evolutionary Biology*, Vol. 27, eds M.K. Hecht, R.J. MacIntyre & M.T. Clegg. Plenum Press, New York.

Rimm, D.L. & Morrow, J.S. 1994. Molecular cloning of human E-cadherin suggests a novel subdivision of the cadherin superfamily. *Biochem. Biophys. Res. Comm.*, **200**, 1754–1761.

Rivera-Pomar, R., Lu, X., Perrimon, N., Taubert, H. & Jackle, H. 1995. Activation of posterior gap gene expression in the *Drosophila* blastoderm. *Nature*, **376**, 253–256.

Roelink, H., Augsburger, A., Heemskerk, J., Korzh, V., Norlin, S., Ruiz i Ataba, A., Tanabe, Y., Placzek, M., Edlund, T., Jessell, T.M. & Dodd, J. 1994. Floor plate and motor neuron induction by *vhh-1*, a vertebrate homolog of *hedgehog* expressed by the notochord. *Cell*, **76**, 761–775.

Roelink, H., Porter, J.A., Chiang, C., Tanabe, Y., Chang, D., Beachy, P.A. & Jessell, T.M. 1995. Floor plate and motor neuron induction by different concentrations of the amino-terminal cleavage product of Sonic hedgehog autoproteolysis. *Cell*, **81**, 445–455.

Rosen, D. 1984. Hierarchies and history. In: *Evolutionary Theory: Paths into the Future*, ed. Pollard, J.W. Wiley, Chichester, UK.

Rosenberg, S.M., Harris, R.S. & Torkelson, J. 1995. Molecular handles on adaptive mutation. *Molec. Microbiol.*, **18**, 185–189.

Roth, S., Stein, D. & Nüsslein-Volhard, C. 1989. A gradient of nuclear localization of the *dorsal* protein determines dorsoventral pattern in the *Drosophila* embryo. *Cell*, **59**, 1189–1202.

Roth, V.L. 1988. The biological basis of homology. In: *Ontogeny and Systematics*, ed. C.J. Humphries. British Museum (Natural History), London.

Roux, W. 1894. The problems, methods and scope of developmental mechanics. *Biological Lectures of the Marine Biology Laboratory, Woods Hole*. Ginn, Boston, MA, pp. 149–190.

Ruddle, F.H., Bartels, J.L., Bentley, K.L., Kappen, C., Murtha, M.T. & Pendleton, J.W. 1994a. Evolution of *Hox* genes. *Ann. Rev. Genet.*, **28**, 423–442.

Ruddle, F.H., Bentley, K.L., Murtha, M.T. & Risch, N. 1994b. Gene loss and gain in the evolution of the vertebrates. In: *The Evolution of Developmental Mechanisms*, eds M. Akam, P. Holland, P. Ingham & G. Wray. Company of Biologists, Cambridge, UK.

Ruiz-Trillo, I., Riutort, M., Littlewood, D.T.J., Herniou, E.A. & Baguñà, J. 1999. Acoel flatworms: earliest extant bilaterian metazoans, not members of Platyhelminthes. *Science*, **283**, 1919–1923.

Rushlow, C.A., Han, K., Manley, J.L. & Levine, M. 1989. The graded distribution of the *dorsal* morphogen is initiated by selective nuclear transport in *Drosophila*. *Cell*, **59**, 1165–1177.

Russell, E.S. 1916. *Form and Function: A Contribution to the History of Animal Morphology*. Murray, London.

Salser, S.J. & Kenyon, C. 1996. A *C. elegans Hox* gene switches on, off, on and off again to regulate proliferation, differentiation and morphogenesis. *Development*, **122**, 1651–1661.

Salthe, S.N. 1985. *Evolving Hierarchical Systems: Their Structure and Representation*. Columbia University Press, New York.

Sang, J. 1984. *Genetics and Development*. Longman, London.

Sano, K., Tanihara, H., Heimark, R.L., Obata, S., Davidson, M., St John, T., Taketani, S. & Suzuki, S. 1993. Protocadherins: a large family of cadherin-related molecules in central nervous system. *EMBO J.*, **12**, 2249–2256.

Sanson, B., White, P. & Vincent, J.-P. 1996. Uncoupling cadherin-based adhesion from *wingless* signalling in *Drosophila*. *Nature*, **383**, 627–630.

Sasai, Y., Lu, B., Steinbeisser, H., Geissert, D., Gont, L.K. & De Robertis, E.M. 1994. *Xenopus chordin*: a novel dorsalizing factor activated by organizer-specific homeobox genes. *Cell*, **79**, 779–790.

Saunders, P.T. & Ho, M.W. 1976. On the increase in complexity in evolution. *J. Theor. Biol.*, **63**, 375–384.

Schmalhausen, I.I. 1949. *Factors of Evolution: The Theory of Stabilizing Selection*. Blakiston, Philadelphia, PA.

Schoener, T.W. 1974. Resource partitioning in ecological communities. *Science*, **185**, 27–39.

Schram, F.R. 1991. Cladistic analysis of metazoan phyla and the placement of fossil problematica. In: *The Early Evolution of Metazoa and the Significance of Problematic Taxa*, eds A.M. Simonetta & S. Conway Morris. Cambridge University Press, Cambridge, UK.

Schulz, C. & Tautz, D. 1995. Zygotic caudal regulation by *hunchback* and its role in abdominal segment formation of the *Drosophila* embryo. *Development*, **121**, 1023–1028.

Scotland, R.W. 1992. Cladistic theory. In: *Cladistics, A Practical Course in Systematics*, eds P.L. Forey, C.J. Humphries, I.L. Kitching, R.W. Scotland, D.J. Siebert & D.M. Williams. Clarendon Press, Oxford, UK.

Scott, M.P. 1992. Vertebrate homeobox gene nomenclature. *Cell*, **71**, 551–553.

Sedkov, Y., Tillib, S., Mizrokhi, L. & Mazo, A. 1994. The Bithorax complex is regulated by *trithorax* earlier during *Drosophila* embryogenesis than is the Antennapedia complex, correlating with a bithorax-like expression pattern of distinct early *trithorax* transcripts. *Development*, **120**, 1907–1917.

Seilacher, A. 1984. Late Precambrian Metazoa: Preservational or real extinctions? In: *Patterns of Change in Earth Evolution*, eds H.D. Holland and A.F. Trendall. Springer-Verlag, Berlin.

Seilacher, A. 1985. Discussion of Precambrian metazoans. *Phil. Trans. Roy. Soc. Lond., B*, **311**, 47–48.

Seilacher, A. 1989. Vendozoa: organismic construction in the Proterozoic biosphere. *Lethaia*, **22**, 229–239.

Shenk, M.A., Bode, H.R. & Steele, R.E. 1993. Expression of *Cnox*-2, a HOM/HOX homebox gene in *Hydra*, is correlated with axial pattern formation. *Development*, **117**, 657–667.

Shimoyama, Y., Gotoh, M., Terasaki, T., Kitajima, M. & Hirohashi, S. 1995. Isolation and sequence analysis of human cadherin-6 complementary DNA for the full coding sequence and its expression in human carcinoma cells. *Cancer Res.*, **55**, 2206–2211.

Shu, D., Zhang, X. & Chen, L. 1996. Reinterpretation of *Yunnanozoon* as the earliest known hemichordate. *Nature*, **380**, 428–430.

Shu, D.-G., Conway Morris, S. & Zhang, X.-L. 1996. A *Pikaia*-like chordate from the Lower Cambrian of China. *Nature*, **384**, 157–158.

Shubin, N., Tabin, C. & Carroll, S. 1997. Fossils, genes and the evolution of animal limbs. *Nature*, **388**, 639–648.

Shumway, W. 1932. The recapitulation theory. *Q. Rev. Biol.*, **7**, 93–99.

Simmonds, A.J., Brook, W.J., Cohen, S.M. & Bell, J.B. 1995. Distinguishable functions for *engrailed* and *invected* in anterior–posterior patterning of the *Drosophila* wing. *Nature*, **376**, 424–427.

Simon, J., Chiang, A. & Bender, W. 1992. Ten different Polycomb group genes are required for spatial control of the abdA and AbdB homeotic products. *Development*, **114**, 493–505.

Simonneau, L., Broders, F. & Thiery, J.P. 1992. N-cadherin transcripts in *Xenopus laevis* from early tailbud to tadpole. *Dev. Dynamics*, **194**, 247–260.

Simpson, G.G. 1944. *Tempo and Mode in Evolution*. Columbia University Press, New York.

Simpson, G.G. 1951. *Horses*. Oxford University Press, New York.

Sinnott, E.W., Dunn, L.C. & Dobzhansky, T. 1958. *Principles of Genetics*, 5th edition. McGraw-Hill Kogakusha, Tokyo.

Skelton, P. 1991. Morphogenetic versus environmental cues for adaptive radiations. In: *Constructional Morphology and Evolution*, eds Schmidt-Kittler, N. & Vogel, K. Springer-Verlag, Heidelberg.

Slack, J.M.W. 1983. *From Egg to Embryo: Determinative Events in Early Development*. Cambridge University Press, Cambridge, UK.

Slack, J.M.W., Holland, P.W.H. & Graham, C.F. 1993. The zootype and the phylotypic stage. *Nature*, **361**, 490–492.

Slatkin, M. 1980. Ecological character displacement. *Ecology*, **61**, 163–177.

Smith, A.B. 1994. *Systematics and the Fossil Record*. Blackwell, Oxford, UK.

Sneath, P.H.A. & Sokal, R.R. 1973. *Numerical Taxonomy: The Principles and Practice of Numerical Classification*. Freeman, San Francisco.

Sokal, R.R. 1986. Phenetic taxonomy: theory and methods. *Ann. Rev. Ecol. Syst.*, **17**, 423–442.

Sommer, R. & Tautz, D. 1991. Segmentation gene expression in the housefly *Musca domestica*. *Development*, **113**, 419–430.

Sordino, P., van der Hoeven, F. & Duboule, D. 1995. *Hox* gene expression in teleost fins and the origin of vertebrate digits. *Nature*, **375**, 678–681.

Spemann, H. 1938. *Embryonic Development and Induction.* Yale University Press, New Haven, CT.

Sprigg, R.C. 1947. Early Cambrian (?) jellyfishes from the Flinders Ranges, South Australia. *Trans. Roy. Soc. S. Aust.*, **71**, 212–224.

Sprigg, R.C. 1949. Early Cambrian 'Jellyfishes' of Ediacara, South Australia, and Mount John, Kimberley District, Western Australia. *Trans. Roy. Soc. S. Aust.*, **73**, 72–99.

Springer, M. & Krajewski, C. 1989. DNA hybridization in animal taxonomy: a critique from first principles. *Q. Rev. Biol.*, **64**, 291–318.

St.Johnston, D. & Nüsslein-Volhard, C. 1992. The origin of pattern and polarity in the *Drosophila* embryo. *Cell*, **68**, 201–219.

Stanley, S.M. 1975. A theory of evolution above the species level. *Proc. Nat. Acad. Sci. USA*, **72**, 646–650.

Stanley, S.M. 1977. Trends, rates and patterns of evolution in the Bivalvia. In: *Patterns of Evolution as Illustrated by the Fossil Record*, ed. A. Hallam. Elsevier, Amsterdam.

Stanley, S.M. 1979. *Macroevolution: Pattern and Process.* Freeman, San Francisco, CA.

Stanley, S.M. 1987. *Extinction.* Freeman, San Francisco, CA.

Stearns, L.W. 1974. *Sea Urchin Development: Cellular and Molecular Aspects.* Dawden, Hutchinson & Ross, Stroudsburg, PA.

Stearns, S.C. 1992. *The Evolution of Life Histories.* Oxford University Press, Oxford, UK.

Stern, D.L. 1998. A role of *Ultrabithorax* in morphological differences between *Drosophila* species. *Nature*, **396**, 463–466.

Stern, M.J. & DeVore, D.L. 1994. Extending and connecting signaling pathways in *C. elegans. Dev. Biol.*, **166**, 443–459.

Sternberg, P.W. 1993. Intercellular signaling and signal transduction in *C. elegans. Ann. Rev. Genet.*, **27**, 497–521.

Steward, R. 1989. Relocalization of the *dorsal* protein from the cytoplasm to the nucleus correlates with function. *Cell*, **59**, 1179–1188.

Stewart, C.-B. 1993. The powers and pitfalls of parsimony. *Nature*, **361**, 603–607.

Stone, B.L. & Thummel, C.S. 1993. The *Drosophila* 78C early late puff contains E78, an ecdysone-inducible gene that encodes a novel member of the nuclear hormone receptor superfamily. *Cell*, **75**, 307–320.

Stone, D.M., Hynes, M., Armanini, M., Swanson, T.A., Gu, Q., Johnson, R.L., Scott, M.P., Pennica, D., Goodard, A., Phillips, H., Noll, M., Hooper, J.E., de Sauvage, F. & Rosenthal, A. 1996. The tumour-suppressor gene *patched* encodes a candidate receptor for Sonic hedgehog. *Nature*, **384**, 129–134.

Stone, J.R. 1996. Computer-generated shell size and shape variation in the Caribbean land snail genus *Cerion*: a test of geometrical constraints. *Evolution*, **50**, 341–347.

Strachan, T. & Read, A.P. 1996. *Human Molecular Genetics.* BIOS Scientific Publishers, Oxford, UK.

Strathmann, R.R. 1985. Feeding and nonfeeding larval development and life-history evolution in marine invertebrates. *Ann. Rev. Ecol. Syst.*, **16**, 339–361.

Strobeck, C. 1974. Sufficient conditions for polymorphism with $n$ niches and $m$ mating groups. *Am. Nat.*, **108**, 152–156.

Struhl, G. 1981. A gene product required for correct initiation of segmental determination in *Drosophila. Nature*, **293**, 36–41.

Struhl, G., Johnston, P. & Lawrence, P.A. 1992. Control of *Drosophila* body pattern by the *hunchback* morphogen gradient. *Cell*, **69**, 237–249.

Sturtevant, A.H. 1923. Inheritance of direction of coiling in *Limnaea. Science*, **58**, 269–270.

Sturtevant, A.H. 1965. *A History of Genetics.* Harper & Row, New York.

Subramaniam, V., Bomze, H.M. & Lopez, A.J. 1994. Functional differences between Ultrabithorax protein isoforms in *Drosophila melanogaster*: evidence from elimination, substitution and ectopic expression of specific isoforms. *Genetics*, **136**, 979–991.

Südhof, T.C., Goldstein, J.L., Brown, M.S. & Russell, D.W. 1985. The LDL receptor gene: a mosaic of exons shared with different proteins. *Science*, **228**, 815–822.

Swofford, D.L. 1985. PAUP. Computer program and manual. Illinois Natural History Survey, Urbana, IL.

Tabata, T. & Kornberg, T.B. 1994. Hedgehog is a signaling protein with a key role in patterning *Drosophila* imaginal discs. *Cell*, **76**, 89–102.

Tabata, T., Eaton, S. & Kornberg, T.B. 1992. The *Drosophila hedgehog* gene is expressed specifically in posterior compartment cells and is a target of *engrailed* regulation. *Genes Dev.*, **6**, 2635–2645.

Tabin, C.J. 1992. Why we have (only) five fingers per hand: *Hox* genes and the evolution of paired limbs. *Development*, **116**, 289–296.

Takeichi, M. 1991. Cadherin cell adhesion receptors as a morphogenetic regulator. *Science*, **251**, 1451–1455.

Takeichi, M., Watabe, M., Shibamoto, S., Ito, F., Oda, H., Uemura, T. & Shimamura, K. 1993. Dynamic control of cell–cell adhesion for multicellular organization. *Comptes Rendus de l'Academie des Sciences – Serie Iii, Sciences de la Vie*, **316**, 813–821.

Tamkun, J.W., DeSimone, D.W., Fonda, D., Patel, R.S., Buck, C., Horwitz, A.F. & Hynes, R.O. 1986. Structure of integrin, a glycoprotein involved in transmembrane linkage between fibronectin and actin. *Cell*, **46**, 271–282.

Tardent, P. & Tardent, R. (eds) 1980. *Developmental and Cellular Biology of Coelenterates.* Elsevier/North Holland, Amsterdam.

Taylor, A.M., Nakano, Y., Mohler, J. & Ingham, P.W. 1993. Contrasting distributions of patched and hedgehog proteins in the *Drosophila* embryo. *Mech. Dev.*, **42**, 89–96.

Telford, M.J. & Holland, P.W. 1993. The phylogenetic affinities of the chaetognaths: a molecular analysis. *Molec. Biol. Evol.*, **10**, 660–676.

Thompson, D'A.W. 1917. *On Growth and Form.* Cambridge University Press, Cambridge, UK.

Thomson, K. 1988. *Morphogenesis and Evolution*. Oxford University Press, New York.

Thorpe, W.H. 1966. Molecules and evolution. (Review of *Internal Factors in Evolution* by L.L. Whyte.) *Nature*, **210**, 663–664.

Thorpe, W.H. 1978. *Purpose in a World of Chance*. Oxford University Press, Oxford, UK.

Thummel, C.S. 1996. Flies on steroids – *Drosophila* metamorphosis and the mechanisms of steroid hormone action. *TIG*, **12**, 306–310.

Thuringer, F., Cohen, S.M. & Bienz, M. 1993. Dissection of an indirect auto-regulatory response of a homeotic *Drosophila* gene. *EMBO J.*, **12**, 2419–2430.

Tomlinson, A. & Ready, D.F. 1986. *Sevenless*: a cell-specific homeotic muta-tion of the *Drosophila* eye. *Science*, **231**, 400–402.

Tooi, O., Fujii, G., Tashiro, K. & Shiokawa, K. 1994. Molecular cloning of cDNA for XTCAD-1, a novel *Xenopus* cadherin, and its expression in adult tissues and embryos of *Xenopus laevis*. *Biochimica et Biophysica Acta*, **1219**, 121–128.

Torrey, T.W. & Feduccia, A. 1979. *Morphogenesis of the Vertebrates*, 4th edi-tion. Wiley, New York.

Turing, A.M. 1952. The chemical basis of morphogenesis. *Phil. Trans. R. Soc. Lond. B.*, **237**, 37–72.

Turner, J.R.G. 1977. Butterfly mimicry: the genetical evolution of an adapta-tion. *Evol. Biol.*, **10**, 163–226.

Tycowski, K.T., Shu, M.-D. & Steitz, J.A. 1996. A mammalian gene with introns instead of exons generating stable RNA products. *Nature*, **379**, 464–466.

Vaahtokari, A., Aberg, T. & Thesleff, I. 1996. Apoptosis in the developing tooth: association with an embryonic signaling center and suppression by EGF and FGF-4. *Development*, **122**, 121–129.

Vacelet, J. & Boury-Esnault, N. 1995. Carnivorous sponges. *Nature*, **373**, 333–335.

Vachon, G., Cohen, B., Pfeifle, C., McGuffin, M.E., Botas, J. & Cohen, S.M. 1992. Homeotic genes of the Bithorax complex repress limb development in the abdomen of the *Drosophila* embryo through the target gene *Distal-less*. *Cell*, **71**, 437–450.

Valentine, J.W. 1986. Fossil record of the origin of Baupläne and its implica-tions. In: *Patterns and Processes in the History of Life*, eds D.M. Raup & D. Jablonski. Springer-Verlag, Berlin.

Valentine, J.W. 1991. Major factors in the rapidity and extent of the metazoan radiation during the Proterozoic–Phanerozoic transition. In: *The Early Evolution of Metazoa and the Significance of Problematic Taxa*, eds A.M. Simonetta & S. Conway Morris. Cambridge University Press, Cambridge, UK.

Valentine, J.W. 1994. Late Precambrian bilaterians: Grades and clades. *Proc. Nat. Acad. Sci. USA*, **91**, 6751–6757.

Valentine, J.W. & Erwin, D.H. 1987. Interpreting great developmental experiments: the fossil record. In: *Development as an Evolutionary Process*, eds R.A. Raff & E.C. Raff. Liss, New York.

Valentine, J.W., Collins, A.G. & Meyer, C.P. 1994. Morphological complexity increase in metazoans. *Paleobiology*, **20**, 131–142.

Valentine, J.W., Erwin, D.H. & Jablonski, D. 1996. Developmental evolution of metazoan bodyplans: the fossil evidence. *Dev. Biol.*, **173**, 373–381.

Van Batenburg, F.H.D. & Gittenberger, E. 1996. Ease of fixation of a change in coiling: computer experiments on chirality in snails. *Heredity*, **76**, 278–286.

Van Dijk, M.A. & Murre, C. 1994. *extradenticle* raises the DNA binding specificity of homeotic selector gene products. *Cell*, **78**, 617–624.

Van Noordwijk, A.J. 1984. Quantitative genetics in natural populations of birds, illustrated with examples from the great tit, *Parus major*. In: *Population Biology and Evolution*, eds K. Wöhrmann & V. Loeschcke. Springer-Verlag, Berlin.

Van Valen, L. 1974. A natural model for the origin of some higher taxa. *J. Herpetol.*, **8**, 109–121.

Vermeij, G.J. 1996. Animal origins. *Science*, **274**, 525–526.

Von Allmen, G., Hogga, I., Spierer, A., Karch, F., Bender, W., Gyurkovics, H. & Lewis, E. 1996. Splits in fruitfly *Hox* gene complexes. *Nature*, **380**, 116.

Von Baer, K.E. 1828. *Uber Entwicklungsgeschichte der Tiere: Beobachtung und Reflexion*. Bornträger, Königsberg.

Von Neumann, J. 1966. *Theory of Self-reproducing Automata*, ed. A.W. Burks. University of Illinois Press, Urbana, IL.

Vorobyeva, E. & Hinchliffe, R. 1996. From fins to limbs: developmental perspectives on palaeontological and morphological evidence. *Evol. Biol.*, **29**, 263–311.

Vortkamp, A., Lee, K., Lanske, B., Segre, G.V., Kronenberg, H.M. & Tabin, C.J. 1996. Regulation of rate of cartilage differentiation by Indian Hedgehog and PTH-related protein. *Science*, **273**, 613–622.

Waddington, C.H. 1940. *Organizers and Genes*. Cambridge University Press, Cambridge, UK.

Waddington, C.H. 1953. The genetic assimilation of an acquired character. *Evolution*, **7**, 118–126.

Waddington, C.H. 1956. Genetic assimilation of the bithorax phenotype. *Evolution*, **10**, 1–13.

Waddington, C.H. 1957. *The Strategy of the Genes*. Allen & Unwin, London.

Waddington, C.H. 1975. *The Evolution of an Evolutionist*. Edinburgh University Press, Edinburgh, UK.

Walcott, C.D. 1911a. Middle Cambrian Merostomata. Cambrian Geology and Paleontology, II. *Smithsonian Miscellaneous Collections*, **57**, 17–40.

Walcott, C.D. 1911b. Middle Cambrian holothurians and medusae. Cambrian Geology and Paleontology, II. *Smithsonian Miscellaneous Collections*, **57**, 41–68.

Walcott, C.D. 1911c. Middle Cambrian annelids. Cambrian Geology and Paleontology, II. *Smithsonian Miscellaneous Collections*, **57**, 109–144.

Walcott, C.D. 1912. Middle Cambrian Branchiopoda, Malacostraca, Trilobita and Merostomata. Cambrian Geology and Paleontology, II. *Smithsonian Miscellaneous Collections*, **57**, 145–228.

Walker, M.H. 1992. Scanning electron microscope observations of embryonic development in *Opisthopatus cinctipes* Purcell (Onychophora, Peripatopsidae). In: *Proc. 8th Int. Cong. Myriapodol; Berichte des Naturwissenschaftlich – Medizinischen Verein in Insbruck, Suppl.,* **10**, 459–646, eds E. Meyer & K. Thaler.

Walker, M.H. 1995. Relatively recent evolution of an unusual pattern of early embryonic development (long germ band?) in a South African onychophoran, *Opisthopatus cinctipes* Purcell (Onychophora: Peripatopsidae). *Zool. J. Linn. Soc.,* **114**, 61–75.

Wallace, A.R. 1870. *Contributions to the Theory of Natural Selection: A Series of Essays.* Macmillan, London.

Walossek, D. & Müller, K.J. 1994. Pentastomid parasites from the Lower Palaeozoic of Sweden. *Trans. Roy. Soc. Edin.: Earth Sciences*, **85**, 1–37.

Walter, M.R., Du, R. & Horodyski, R.J. 1990. Coiled carbonaceous megafossils from the middle Proterozoic of Jixian (Tianjin) and Montana. *Am. J. Sci.,* **290A**, 133–148.

Wang, B.B., Müller-Immergluck, M.M., Austin, J., Robinson, N.T., Chisholm, A. & Kenyon, C. 1993. A homeotic gene cluster patterns the anteroposterior body axis of *C. elegans. Cell*, **74**, 29–42.

Wang, C. & Lehmann, R. 1991. Nanos is the localized posterior determinant in *Drosophila. Cell*, **66**, 637–647.

Wang, D. Y.-C., Kumar, S. & Hedges, S.B. 1999. Divergence time estimates for the early history of animal phyla and the origin of plants, animals and fungi. *Proc. Roy. Soc. Lond. B*, **266**, 163–171.

Wang, S. & Hazelrigg, T. 1995. Implications for *bcd* mRNA localization from spatial distribution of *exu* protein in *Drosophila* oogenesis. *Nature*, **369**, 400–403.

Warren, R.W., Nagy, L., Selegue, J., Gates, J. & Carroll, S. 1994. Evolution of homeotic gene regulation and function in flies and butterflies. *Nature*, **372**, 458–461.

Watrous, L.E. & Wheeler, Q.D. 1981. The out-group comparison method of character analysis. *Syst. Zool.,* **30**, 1–11.

Weismann, A. 1904. *The Evolution Theory*, vols I & II, translated by J.A. Thomson & M.R. Thomson. Edward Arnold, London.

Weiss, P. 1940. The problem of cell individuality in development. *Am. Nat.,* **74**, 34–46.

Weldon, W.F.R. 1901. A first study of natural selection in *Clausilia laminata* (Montagu). *Biometrika*, **1**, 109–124.

Werner, E.E. 1988. Size, scaling and the evolution of complex life cycles. In: *Size-Structured Populations: Ecology and Evolution*, eds B. Ebenman & L. Persson. Springer-Verlag, Berlin.

West-Eberhard, M.J. 1989. Phenotypic plasticity and the origins of diversity., *Ann. Rev. Ecol. Syst.,* **20**, 249–278.

Whalen, A.M. & Steward, R. 1993. Dissociation of the dorsal-cactus complex and phosphorylation of the dorsal protein correlate with the nuclear localization of Dorsal. *J. Cell Biol.*, **123**, 523–534.

Wheeler, Q.D. 1990. Ontogeny and character phylogeny. *Cladistics*, **6**, 225–268.

Wheeler, W.C., Cartwright, P. & Hayashi, C.Y. 1993. Arthropod phylogeny: a combined approach. *Cladistics*, **9**, 1–39.

Whewell, W. 1832. Review of *Principles of Geology*, volume 2, by Charles Lyell. *Quart. Rev.*, **47**, 103–132.

White, M.J.D. 1978. *Modes of Speciation*. Freeman, San Francisco.

Whiting, M.F. & Wheeler, W.C. 1994. Insect homeotic transformation. *Nature*, **368**, 696.

Whittington, H.B. 1974. *Yohoia* Walcott and *Plenocaris* n.gen., arthropods from the Burgess Shale, Middle Cambrian, British Columbia. *Geological Survey of Canada Bulletin*, **231**, 1–21.

Whittington, H.B. 1975. The enigmatic animal *Opabinia regalis*, Middle Cambrian, Burgess Shale, British Columbia. *Phil. Trans. Roy. Soc. Lond. B*, **271**, 1–43.

Whittington, H.B. 1985. *The Burgess Shale*. Yale University Press, New Haven, CT.

Whittington, H.B. & Briggs, D.E.G. 1985. The largest Cambrian animal, *Anomalocaris*, Burgess Shale, British Columbia. *Phil. Trans. Roy. Soc. Lond. B*, **309**, 569–609.

Whyte, L.L. 1960a. Developmental selection of mutations. *Science*, **132**, 954.

Whyte, L.L. 1960b. Developmental selection of mutations. (Reply to Lewontin & Caspari). *Science*, **132**, 1692–1694.

Whyte, L.L. 1964. Internal factors in evolution. *Acta Biotheoretica*, **17**, 33.

Whyte, L.L. 1965. *Internal Factors in Evolution*. Tavistock Publications, London.

Wiegmann, B.M., Mitter, C. & Thompson, F.C. 1993. Evolutionary origin of the Cyclorrhapha (Diptera): tests of alternative morphological hypotheses. *Cladistics*, **9**, 41–81.

Wieschaus, E. & Gehring, W. 1976. Clonal analysis of primordial disc cells in the early embryo of *Drosophila melanogaster. Dev. Biol.*, **50**, 249–263.

Wigglesworth, V.B. 1935. Functions of the corpus allatum of insects. *Nature*, **136**, 338–339.

Wilbur, H.M. 1980. Complex life cycles. *Ann. Rev. Ecol. Syst.*, **11**, 67–93.

Williams, G.C. 1992. *Natural Selection: Domains, Levels, and Challenges*. Oxford University Press, New York.

Williamson, M. 1957. An elementary theory of interspecific competition. *Nature*, **180**, 422–425.

Williamson, P.G. 1981. Palaeontological documentation of speciation in Cenozoic molluscs from Turkana Basin. *Nature*, **293**, 437–443.

Willis, J.C. 1940. *The Course of Evolution by Differentiation or Divergent Mutation rather than by Selection*. Cambridge University Press, Cambridge, UK.

Willmer, P. 1990. *Invertebrate Relationships: Patterns in Animal Evolution.* Cambridge University Press, Cambridge, UK.

Wills, M.A., Briggs, D.E.G. & Fortey, R.A. 1994. Disparity as an evolutionary index: a comparison of Cambrian and Recent arthropods. *Paleobiology*, **20**, 93–130.

Wilson, R. *et al.* (54 others), 1994. 2.2 Mb of contiguous nucleotide sequence from chromosome II of *C. elegans. Nature*, **368**, 32–38.

Wimsatt, W.C. 1986. Developmental constraints, generative entrenchment, and the innate-acquired distinction. In: *Integrating Scientific Disciplines*, ed. W. Bechtel. Martinus-Nijhoff, Dordrecht.

Wolpert, L. 1968. The French flag problem: a contribution to the discussion on pattern formation and regulation. In: *Towards a Theoretical Biology. I. Prolegomena*, ed C.H. Waddington. Edinburgh University Press, Edinburgh, UK.

Wolpert, L. 1969. Positional information and the spatial pattern of cellular differentiation. *J. Theor. Biol.*, **25**, 1–47.

Wolpert, L. 1971. Positional information and pattern formation. *Curr. Top. Dev. Biol.*, **6**, 183–224.

Wolpert, L. 1990. Signals in limb development: STOP, GO, STAY and POSITION. In: *Growth Factors in Cell and Developmental Biology*, ed. M.D. Waterfield. Company of Biologists, Cambridge, UK.

Wray, G.A. 1994. The evolution of cell lineage in echinoderms. *Amer. Zool.*, **34**, 353–363.

Wray, G.A. & Bely, A.E. 1994. The evolution of echinoderm development is driven by several distinct factors. In: *The Evolution of Developmental Mechanisms*, eds M. Akam, P. Holland, P. Ingham & G. Wray. Company of Biologists, Cambridge, UK.

Wray, G.A. & McClay, D.R. 1989. Molecular heterochronies and heterotopies in early echinoid development. *Evolution*, **43**, 803–813.

Wray, G.A. & Raff, R.A. 1991. The evolution of developmental strategy in marine invertebrates. *TREE*, **6**, 45–50.

Wray, G.A., Levinton, J.S. & Shapiro, L.H. 1996. Molecular evidence for deep pre-Cambrian divergences among metazoan phyla. *Science*, **274**, 568–573.

Wright, S. 1931. Evolution in Mendelian populations. *Genetics*, **16**, 97–159.

Wright, S. 1932. The roles of mutation, inbreeding, crossbreeding and selection in evolution. *Proc. 6th Int. Congr. Genet.*, **1**, 356–366.

Wynne-Edwards, V.C. 1962. *Animal Dispersion in Relation to Social Behaviour.* Oliver and Boyd, London.

Xiang, Y.Y., Tanaka, M., Suzuki, M., Igarashi, H., Kiyokawa, E., Naito, Y., Ohtawara, Y., Shen, Q., Sugimura, H. & Kino, I. 1994. Isolation of complementary DNA encoding κ-cadherin, a novel rat cadherin preferentially expressed in fetal kidney and kidney carcinoma. *Cancer Res.*, **54**, 3034–3041.

Yamada, T., Placzek, M., Tanaka, H., Dodd, J. & Jessell, T.M. 1991. Control of cell pattern in the developing nervous system: polarizing activity of the floor plate and notochord. *Cell*, **64**, 635–647.

Yamamoto, D. 1994. Signaling mechanisms in induction of the R7 photore-ceptor in the developing *Drosophila* retina. *Bioessays*, **16**, 237–244.

Yao, T.P., Segraves, W.A., Oro, A.E., McKeown, M. & Evans, R.M. 1992. *Drosophila ultraspiracle* modulates ecdysone receptor function via hetero-dimer formation. *Cell*, **71**, 63–72.

Yao, T.P., Forman, B.M., Jiang, Z., Cherbas, L., Chen, J.D., McKeown, M., Cherbas, P. & Evans, R.M. 1993. Functional ecdysone receptor is the product of *EcR* and *ultraspiracle* genes. *Nature*, **366**, 476–479.

Yockey, H.P. 1992. *Information Theory and Molecular Biology*. Cambridge University Press, Cambridge, UK.

Yokoyama, S. & Schaal, B.A. 1985. A note on multiple-niche polymorphisms in plant populations. *Am. Nat.*, **125**, 158–163.

Young, C.M., Vázquez, E., Metaxas, A. & Tyler, P.A. 1996. Embryology of vestimentiferan tube worms from deep-sea methane/sulphide seeps. *Nature*, **381**, 514–516.

Young, J.Z. 1962. *The Life of Vertebrates*, 2nd edition. Clarendon Press, Oxford, UK.

Zhang, C.C., Muller, J., Hoch, M., Jackle, H. & Bienz, M. 1991. Target sequences for hunchback in a control region conferring *Ultrabithorax* expression boundaries. *Development*, **113**, 1171–1179.

Zhang, Z. 1986. Clastic facies microfossils from the Chuanlinggou Formation (1800 Ma) near Jixtan, North China. *J. Micropalaeont.*, **5**, 9–16.

Zou, H. & Niswander, L. 1996. Requirement for BMP signaling in interdigital apoptosis and scale formation. *Science*, **272**, 738–741.

# Index

Notes on the indexing of names: *Animal names* - names of genera and phyla are indexed; common names and names of intermediate-level taxa are not. *Gene names* - individual genes (e.g. *bicoid*) are listed alphabetically, not under 'gene'; names of gene groups (e.g. pair-rule) and of types of genes (e.g. housekeeping) are listed as sub-entries under 'gene'. *Author and place names* - these are not indexed except where they have become part of a phrase (e.g. Von Baer's laws, Ediacaran fauna).

Printed in the United States
By Bookmasters